Mathematical Principle and Fractal Analysis of Mesoscale Eddy

Shu-Tang Liu · Yu-Pin Wang · Zhi-Min Bi ·
Yin Wang

Mathematical Principle and Fractal Analysis of Mesoscale Eddy

Springer

Shu-Tang Liu
College of Control Science and Engineering
Shandong University
Jinan, Shandong, China

Yu-Pin Wang
Institute of Marine Science and Technology
Shandong University
Qingdao, Shandong, China

Zhi-Min Bi
School of Control Science and Engineering
Shandong University
Jinan, Shandong, China

Yin Wang
Institute of Marine Science and Technology
Shandong University
Qingdao, Shandong, China

ISBN 978-981-16-1841-3 ISBN 978-981-16-1839-0 (eBook)
https://doi.org/10.1007/978-981-16-1839-0

This Springer imprint is published by the registered company Springer Nature Singapore Pte Ltd.
The registered company address is: 152 Beach Road, #21-01/04 Gateway East, Singapore 189721, Singapore

Preface

Throughout the ages, the research on mesoscale eddies in the ocean is based on the statistical analysis of data formed by the observation of satellite altimeters and Argo buoy profiles. The abundant numerical simulation of the Navier-Stokes system in Euler form has been obtained. Unfortunately, so far, there has not been a universal mathematical model of mesoscale eddies in the analytical form.

Mesoscale eddy is a large-scale water body from 100 to 200 nautical miles, which is bounded and continuous. And any water particle tends to have the same dynamic behavior. To understand the basic properties and mechanism of the mesoscale eddy, we must reveal its essential law from the perspective of Lagrangian particle dynamics. Starting from the movement of the Lagrangian water particle, this work establishes a nonlinear dynamic model for the universality of mesoscale eddies using limit cycle theory, fractal theory, chaos theory, fractal dimension theory, nonlinear distributed parameter system theory, and strange attractor theory, then obtains a universal mathematical model of mesoscale eddy, its motion, and spatiotemporal motion laws, self-similar fractal structure characteristics, results of the model are consistent with the objective reality, complex characteristics analysis, the mathematical relationship between two-dimensional geostrophic balance equations and mesoscale eddies, the same solution between the Navier-Stokes equation of the Euler flow field and the mathematical model of mesoscale eddy, etc.

The results of the establishment of the universal model and experimental verification not only verify that the relevant results obtained by the Euler form and different measurement statistics are consistent, but more importantly, the results of the system theory are more abundant, complete, and analytical due to the universality of the model.

On the other hand, the results of this work are not only the direct and typical applications of limit cycle theory and fractal theory in pure mathematical theories but also the classic combination of nonlinear dynamic systems in mathematics and physical oceanography. Although this work mainly focuses on mesoscale processes, all the results are also applicable to vortex behavior at other scales.

The authors are very grateful for the generous help of the Institute of Oceanography, Chinese Academy of Sciences, especially Researcher Shan Gao and Dr. Jun Zhao; Thanks to the relevant teachers from the Department of Oceanography, School of Ocean and Atmosphere, and the Department of Computer Science and Technology, School of Information Science and Engineering, Ocean University of China. The entire contents of this book are innovative achievements of our research team. We would like to express our appreciation for the great efforts of Drs. Zhibin Liu, Xingao Zhu, Xiang Wu, and Miao Ouyang, and graduate students Changxu Shao and Tong Shen. In particular, we would like to thank Dr. Lingxia Ran for her work in the field of literature collection and some mathematical transformation of particle motion.

Finally, we are grateful for the sequential support of the National Natural Science Foundation of China-Shandong joint fund (No. U1806203) and the Key Program of the National Natural Science Foundation of China (No. 61533011), the Foundation for the Author of National Excellent Doctoral Dissertation of China (No. 200444), the National Natural Science Foundation of China (No. 60472112, No. 60372028, No. 600874009, No. 10971120, No. 61273088), the Shandong Province Natural Science Foundation (No. ZR2010FM010, No. Y98A02005), the China Postdoctoral Science Foundation (No. 2004035036), Project supported by the Scientific Research Staring Foundation for the Returned Overseas Chinese Scholars, Ministry of Education of China, and the K. C. Wong Education Foundation, Hong Kong, China, which ensure the smooth developments of this research project.

Due to the influence of scientific research level and the limitations, there are inevitably omissions and deficiencies in the book, and readers are welcome to put forward valuable suggestions.

Jinan, China	Shu-Tang Liu
Qingdao, China	Yu-Pin Wang
Jinan, China	Zhi-Min Bi
Qingdao, China	Yin Wang

Contents

Abbreviations

a.e.	almost everywhere
a.s.	almost surely
cos	cosine function
cov	covariance of a pair of random variables
deg	degree of a polynomial
dim	dimension of a vector space
div	divergence of a vector field
Im	imaginary part of a complex number
inf	infimum of a set
lim	limit of a sequence, or of a function
lim inf	limit inferior
lim sup	limit superior
max	maximum of a set
min	minimum of a set
No.	number
rank	rank
Re	real part of a complex number
rot	rotor of a vector field
s.t.	such that or so that
sin	sine function
sup	supremum of a set
tan	tangent function

Symbols

$\mathbb{R}, \mathbb{R}^n$	set of real numbers, set of n-dimensional real vectors
$\mathbb{C}, \mathbb{C}^n$	set of complex numbers, set of n-dimensional complex vector
$\mathbb{R}^+$	set of non-negative real numbers
$\mathbb{Z}^+$	set of non-negative integers
$\forall$	for all
$\exists$	there exists
$\mathrm{Re}(s)$	real part of $s \in \mathbb{C}$
$\mathrm{Im}(s)$	imaginary part of $s \in \mathbb{C}$
I_n	$n \times n$ identify matrix (the subscript is omitted if no confusion will occur)
A^{T}	transpose of matrix A
A^{-1}	inverse of matrix A
$A > 0 (< 0)$	symmetric positive (negative) definite matrix
$A \geq 0 (\leq 0)$	symmetric positive (negative) semi-definite matrix
$\mathrm{diag}\{A_1, \ldots, A_n\}$	diagonal matrix with A_i as its i th diagonal element
$\det(A)$	determinant of matrix A
$\mathrm{Tr}(A)$	trace of matrix A
$\lambda(A)$	eigenvalue of matrix A
$\lambda_{max}(A)$	largest eigenvalue of matrix A
$\lambda_{min}(A)$	smallest eigenvalue of matrix A
$\sigma_{max}(A)$	largest singular value of matrix A
$\lvert \cdot \rvert$	absolute value (or modulus)
$\lVert \cdot \rVert$	Euclidean norm of a vector or spectral norm of a matrix
$\in$	belongs to
$\subseteq$	is a subset of
$\cup$	union
$\cap$	intersection
$\rightarrow$	tends toward or is mapped into (case sensitive)
$\frac{\partial V}{\partial t}$	partial derivative of function V
$\mathrm{grad}\, V$	gradient of function V
$\prod_{i=1}^{n} \lambda_i$	product of $\lambda_1, \lambda_2, \ldots, \lambda_n$
$\square$	end of proof

Chapter 1
Introduction

Throughout the research history of relevant mesoscale eddies in the physical oceanography, for hundreds of years, Navier-Stokes equation in the Euler form and statistics of observation data have been mainly used in the study of mesoscale eddies. Although rich results have been obtained [1]–[190], the progress is relatively slow.

To a certain extent, it is obtained that the statistical laws of temperature, density, salinity, pressure, seawater velocity, and other related issues of mesoscale eddies, as well as geometric evolution behavior of mesoscale eddies, from the measurement data, statistical methods, and numerical simulations of the flow field in the Euler form. The noticed behavior, in many cases, only the geometric gradual behavior of mesoscale eddies on the sea surface is demonstrated, and the universal mathematical model of mesoscale eddies is truly established from the systematic law. The internal mechanism reveals its essence. At present, there is no universal mathematical model or qualitative analysis of the internal mechanism of the mathematical system. The current bottleneck of the mesoscale eddy is the key problem that this book urgently solves.

To the formation of the geometric shapes of the mesoscale eddy field on the sea surface, from a microscopic point of view, the fundament is the movement of water particles of the ocean continuum with uniform characteristics. Therefore, the study of mesoscale eddies must be solved fundamentally. The problem is perhaps that the fate of any water particles in the flow field must be considered, which means that the problem must be explored from the movement of the Lagrangian mass point to break the framework constraints of the Navier-Stokes equation in Euler form.

The Lagrangian form has not been seen in previous studies on mesoscale eddies. Although mathematical systems of Lagrangian particle dynamics and Euler flow field are theoretically mutually transformable, the conversion from Euler form to Lagrangian form requires integral operations, and the conversion from Lagrangian form to Euler form requires differential operations. In most problems, integral

S.-T. Liu et al., *Mathematical Principle and Fractal Analysis of Mesoscale Eddy*,
https://doi.org/10.1007/978-981-16-1839-0_1

operations are generally difficult. It can be transformed with each other only for individual special forms. That is to say, the analytical process of the conversion between the two, that is, the conversion from Euler form to Lagrangian form is generally impossible under normal circumstances.

Regardless of mathematical or physical considerations, the nature of mesoscale eddies is a macroscopic trend behavior in which all water particles in a certain area of the ocean tend to be unified. It can also be said that each water particle in the Euler flow field has uniform characteristics—the mighty trend of the mesoscale eddy is formed in the ocean. In this tendency, every water particle has a consistent behavior. This feature makes any particle in the field must conform to a definite attraction rule. This unified and consistent attraction rule makes arbitrary particle is closely related to a nonlinear particle dynamic system described in Lagrangian. Therefore, the motion of the infinite particle of the mesoscale eddy can be regarded as the motion of a particle in the Euler field, which is also a local system in the Euler flow field. And this happens to be Lagrangian's nonlinear particle motion.

Starting from the movement of Lagrangian particles, this work establishes a universal nonlinear dynamics model of mesoscale eddies, using limit cycle theory, fractal theory, chaos theory, fractal dimension theory, nonlinear distributed parameter system theory, and strange attractor theory, then obtains the universal mathematical model of mesoscale eddy spatiotemporal motion law, self-similar fractal structure characteristics, results of the model are consistent with the objective reality, analysis of the complex characteristics of the mesoscale eddy, the mathematical relationship between the two-dimensional geostrophic balance equations and the mesoscale eddies, the same solution between the Navier-Stokes equation of the Euler flow field and the mathematical model of mesoscale eddy, etc.

In fact, the universal nonlinear system model must meet certain conditions, that is, the strange attractor or semi-strange attractor that produces a semi-stable limit cycle is a region where any particle in the flow field tends to be consistent, and this point or region maintaining the unique energy is also that can be maintained by the minimum energy described in [43, 191].

According to the nonlinear theory, the initial point in each basin of attraction is based on its characteristic attractive movement. And the process of movement tendency is the process of generating a semi-strange attractor. For each semi-strange attractor, there is a field of attraction. This field of attraction is also the fractal basin of attraction, which is the actual mesoscale eddy.

The results of this work are not only the direct type applications of pure mathematical limit cycle theory and fractal theory in practice but also the classic combination of nonlinear dynamic systems in mathematics and physical oceanography. The universal model and experimental verification not only verify the relevant results that are obtained by Euler's form but also, more importantly, are consistent with observational numerical statistics. Due to the universality of the model, the consequences of the system are richer and more complete. From the study of the Lagrangian particle movement, it will open up original and subversive new ideas for related issues such as temperature, density, salinity, pressure, seawater velocity, etc. New methods and new technologies will also be a question of gradual research.

In addition, all the results of this work are equally applicable to vortex behaviors in large, meso-, and small scales in the ocean.

Chapter 2
Preliminaries

2.1 Limit Cycle and Semi-stable Limit Cycle

Consider the nonlinear dynamic system

$$\begin{cases} \frac{dx}{dt} = P(x, y), \\ \frac{dy}{dt} = Q(x, y), \end{cases} \tag{2.1.1}$$

where $P(x, y)$ and $Q(x, y)$ are the nonlinear functions. From [192–194], we have

Definition 2.1 Suppose system (2.1.1) has a closed trajectory C, if in the sufficiently small neighborhood of C, except for C, the trajectory is not closed, and non-closed trajectories tend to C when $t \to +\infty$, it is said that C is isolated, and the closed trajectory at this time is called a limit cycle of system (2.1.1).

It can be seen that C divides the phase plane (phase space, if it is high-dimensional) into two regions: the inner domain and the outer domain.

Definition 2.2 In C,

1. When $t \to +\infty(-\infty)$ is hovering close to C, it is said that C is internally stable (internally unstable).

S.-T. Liu et al., *Mathematical Principle and Fractal Analysis of Mesoscale Eddy*,
https://doi.org/10.1007/978-981-16-1839-0_2

2. In the outer domain of C, close to the trajectory of C, when $t \to +\infty(-\infty)$ circlingly approaches C, it is said that C is externally stable (externally unstable).

3. When $t \to +\infty(-\infty)$, the internal and external trajectories of C close to C spirally approach C, then C is said to be stable (unstable).

4. When $t \to +\infty(-\infty)$, the internal and external stability of C is opposite, then C is semi-stable.

Besides, notice that the closed trajectory C, especially the stable closed trajectory C, because the internal and external trajectories of C tend to be C so that C is attractive to other trajectories. Then **the stable limit cycle is called limit cycle attractor**.

In particular, the determination of the existence, stability, and instability of limit cycles of nonlinear dynamic systems is a very mature branch of the theory of differential dynamic systems, and there are very rich theoretical and application results [192–194].

2.2 Criterion of Semi-stable Limit Cycle

2.2.1 *Limit Cycles of Oscillatory Approach and Monotone Approach*

For a nonlinear system (2.1.1), usually limit cycles can be divided into two types, namely, odd and even. According to the stability theory of A. M. Libons, both sides of the limit cycle of odd weight are both stable or both unstable, and the limit cycle of even weight is stable on one side and unstable on the other side. So it is called semi-stable or stable with conditions.

According to [192–194], we first introduce the following content and notation. Let M be any non-singular point of system (2.1.1), (x, y) be the rectangular coordinates of M. then

$$P^2(x, y) + Q^2(x, y) \neq 0.$$

γ represents any integral curve of system (2.1.1). The integral curve past M is recorded as $\gamma(M)$. If $\gamma(M)$ is not a simple closed curve, then M divides $\gamma(M)$ into two parts. We use $\gamma^+(M)$ to represent the part of t increasing to positive infinity and $\gamma^-(M)$ to represent the part of t that decreases toward negative infinity. Set

$$\Lambda^+(\gamma) = \bigcap_{M\in\gamma} \overline{\gamma^+(M)} \quad \text{and} \quad \Lambda^-(\gamma) = \bigcap_{M\in\gamma} \overline{\gamma^-(M)},$$

where γ means along the positive direction or along the negative direction to approach limit cycle C, depending on

$$\Lambda^+(\gamma) = C \quad \text{or} \quad \Lambda^-(\gamma) = C.$$

Take $\rho(L, K)$, $\rho(L, C)$, $\rho(C_1, C_2)$ as distances between two points L and K, between one point L and one point set C and between the two point sets C_1 and C_2. Here, $\rho(L, K)$ are all set to non-negative real numbers.

$\gamma^+(M)$ or $\gamma^-(M)$ is called monotonically close to C, if the following two conditions are met:

1. $\Lambda^+[\gamma(M)] = C$ (or $\Lambda^-[\gamma(M)] = C$);
2. In $\gamma^+(M)$ (or $\gamma^-(M)$), take two points K_1, K_2. The corresponding parameters are t_1, t_2. If $t_1 < t_2$ (or $t_1 > t_2$), then there is

$$\rho(K_1, C) > \rho(K_2, C).$$

Then we introduce the following definitions.

Limit cycles of oscillatory approach: If γ satisfies the above condition (1) but does not meet the condition (2), then $\gamma^+(M)$ (or $\gamma^-(M)$) is called vibration close to C.

Limit cycles of monotone approach: If there is an open set S, $S \supset C$, for any point M in $S - C$, then one of $\gamma^+(M)$ and $\gamma^-(M)$ is monotonously close to C.

In particular, the limit cycle without the above-mentioned open set S is called oscillatory approach. Then C is called limit cycle of monotonic proximity.

2.2.2 *Criterions*

For nonlinear system (2.1.1), we review the following criterion [194] for the semi-stable limit cycle:

2.2.2.1 Necessary Conditions for the Existence of Semi-stable Limit Cycles

Theorem 2.1 *For nonlinear system (2.1.1), each of the following conditions is necessary for that limit cycle C is semi-stable:*

1. *Suppose P and Q have second-order partial derivatives. If system (2.1.1) has a monotonically close semi-stable limit cycle C, then the coordinates of any point L on C (x, y) must satisfy equation*

$$F(x, y) \equiv P^2 \frac{\partial Q}{\partial y} - PQ \left(\frac{\partial P}{\partial y} + \frac{\partial Q}{\partial x} \right) + Q^2 \frac{\partial P}{\partial x} = 0.$$

2. *Assuming the same as (1). If system (2.1.1) has a monotonically close semi-stable limit cycle C, then for any point L on C, the curvature $\Gamma(L)$ is zero.*

From Theorem 2.1, we know that for the monotonic even-weight limit cycle line C of system (2.1.1), only the trajectory of $F = 0$ needs to be studied. In particular, when P and Q are a polynomial of x and y, there are the following theorems.

Theorem 2.2 *Suppose P and Q are polynomials of x and y, then all monotonic semi-stable limit cycles of system (2.1.1) are algebraic curves, and the number of them is finite.*

In fact, the number of C is far less than N. Also, P and Q in Theorem 2.2 can also be set as algebraic functions, and the result is the same.

Theorem 2.1 can also be used to determine when a monotonically close semi-stable limit cycle does not exist.

Theorem 2.3 *Suppose there is no single closed curve in the trajectory of $F = 0$, then system (2.1.1) does not have a monotonically close semi-stable limit cycle.*

If a single closed curve appears in the trajectory of $F = 0$, we still need to know whether it is a periodic solution, and if it is a periodic solution, we also need to know its stability.

Sufficient Conditions for the Existence of Semi-stable Limit Cycle

Theorem 2.4 *Let C be a periodic solution of system (2.1.1). If there is an open set S satisfies conditions*

1. *There is no singularity in S;*
2. $S \supset C$;
3. *Take two points K_1 and K_2 from $S - C$. It always has*

$$(\sin\alpha(K_1))\,(\sin\alpha(K_2)) > 0.$$

Then C is one of the semi-stable monotonically close limit cycles of Eq. (2.1.1).

Theorem 2.5 *The condition (3) in Theorem 2.4 is changed to the following condition:*
Any two points K_1 and K_2 on the same side of C in $S - C$ always have

$$(\sin\alpha(K_1))\,(\sin\alpha(K_2)) > 0;$$

Any two points K_1 and K_2 on the opposite side of C in $S - C$ always have

$$(\sin\alpha(K_1))\,(\sin\alpha(K_2)) < 0.$$

Then C is a monotonically close limit cycle line, but not semi-stable. C is stable or unstable still according to the opposite sign or the same sign of $\sin\alpha$ *and* ρ.

Theorem 2.6 *Let C be a periodic solution of system (2.1.1). For arbitrarily small positive numbers d, there are always two points K_1 and K_2 in $\bigcup_{|\delta|\le d} C_\delta$, such that $(\sin\alpha(K_1))\,(\sin\alpha(K_2)) < 0$. Then C is a semi-stable limit cycle that is not monotonically close.*

According to [194], in order to further study the semi-stability of the limit cycle, the assume that P and Q on the right side of system (2.1.1) has any order of continuous partial derivatives we need. And there is a closed trajectory Γ, the negative direction equation are

$$x = f(t) \quad \text{and} \quad y = g(t),$$

where f and g is a periodic function of t with a period of T. The curve coordinates (s, n) are introduced into the small enough neighborhood of Γ, where s represents the arc length of Γ, measured from a certain point on Γ, clockwise is positive, that is, the increasing direction of s is the same as that of t; n represents the normal length of Γ, which is positive outwards. Let the equation of Γ take the arc length as the parameter

$$x = \varphi(s) \quad \text{and} \quad y = \psi(s).$$

For a point A near Γ, suppose it is located on the normal line Γ near B (see Fig. 2.1).

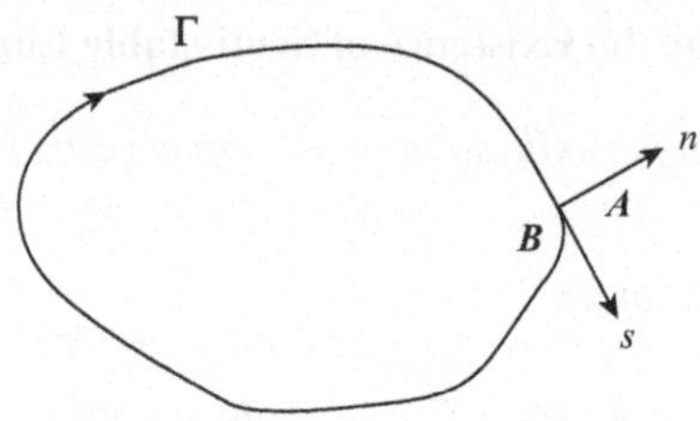

Fig. 2.1 Normal line Γ

The rectangular coordinates of point B are $(\varphi(s), \psi(s))$, so relationship between the rectangular coordinates (x, y) of A and the curve coordinates (s, n) are

$$x = \varphi(s) - n\psi'(s), \quad y = \psi(s) + n\varphi'(s), \tag{2.2.1}$$

where

$$\varphi'(s) = \left.\frac{\mathrm{d}x}{\mathrm{d}s}\right|_B = \frac{Q_0}{\sqrt{P_0^2 + Q_0^2}}, \quad \psi'(s) = \left.\frac{\mathrm{d}y}{\mathrm{d}s}\right|_B = \frac{Q_0}{\sqrt{P_0^2 + Q_0^2}},$$

P_0 and Q_0 represents the value of P and Q at point B, namely

$$P_0 = P(\varphi(s), \psi(s)) \quad \text{and} \quad Q_0 = Q(\varphi(s), \psi(s)).$$

Substituting (2.2.1) into system (2.1.1), we get

$$\begin{aligned}\frac{\mathrm{d}y}{\mathrm{d}x} &= \frac{\psi'(s) + \varphi'(s)\frac{\mathrm{d}n}{\mathrm{d}s} + n\varphi''(s)}{\varphi'(s) - \psi'(s)\frac{\mathrm{d}n}{\mathrm{d}s} - n\psi''(s)} \\ &= \frac{Q(\varphi(s) - n\psi'(s), \psi(s) + n\varphi'(s))}{P(\varphi(s) - n\psi'(s), \psi(s) + n\varphi'(s))},\end{aligned}$$

then

$$\frac{\mathrm{d}n}{\mathrm{d}s} = \frac{Q\varphi' - P\psi' - n(P\varphi'' + Q\psi'')}{P\varphi' + Q\psi'} = F(s, n). \tag{2.2.2}$$

Let

$$F_n(s, n)|_{n=0} = \frac{P_0 Q_{y0} - P_0 Q_0(P_{y0} + Q_{x0}) + Q_0 P_{x0}}{({P_0}^2 + {Q_0}^2)^{\frac{3}{2}}} = H(s),$$

where Q_{y0}, P_{y0}, Q_{x0}, P_{x0} are the partial derivatives of P and Q at $n = 0$. $H(s)$ represents the curvature of the orthogonal trajectory of system (2.1.1) at point B. Let the arc length of the closed trajectory C of system (2.1.1) be l. Take

$$\int_0^l H(s)\mathrm{d}s < 0 \quad (> 0). \tag{2.2.3}$$

Then

$$\int_0^T \left(\frac{\partial P}{\partial x} + \frac{\partial Q}{\partial y}\right)dt < 0 \quad (> 0) \tag{2.2.4}$$

and

$$\int_0^T \left(\frac{\partial P}{\partial x} + \frac{\partial Q}{\partial y}\right)dt = \int_0^l H(s)\mathrm{d}s = 0.$$

Definition 2.3 When the conditions (2.2.2) and (2.2.3) are satisfied, Γ is called a single cycle or a rough cycle, and when the condition (2.2.4) is established, Γ is called a multiple cycle or a non-rough cycle.

In order to study the stability of multiple cycles, we put (2.2.2) from $s = 0$ to $s = l$ integral, and get

$$\psi(n_0) = n(l, n_0) - n_0 - \int_0^l F(s, n(s, n_0))\mathrm{d}s,$$

where $\psi(n_0)$ is called successor function, which means the difference between the coordinates that starts point $(0, n_0)$ in the neighborhood of Γ goes in the direction of increasing t (or s) and reaches point

$$(l, n(l, n_0)) = (0, n(l, n_0)).$$

Then

$$\psi'(n_0) = 0$$

is the necessary and sufficient condition for the trajectory passing $(0, n_0)$ to be a closed trajectory;

$$n_0\psi'(n_0) < 0 \quad (> 0)$$

for everything small enough $|n_0|$ is a necessary and sufficient condition for Γ to be stable (unstable);

$$\psi'(n_0) < 0 \quad (> 0)$$

for everything small enough $|n_0|$ is a necessary and sufficient condition for Γ to be internally unstable and externally stable (internally stable and externally unstable). Since $\psi'(n_0) = 0$.

Theorem 2.7 *If*

$$\psi'(0) < 0 \quad (> 0),$$

then Γ is a stable (unstable) limit cycle; if

$$\psi'(0) = 0$$

and

$$\psi''(0) \neq 0,$$

then Γ *is a semi-stable limit cycle.*

According to the well-known theorem in differential calculus, the following theorem is generally true.

Theorem 2.8 *For the closed track* Γ, *if*

$$\psi'(0) = \psi''(0) = \cdots = \psi^{(k-1)}(0) = 0, \quad \psi^{(k)}(0) < 0 \quad (> 0),$$

where k is an odd number, then Γ *is a stable (unstable) limit cycle; if*

$$\psi'(0) = \psi''(0) = \cdots = \psi^{(k-1)}(0) = 0, \quad \psi^{(k)}(0) \neq 0,$$

where k is an even number, then Γ *is a semi-stable limit cycle.*

Theorem 2.9 *If Theorem 2.8 holds on the closed trajectory* Γ *of system (2.1.1), then* Γ *is a stable (unstable) limit cycle.*

2.3 Feature Scale, Scale-Free Domain, Fractal, Random Fractal, Dimension

Feature scale It is not appropriate to use an ordinary ruler to measure the Great Wall of China, or to measure cells. "Kill a chicken with a sledgehammer" refers to the unreasonable use of scales. This common saying contains the profound meaning that everything has its a characteristic scale.

Scale-free domain The scale is the unit of measurement. Scale-free means that the research object has nothing to do with the scale. No matter how the measurement unit changes, the nature of the research object will not. For example, phase transitions and turbulence in physics are things that have nothing to do with scale.

Since the publication of Mandelbrot's monographs [195, 196] in the 1970s, there are rich theoretical and applied researches on fractal theory and engineering technology, medicine, biology, informatics, geophysics, marine science, etc. Please see [197]–[467]. From [468]–[472] the following concepts can be obtained:

Fractal Irregular geometry with scale-free and self-similarity, whose Hausdorff dimension is generally greater than its topological dimension, is a fractal.

Random fractal A pattern with an approximate or statistically self-similar structure is called a random fractal.

Fractal dimension There are many definitions of fractal dimension, usually, we mainly use the following definition.

Definition 2.4 Let the set S be a subset of n-dimensional space, and $N(a)$ is the minimum number of n-dimensional hypercubes that cover the set S with a side length of a. Define

$$d_H = \lim_{a \to 0} \frac{\ln N(a)}{\ln 1/a},$$

which is called Hausdorff dimension.

2.4 Iterative Function System and Fractal

Set $f(x) = a + x$. Then

$$\begin{aligned} f^1(x) &= f(x) = a + x, \\ f^2(x) &= f(f(x)) = f(a + x) = 2a + x, \\ f^3(x) &= f(f^2(x)) = f(2a + x) = 3a + x. \end{aligned}$$

The general the n-th iteration is

$$f^n(x) = na + x.$$

This formula is called a iterative operation of the function. If $f^0 = I$ (unit map), then

$$\begin{cases} F_0(x) = f^0(x) = I(x) = x, \\ F_{n+1} = f(F_n(x)), \quad n = 0, 1, 2 \dots. \end{cases}$$

This formula is called a recurrence relation of the function.

From iterative operations and recurrence relations, we can discuss regular fractals, which is one of the advantages of iterative function systems. From [472], the concept of compression map is first given.

Definition 2.5 Let D be a closed subset of $\mathbb{R}^n$. Map $S : D \to D$ is called D a contraction, if for any $x, y \in D$, there is a number c satisfying $0 < c < 1$, such that

$$|S(x) - S(y)| \leq c\,|x - y|.$$

Obviously, any compression map is continuous. If the equal sign holds, that is, if $|S(x) - S(y)| = c\,|xy|$, then S turns the set into a geometric similarity set. At this time, map S is called contracting similarity.

A finite set of compressed map families $\{S_1, \dots, S_m\}$, among which $m \geq 2$, is called an iterative function system (IFS for short). The non-empty compact subset $F \subset D$ is called attractor (or invariant set) of IFS, if there is

$$F = \bigcup_{i=1}^{m} S_i(F).$$

The basic property of the iterative function system is that it determines only an attraction. It is usually fractal. For example, F is a three-point Cantor set. Suppose that S_1 and S_2 are all transformations with $\mathbb{R} \to \mathbb{R}$:

$$S_1(x) = \frac{1}{3}x \quad \text{and} \quad S_2(x) = \frac{1}{3}x + \frac{2}{3}.$$

Then $S_1(F)$ and $S_2(F)$ are exactly half the left and right of F, respectively, and $F = S_1(F) \cup S_2(F)$. Therefore, F is the attractor of the IFS composed of the compressed map $\{S_1, S_2\}$. These two maps are showing the basic self-similarity of the Cantor set.

Theorem 2.10 ([472]) *Consider the iterative function system given by $D \subset \mathbb{R}^n$ compression map family $\{S_1, S_2\}$, that is*

$$|S_i(x) - S_i(y)| \leq c_i |x - y| \quad x, y \in D.$$

For each i, $c_i < 1$ there is a unique attractor, that is, a non-empty compact set F, which satisfies

$$F = \bigcup_{i=1}^{m} S_i(F).$$

If the transformation S is defined on the non-empty compact subset D, such that for $E \in D$, there is

$$S(E) = \bigcup_{i=1}^{m} S_i(E).$$

Let S^k be k iterations of S. Then for any $E \in D$, there is

$$F = \bigcap_{k=0}^{\infty} S^k(E).$$

2.5 Dissipative System

If there is a closed bounded area $\mathcal{D}$ on the phase plane (x, y), such that any closed trajectory line of nonlinear system (2.1.1) enter $\mathcal{D}$ and stay there forever, as long as t is sufficiently large, then the system is called a dissipative system. Otherwise, it is called a non-dissipative system. From [473], we have the following judgment.

Let x', y', z' be $\mathbf{V} = (u, v, w)$. Then the divergence of velocity is

$$\text{div}\mathbf{V} = \frac{\partial u}{\partial x} + \frac{\partial v}{\partial y} + \frac{\partial w}{\partial z}.$$

When div$\mathbf{V} < 0$, system (2.1.1) is dissipative. When div$\mathbf{V} = 0$, system (2.1.1) is a conservative system.

Especially for second-order differential equation

$$x'' + f(x, x')x' + g(x) = p(t)$$

or

$$\begin{cases} x' = y, \\ y' = -g(x) - f(x, y)y + p(t), \end{cases} \tag{2.5.1}$$

where $f(x, y) \in \mathbb{C}^1(\mathbb{R}, \mathbb{R})$, $g(x) \in \mathbb{C}^1(\mathbb{R}, \mathbb{R})$ and $p(t) \in \mathbb{C}(\mathbb{R}, \mathbb{R})$. Then the dissipation of system (2.5.1) can naturally be judged by the above conclusion.

2.6 Attractor, Attracting Set, Basin of Attraction, Strange Attractor, and Semi-strange Attractor

According to [468, 474], we have the following conclusions:

Attractive The core of the mesoscale eddy is the attraction of particles in the flow field tending to a point or a closed loop. Generally, for $\forall \varepsilon > 0$, initial value t_0 and all disturbances satisfy $|y(t_0)| \leq \delta$, the disturbed motion exists with ε, t_0 and T_0 related to $y(t_0)$ makes $|y(t)| \leq \varepsilon$ when $t > T$, then the disturbed motion $x(t)$ is attractive. If T is only related to ε, it is said that the disturbing motion is uniformly attractive. If δ is arbitrarily large, it is said that the disturbing motion is globally attractive.

Attractor The motivational state of the dynamical system's trajectory after a long period of evolution may be a stable equilibrium point or a periodic trajectory.

Attract set For the invariant set A of the dynamic system flow ϕ_t in the phase space $\mathbb{R}^n$, such as satisfying the invariant set A in a certain field of A at any point x. When $t \to \infty$, $\phi_t \to A$, the set A is the attraction set.

Basin of attraction If point x_0 in the phase space $\mathbb{R}^n$ is such that when $t \to \infty$, the phase trajectory starting from x_0 tends to attract set A, then all $B(A)$ at point x_0 is called basin of attraction of the attraction set A, or called attraction domain.

Strange attractor If the cross section of an attractor is a Cantor-type set and has a sensitive dependence on initial conditions, it is called a strange attractor. Or the dynamic system is fractal, then the attractor is called a strange attractor [473].

Semi-strange attractor In the closed trajectory, the area is divided into inner and outer parts. The attractor requires that the phase "volume" near a point decreases with time. On both sides of the closed trajectory, one side is stable and attractive, and the "volume" decreases with time. The other side is unstable and not attractive, that is, attraction and non-attraction are intertwined. Such attractive features are called semi-strange attractors.

2.7 Relationship between Semi-stable Limit Cycles and Semi-strange Attractors

According to [468, 469], what is composed of points, closed curves or torus is not a fractal attractor, it is called a mediocre attractor, or a trivial attractor.

For example, the equilibrium points of nonlinear dynamic systems and stable limit cycles are all ordinary attractors. The chaotic attractors in [468, 469] are generally strange attractors.

For closed trajectory C, if it is stable since the internal and external trajectories of C tend to C, then C is attractive to other trajectories, and the stable limit cycle is called limit cycle attraction. In particular, the theory of the existence, stability, and instability of limit cycles is already a very mature theoretical branch of differential dynamical systems, and there are very rich theoretical and application results [192–194]. For a semi-stable limit cycle, it attracts on one side and diverges on the other side, and has a weak attraction, so the attraction is intertwined has fractal behaviors. We call this behavior a semi-strange attraction factor.

2.8 Elementary Reaction and Reaction Rate

Definition 2.6 ([475]) Reactions at the molecular level are called elementary reactions (or elementary processes).

Definition 2.7 The reaction rate is proportional to the power product of the reactant concentration, and the power exponent is the measurement coefficient of each reactant in the elementary reaction equation.

Considering the elementary reactions shown in Table 2.1, the reaction rate is given in [476].

Here k and r are the reaction rate factors; a and b are numbers of parts participating in the reaction.

Table 2.1 Elementary reaction and its rate

Elementary reaction	Reaction rate
$A + B \to C$	$\frac{\mathrm{d}u}{\mathrm{d}t} = k[A][B]$
$aA + bB \to \lambda G + \mu H$	$\frac{\mathrm{d}u}{\mathrm{d}t} = r[A]^a[B]^b$

2.9 Lagrangian Particle Dynamic System

First, we introduce the particle motion system in space. Suppose the velocity and time of any particle $Q(x, y, z)$ in $\mathbb{R}^3$ and the coordinates of point (x, y, z). The relationship is

$$\begin{cases} v_x = f_1(f, p(x, y, z), x, y, z, t), \\ v_y = f_2(f, p(x, y, z), x, y, z, t), \\ v_z = f_3(f, p(x, y, z), x, y, z, t), \end{cases}$$

where $f_i(f, p(x, y, z), s, T, \rho, x, y, z)$, $i = 1, 2, 3$, f is the Coriolis-force in geostrophic motion, $p(x, y, z)$ is the pressure factor s, ρ, T is the salinity, density and temperature factors carried by the particle, respectively. And they are all functions of (x, y, z). The mass point passes through point (x_0, y_0, z_0) at time t_0. Therefore, requiring the trajectory of the mass point Q is actually to solve differential equation

$$\begin{cases} \frac{\mathrm{d}x}{\mathrm{d}t} = f_1(f, p(x, y, z), x, y, z, t), \\ \frac{\mathrm{d}y}{\mathrm{d}t} = f_2(f, p(x, y, z), x, y, z, t), \\ \frac{\mathrm{d}z}{\mathrm{d}t} = f_3(f, p(x, y, z), x, y, z, t) \end{cases}$$

with initial conditions $x(t_0) = x_0$, $y(t_0) = y_0$ and $z(t_0) = z_0$.

Similarly, in the coordinate system $O - xyz$, let E be a group of fluid in the ocean. Especially at the moment of $t = 0$, the group of fluid is denoted as E_0. Take a mass point in E_0, mark it as P_0 and set the coordinates as (x_0, y_0, z_0), as shown in Fig. 2.2.

Note that $P_0 = P_0(x_0, y_0, z_0, t)|_{t=0} = P_0(x_0, y_0, z_0, 0)$, then at any time thereafter, as long as the functional relationship φ of the movement of P_0 is given, then the particle P_0 moves to P. Its Lagrangian form of the particle coordinates of $P(x, y, z, t)$ showing in Fig. 2.3 completely certain, that is

$$\begin{cases} x = x(x_0, y_0, z_0, t), \\ y = y(x_0, y_0, z_0, t), \\ z = z(x_0, y_0, z_0, t). \end{cases} \tag{2.9.1}$$

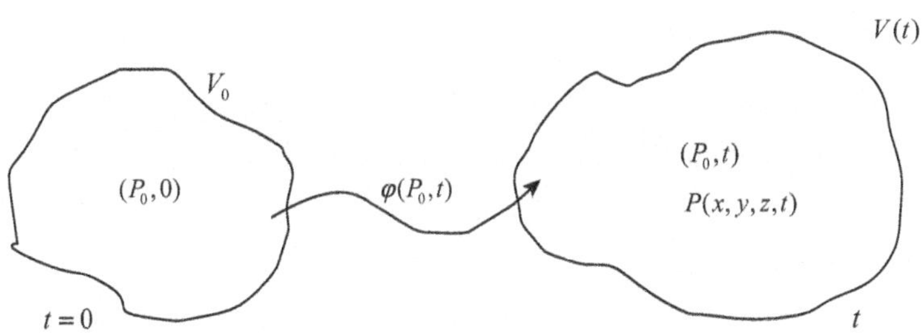

Fig. 2.2 Water particle movement

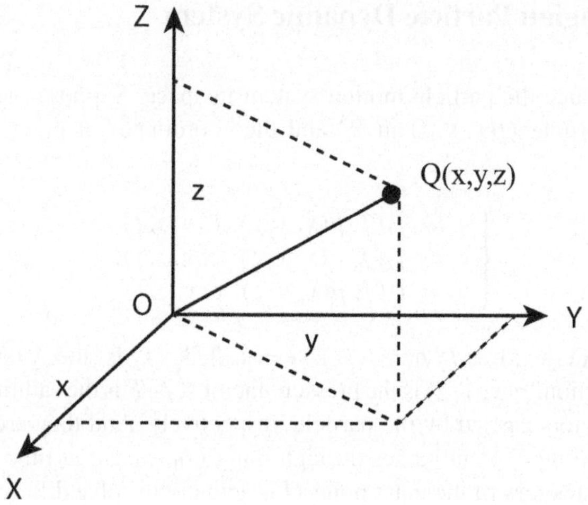

Fig. 2.3 Particle coordinates in Lagrangian form

Assuming that the water mass E is a mesoscale eddy, and any water particle in the mesoscale eddy is recorded as q, that is, $q \in E$, then system (2.9.1) is the mathematical representation of the movement of any water particle in E.

Chapter 3
Universal Mathematical Model of Mesoscale Eddy

This work takes mesoscale eddies as the research object. In fact, the following universal mathematical models are established, including disk eddies and columnar eddies, as well as any other eddies with practical approximate diameter meaning scales [477].

3.1 Mesoscale Eddy

Because of the perturbation of two or more ocean currents, monsoons, and other physical noises, with Coriolis-force, the spiral motion formed by collision with each other is called mesoscale eddy. In particular, the diameter of mesoscale eddies is about 100–300 km, and its life span is 2–10 months. It moves like a typhoon and has great kinetic energy. According to estimates, the kinetic energy of these mesoscale eddies occupies more than 80%–90% of the flow energy of large and medium oceans in the entire ocean. And the maximum value of eddy kinetic energy is not in the center, but in the area with the highest linear velocity of water body. Mesoscale eddies are widely distributed around the world and can have a significant impact on the atmosphere.

1. Mesoscale eddies will change the original water movement through the sea area, causing the direction of the ocean current to vary widely. The flow velocity will increase several to tens of times, and be accompanied by strong vertical movement of the water body. The eddy center has the largest potential energy. And the farther away from the center the smaller the potential energy.
2. The sea surface temperature anomaly caused by the mesoscale eddy changes the turbulent heat flux to cause anomalies in ocean surface wind speed, divergence, cloud cover, and precipitation, and produce abnormal secondary circulations in the vertical direction.

S.-T. Liu et al., *Mathematical Principle and Fractal Analysis of Mesoscale Eddy*,
https://doi.org/10.1007/978-981-16-1839-0_3

3. In the South China Sea, the Kuroshio Extension, and the Southern Ocean, mesoscale eddies can affect the surface wind speed on them by changing the sea surface pressure or the stability of the atmospheric boundary layer, respectively.
4. Mesoscale eddies can change the energy conversion in the atmosphere to affect the storm path and jet stream position, and the weather pattern in the downstream area through teleconnection. In addition, the upper ocean temperature changes caused by the mesoscale eddy will also affect tropical cyclones. Its enhancement and maintenance play an important role.

The mesoscale eddy is the smallest basic scale dominated by the rotational geostrophic characteristics in rotating fluid mechanics. Therefore, strengthening the research on the mesoscale eddy and scientifically controlling its motion mechanism has profound theoretical significance and wide application value. It can highlight the characteristic properties of a series of rotating fluids with geostrophic as the main body, which is the top priority of physical oceanography research.

3.2 Mathematical Model of Mesoscale Eddy

Suppose $E \in \mathbb{R}^3$ is a three-dimensional mesoscale eddy. To establish a universal mathematical model of mesoscale eddies, it is necessary to find the formation of mesoscale eddies' basic element, that is, the generator of mesoscale eddies.

To this end, we first observe the change in the radius of a mesoscale eddy from its beginning to the extinction in Mesoscale Eddies in Altimeter Observations of SSH. Take the mesoscale eddy data in Table 3.1 as an example. From the beginning to the extinction, there are 28 radius data points. The changing trend is shown in Fig. 3.1

It can be seen that the radius of the mesoscale eddy is an important characteristic feature of the motion behavior of the mesoscale eddy. The radius is the distance from the approximate circumference C of the maximum edge of the mesoscale eddy to its center.

Table 3.1 Radius observations from the mesoscale eddy database

Time	Radius	Time	Radius	Time	Radius	Time	Radius
1	146.74	8	128.29	15	115.05	22	85.37
2	138.67	9	128.29	16	114.05	23	89.28
3	137.84	10	130.93	17	112.01	24	79.85
4	136.15	11	118.04	18	110.97	25	75.43
5	136.12	12	125.54	19	100.16	26	62.19
6	132.70	13	124.63	20	99.00	27	65.76
7	130.08	14	121.82	21	88.00	28	45.24

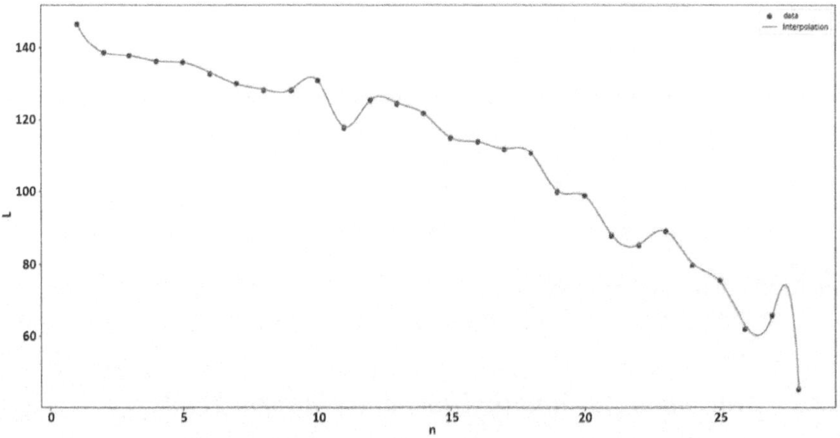

Fig. 3.1 The radius extinction curve of the mesoscale eddy database

For the maximum approximate circle C of the mesoscale eddy, it is usually the change function of time t. Let $C(t)$ be an approximate ellipse. So for the change of t at any time, it can be expressed as

$$C(t) : a(t)x^2 + b(t)y^2 \approx R^2(t), \tag{3.2.1}$$

where $R(t)$ is the radius. In particular, the approximate ellipse satisfying (3.2.1) is called a **stochastic ellipse** that is the basic element of the mesoscale eddy we are looking for. For any t, we have

$$\bar{C}(t) : a(t)x^2 + b(t)y^2 = R^2(t).$$

Then $\bar{C}(t)$ is a **standard ellipse**.

In the coordinate system $O - xy$, analyze the radius data of the set of mesoscale eddies and obtain the changes in the radius. This reflects the overall change of the stochastic ellipse, and the size of the overall change is random quantified. The coordinates of ellipse (x, y) are closely related, which causes the overall change of the stochastic ellipse to be related to the **change rate** of the coordinates, that is, it is a proportional relationship in quantity.

In order to obtain the more prominent characteristics of the mesoscale eddy, we further decompose the mesoscale eddy. Roughly speaking, the mesoscale eddy is a disk-shaped geometric body that continuously rotates in the ocean, usually called disk eddy and columnar eddy [477]. Whether it is a disk eddy or a columnar eddy, it is a moving body in the ocean. It can be seen from Sect. 3.1 that the physical characteristics of the mesoscale eddy have the following characteristics.

3.2.1 Bounded Motion

Note that mesoscale eddies are generally disk eddies. For a specific mesoscale disk eddy E, it is regionally bounded. This is because C is taken as the disk eddy E on the sea surface. Let $U(C)$ be set to the sea area enclosed by C. Then

$$|U(C) \cup C| \leq M,$$

where M is a positive number.

3.2.2 Movement Asymptotic Unity and Uniform Tendency

For the mesoscale eddy E, set $p(x, y, z) \in E$. Then the motion behavior of p is an asymptotically unified and consistent motion form that tends to rotate. Note that point p is the mass point on the mesoscale eddy. According to this arbitrariness feature, we know that the motion of any point on the mesoscale eddy conforms to the uniform characteristics. This determines its motion behavior is regular. See Fig. 3.2.

The above two conditions are enough to determine the existence of its universal mathematical model.

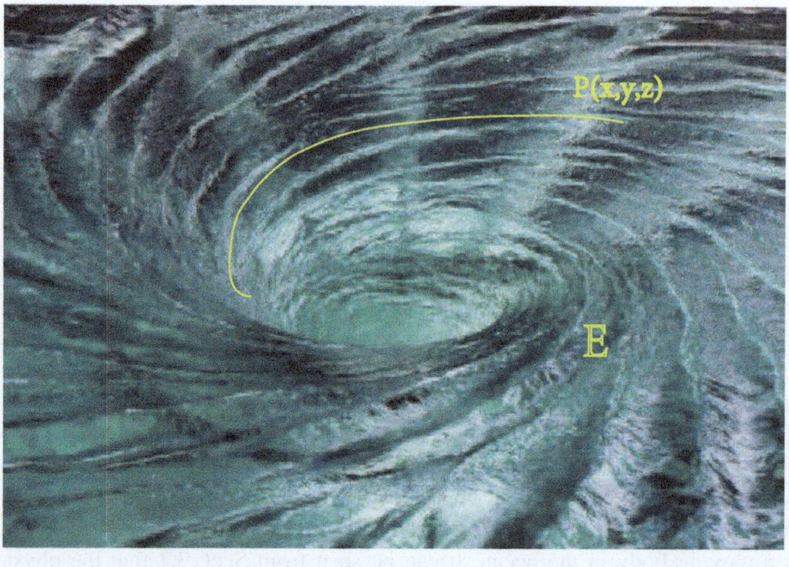

Fig. 3.2 The boundedness of the mesoscale eddy and the asymptotic uniformity of the swirling motion of any water particle in the mesoscale eddy

3.3 Universal Mathematical Model of Mesoscale Eddy

First, we discuss the quantitative characteristics of mesoscale eddies from different physical properties.

3.3.1 *Momentum of a Stochastic Ellipse*

On the sea surface, first, consider the ellipse circle C randomly enclosed by an approximately uniform and smooth unit matter curve. C as a whole makes random motions on the sea surface.

1. On the sea surface, establish a rectangular coordinate system $O - xy$. Since the circle of ellipse C is in the coordinate system, C is close to the coordinates (x, y) of the coordinate axes Ox and Oy, that is, stochastic ellipse C and the coordinates (x, y) are a balanced relationship bound together:

$$(x, y) \leftrightarrow C.$$

C can be expressed as a function of coordinates, $C : ax^2 + by^2 \approx R^2$. Since it is a random change of t over time on the sea surface, it is generally expressed as

$$a(t)x^2 + b(t)y^2 - R^2(t) \approx 0$$

or

$$a(t)x^2 + b(t)y^2 \approx R^2(t).$$

Due to the change of time t, C is a stochastic ellipse. Without loss of generality, we denote it as

$$C(t) \cong a(t)x^2 + b(t)y^2.$$

This can be regarded as the speed of the change of stochastic ellipse C as a whole. Let us suppose that $v(t) = C(t)$. From Sect. 3.2.1, we have

$$|C(t)| \leq M. \tag{3.3.1}$$

The variation of the numerical difference is bounded, so for the momentum of unit mass $m = 1$, stochastic ellipse C is

$$mv(t) = a(t)x^2 + b(t)y^2 - R^2(t). \tag{3.3.2}$$

Noting that Eqs. (3.3.1) and (3.3.2) reflects the numerical difference of the stochastic ellipse of the sea surface. Due to the constant change of the time t, there are very few cases of the standard ellipse, usually irregular stochastic ellipse. In order to ensure that the edge point of a certain maximum diameter arc segment and a certain minimum diameter arc segment of the mesoscale eddy, stochastic ellipse cannot be missing, that is, the arc segment with large deviation is also stored with small deviation. So, we consider the power of formula (3.3.2) as

$$(a(t)x^2 + b(t)y^2 - R^2(t))^m, \tag{3.3.3}$$

where $m \geq 0$ is an integer, and formula (3.3.3) is called generalized stochastic ellipse.

2. In particular, the difference between points on a certain arc $a(t)x^2 + b(t)y^2 - R^2(t) > 0$, it should be understood that the arc segment, where the point on the stochastic ellipse is located is convex outward on the arc segment. For formula (3.3.3), it is also understood that the arc where the point on the stochastic ellipse is located on the arc is convex outward. And, when $m = 2$, if the difference of (3.3.3) is greater than 1, the power is greater; if the difference is less than 1 but greater than zero, the power is smaller. This ensures that the value of the arc segment farthest from the eddy center and the value of the nearest arc segment on the stochastic ellipse. All can be contained in stochastic ellipses.
3. The difference $a(t)x^2 + b(t)y^2 - R^2(t) < 0$ should be understood that the arc of the point on the stochastic ellipse is concave inward. Especially for formula (3.3.3), we use

$$\begin{cases} (a(t)x^2 + b(t)y^2 - R^2(t))^m, & m \text{ is odd}, \\ -(a(t)x^2 + b(t)y^2 - R^2(t))^m, & m \text{ is even}. \end{cases}$$

It still shows that the arc of the point on the stochastic ellipse is concave inward.
4. Similarly, consider the momentum of the overall change motion of the unit mass material straight lines Ox and Oy formed by the coordinates x and y on the coordinate axes Ox and Oy. The momentum of the coordinate x and y of the unit mass material straight line Ox and Oy on the coordinate axes are $\frac{dx}{dt}$ and $\frac{dy}{dt}$. In particular, since the rate of change is positive, the square root of the momentum is taken as

$$\frac{dx}{dt} \text{ and } \frac{dy}{dt}. \tag{3.3.4}$$

5. In coordinate system $O - xy$, the mesoscale eddy has a relatively balanced motion period, that is, the momentum of the unit mass stochastic ellipse C and the momentum of the unit mass straight lines Ox and Oy. The two are in a relatively balanced state in the same coordinate system $O - xy$, that is, the components of their respective coordinates are relatively balanced. Therefore, combining formulas (3.3.3) and (3.3.4), according to the law of conservation of momentum and the regulation of momentum conservation factors $\alpha_1(t)$ and $\beta_1(t)$, we get

$$\begin{cases} \left[\frac{\mathrm{d}x}{\mathrm{d}t}\right]^{\frac{1}{2}} = \alpha_1(t)(a(t)x^2 + b(t)y^2 - R^2(t))^m, \\ \left[\frac{\mathrm{d}y}{\mathrm{d}t}\right]^{\frac{1}{2}} = \beta_1(t)(a(t)x^2 + b(t)y^2 - R^2(t))^m. \end{cases} \tag{3.3.5}$$

6. In a similar physical sense, consider the momentum of the unit material lines Ox and Oy, the unit mass material line on the coordinate axes, as well as the momentum of the coordinates x and y of straight lines Ox and Oy. From formula (3.3.4), it is still in a relatively balanced state. From the law of conservation of momentum and the regulation of momentum conservation factors $\alpha_2(t)$ and $\beta_2(t)$, we get

$$\begin{cases} \left[\frac{\mathrm{d}x}{\mathrm{d}t}\right]^{\frac{1}{2}} = \alpha_2(t)x, \\ \left[\frac{\mathrm{d}y}{\mathrm{d}t}\right]^{\frac{1}{2}} = \beta_2(t)y. \end{cases} \tag{3.3.6}$$

3.3.2 Elementary Reaction Rate

It can be seen from Sect. 3.1 that in the coordinate system $O-xy$, the generation of mesoscale eddy E is related to the coordinate (x, y) and generalized stochastic ellipse (3.3.3) is closely related. In particular, the change (or reaction) rate of the movement (or reaction) of stochastic ellipse C of the largest outer the eddy edge E can be regarded as a primitive reaction of coordinates (x, y) and generalized stochastic ellipse. Since the reaction rate is proportional to the product of the power of the reactant concentration, the reaction rate is the product of the coordinates (x, y) and the generalized stochastic ellipse C as the power of the concentration of the reactant. Thus, we have

$$\begin{cases} \frac{\mathrm{d}x}{\mathrm{d}t} = \alpha(t)x(a(t)x^2 + b(t)y^2 - R^2(t))^m, \\ \frac{\mathrm{d}y}{\mathrm{d}t} = \beta(t)y(a(t)x^2 + b(t)y^2 - R^2(t))^m. \end{cases} \tag{3.3.7}$$

According to the elementary reaction kinetics, where $\alpha(t)$ and $\beta(t)$ are the reaction factors, m is the number of reaction molecules in the reactant. In this reactant, the number of reactions at the coordinates x and y are both 1, and number of reactions of the stochastic ellipse is m.

Remark 3.1 In system (3.3.7) particularly, $\alpha(t)$ and $\beta(t)$ are closely related to the temperature, density, salinity, flow velocity, water pressure, and other indicators of seawater. This impact is also a very meaningful research topic, which will be carried out in the future.

3.3.3 Basic Mathematical Model of Mesoscale Eddy

From Sects. 3.3.1 and 3.3.2, it can be seen that regardless of the momentum of the stochastic ellipse on the sea surface or the reaction rate of the elementary reaction, the two sides are actually the same. Since the mesoscale eddy is in relative equilibrium in coordinate system $O - xy$, Eqs. (3.3.5) and (3.3.6) are equal at each time. Therefore, multiply the two sides of the Eqs. (3.3.5) and (3.3.6) to obtain

$$\begin{cases} \frac{dx}{dt} = \alpha_1(t)\alpha_2(t)x(x^2 + b(t)y^2 - R^2(t))^m, \\ \frac{dy}{dt} = \beta_1(t)\beta_2(t)y(x^2 + b(t)y^2 - R^2(t))^m. \end{cases} \tag{3.3.8}$$

Set $\alpha(t) = \alpha_1(t)\alpha_2(t)$ and $\beta(t) = \beta_1(t)\beta_2(t)$. Then system (3.3.8) becomes system (3.3.7).

In addition, we pay attention to the rate of elementary reaction $A + B \to C$ in Table (2.1), in which one is a stochastic ellipse and the other is variable quantities x and y on the axis Ox and Oy. Therefore, from the calculation methods of reaction rate of elementary reaction kinetics in [475, 476], we also get system (3.3.7), that is, the results of the two methods are consistent.

3.3.4 Universal Mathematical Model of Mesoscale Eddy

Excluding the physical meaning and the meaning of elementary reaction dynamics in Sects. 3.3.1 and 3.3.2, we can obtain a universal nonlinear dynamic system model of mesoscale eddies.

In the following, the definition of mesoscale eddy is introduced from a purely mathematical point of view. In coordinate system $O - xy$, the generation of mesoscale eddy E is related to coordinates (x, y) and generalized stochastic ellipse (3.3.3) closely related. According to the conclusions discussed in this chapter, system (3.3.7) we have obtained is the most basic Lagrangian particle dynamics mathematical model for mesoscale eddies.

Considering the external perturbation or noise $p(x, y)$ and $q(x, y)$, nonlinear system (3.3.7) becomes

$$\begin{cases} \frac{dx}{dt} = \alpha(t)x(x^2 + b(t)y^2 - R^2(t))^m + \omega p(x, y), \\ \frac{dy}{dt} = \beta(t)y(x^2 + b(t)y^2 - R^2(t))^m + \omega q(x, y), \end{cases} \tag{3.3.9}$$

where ω is the perturbation coefficient related to the Coriolis-force, usually taken as $\omega = \omega_0 f$, ω_0 is the perturbation factor, and f is the Coriolis-force. The physical meaning of the noise functions $p(x, y)$ and $q(x, y)$ comes from various disturbances of the objective environment, such as atmospheric circulation, seawater temperature, density, salinity, etc. In short, they are two very complex nonlinear functions, or one of the solution of nonlinear partial differential equation:

$$F\left(\frac{\partial u}{\partial x}, \frac{\partial u}{\partial y}, \cdots, \frac{\partial^{\nu} u}{\partial y^{\nu}}\right) = f(a, u(x, y)).$$

The noise terms $p(x, y)$ and $q(x, y)$ are the nonlinear perturbation functions of the two variables x and y. From the Weierstrass polynomial approximation theorem, any function can be derived from a polynomial approximation. So might as well $p(x, y)$ and $q(x, y)$ are polynomials with variable coefficients of x and y, namely,

$$\begin{cases} p(x, y) = \sum\limits_{i=0}^{r} a_i(t)x^i + \sum\limits_{j=0}^{s} b_j(t)y^j + \sum\limits_{k=0}^{u}\sum\limits_{l=0}^{v} c_{ij}(t)x^k y^l, \\ q(x, y) = \sum\limits_{i'=0}^{r'} a'_{i'}(t)x^{i'} + \sum\limits_{j'=0}^{s} b'_{j'}(t)y^{j'} + \sum\limits_{k'=0}^{u'}\sum\limits_{l'=0}^{v'} c'_{i'j'}(t)x^{k'} y^{l'}. \end{cases} \tag{3.3.10}$$

Thus, system (3.3.9) becomes

$$\begin{cases} \frac{dx}{dt} = \alpha(t)x(a(t)x^2 + b(t)y^2 - R^2(t))^m \\ \qquad +\omega\left(\sum\limits_{i=0}^{r} a_i(t)x^i + \sum\limits_{j=0}^{s} b_j(t)y^j + \sum\limits_{k=0}^{u}\sum\limits_{l=0}^{v} c_{ij}(t)x^k y^l\right), \\ \frac{dy}{dt} = \beta(t)y(a(t)x^2 + b(t)y^2 - R^2(t))^m \\ \qquad +\omega\left(\sum\limits_{i'=0}^{r'} a'_{i'}(t)x^{i'} + \sum\limits_{j'=0}^{s} b'_{j'}(t)y^{j'} + \sum\limits_{k'=0}^{u'}\sum\limits_{l'=0}^{v'} c'_{i'j'}(t)x^{k'} y^{l'}\right). \end{cases} \tag{3.3.11}$$

This is the universal nonlinear dynamic system of mesoscale eddies we obtained. It is noted that the motion of ocean surface fluids is mainly disturbed by Coriolis-force and pressure gradient force on horizontal circumferential eddies. In particular, in system (3.3.10), for the convenience of studying the problem, we take general terms $a'_1 = -\omega(t)\mu(t)x, b'_1 = \omega(t)\lambda(t)y$ and the other terms are zero, where $\lambda(t)$ and $\mu(t)$ are quantities associated with the geometric variations of the surface of mesoscale eddies. Then system (3.3.10) is turned into

$$\begin{cases} p(x, y) = \omega(t)\lambda(t)y, \\ q(x, y) = -\omega(t)\mu(t)x; \end{cases}$$

system (3.3.11) is turned into

$$\begin{cases} \frac{dx}{dt} = \alpha(t)x(a(t)x^2 + b(t)y^2 - R^2(t))^m + \omega(t)\lambda(t)y, \\ \frac{dy}{dt} = \beta(t)y(a(t)x^2 + b(t)y^2 - R^2(t))^m - \omega(t)\mu(t)x. \end{cases} \tag{3.3.12}$$

In practice, the center of a stochastic ellipse is not necessarily at the origin. In order to make the problem universal, the coordinate of the center of ellipse C is set as (o_x, o_y), so that ellipse C is transformed into

$$C_0 : a(t)(x - o_x)^2 + b(t)(y - o_y)^2 = R(t)^2.$$

Thus, system (3.3.12) develop into

$$\begin{cases} \frac{dx}{dt} = \alpha(t)(x-o_x)(a(t)(x-o_x)^2+b(t)(y-o_y)^2-R^2(t))^m \\ \quad +\omega(t)\lambda(t)(y-o_y), \\ \frac{dy}{dt} = \beta(t)(y-o_y)(a(t)(x-o_x)^2+b(t)(y-o_y)^2-R^2(t))^m \\ \quad -\omega(t)\mu(t)(x-o_x). \end{cases} \tag{3.3.13}$$

It can be seen that system (3.3.13) is a nonlinear ordinary differential dynamic system with a coefficient of variation. Parameters $\alpha(t), \beta(t), R(t), a(t), b(t), \lambda(t), \mu(t)$, and m in system (3.3.13) are all derived from the data fitting in the SSH database on mesoscale eddies.

It is noted that the motion of mesoscale eddies is complex. We only build its plane structure model as the main body. If we consider its vertical structure, its eddy axis is still a function of x, y, z, and other parameters. Because the stochastic ellipses of different depths of mesoscale eddies are different, we use the semi-stability characteristics of the limit cycle. The change of the eddy axis is related to the Coriolis-force (about 10^{-6} m/s) and the variable z. So take the eddy axis as $z=\varepsilon t$, where $\varepsilon \approx \varepsilon(f)$ and ε is a function of Coriolis-force. Then system (3.3.13) becomes

$$\begin{cases} \frac{dx}{dt} = \alpha(t)(x-o_x)(a(t)(x-o_x)^2+b(t)(y-o_y)^2-R^2(t))^m \\ \quad +\omega(t)\lambda(t)(y-o_y), \\ \frac{dy}{dt} = \beta(t)(y-o_y)(a(t)(x-o_x)^2+b(t)(y-o_y)^2-R^2(t))^m \\ \quad -\omega(t)\mu(t)(x-o_x), \\ \frac{dz}{dt} = \varepsilon(f). \end{cases} \tag{3.3.14}$$

Without loss of generality, we choose a system that is convenient for studying problems, that is, all time-varying parameters and time-varying coefficients $\alpha(t)$, $\beta(t)$, $R(t), a(t), b(t), \lambda(t)$ and $\mu(t)$ are recorded as constants $\alpha, \beta, R, a, b, \lambda$, and μ. Then from system (3.3.14), there is

$$\begin{cases} \frac{dx}{dt} = \alpha(x-o_x)(a(x-o_x)^2+b(y-o_y)^2-R^2)^m+\omega\lambda(y-o_y), \\ \frac{dy}{dt} = \beta(y-o_y)(a(x-o_x)^2+b(y-o_y)^2-R^2)^m-\omega\mu(x-o_x), \\ \frac{dz}{dt} = \varepsilon(f). \end{cases} \tag{3.3.15}$$

Furthermore, for the original center of the mesoscale eddy dynamic system (3.3.15) on the sea surface, make a translation transformation of the center, then we get

$$\begin{cases} \frac{dx}{dt} = \alpha x(ax^2+by^2-R^2)^m+\omega\lambda y, \\ \frac{dy}{dt} = \beta y(ax^2+by^2-R^2)^m-\omega\mu x, \\ \frac{dz}{dt} = \varepsilon(f). \end{cases} \tag{3.3.16}$$

Note that system (3.3.16) does not have general coefficients, but it can solve the mainframe and main structure problems. Although system (3.3.14) is general, it is difficult to handle in practice. The possible situation is only what we are doing under numerical simulation at present.

Remark 3.2 The use of system (3.3.16) is inconsistent with the actual situation, but after all, it gives an analytical mathematical model of the mesoscale eddy. Thus it makes up for the gap in this field.

Remark 3.3 The error exists because the mesoscale eddy is 100–200 nautical miles. But the error hardly affects our identification and related predictions.

Chapter 4
Semi-stable Limit Cycle in Mathematical Model of Mesoscale Eddy

System (3.3.16) is the most basic mathematical model to control mesoscale eddies based on Lagrangian particle dynamics. Since the mesoscale eddy is closely related to a stochastic ellipse, its closedness is related to the closed curve on a plane. For a nonlinear system, this closed curve is a limit cycle. So, we first analyze the limit cycle of nonlinear dynamic system (3.3.16). In mathematics, the limit cycle problem has rich theories. Here, we are concerned about:

1. The criteria for the existence of various types of limit cycles;
2. Existence types of limit cycles.

For 1, a series of specific results have been given in Sects. 2.1 and 2.2. For 2, we have the following results on the existing types of limit cycles.

4.1 Analysis of Parameter Distribution of Stable and Unstable Limit Cycles

In order to obtain different types of limit cycles of nonlinear system (3.3.16), it is necessary to adjust α, β, m, a, b, R, λ, μ, ω, and other parameters in detail. From **shooting method**, **screening method** [478, 479], and differential equation stability theory [192, 193], we have the following results:

S.-T. Liu et al., *Mathematical Principle and Fractal Analysis of Mesoscale Eddy*,
https://doi.org/10.1007/978-981-16-1839-0_4

4.1.1 α and β are Positive and m is Odd

Set $m = 3, 5, 7, 9, 11$ and $m \in [12, 13)$. See Figs. 4.1, 4.2, 4.3, 4.4, 4.5, 4.6.

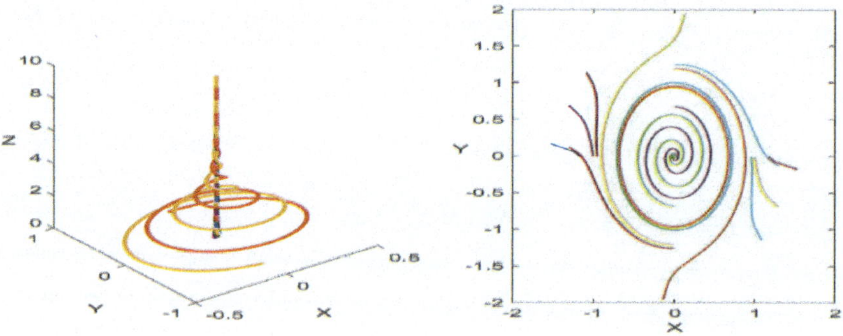

(a) $\alpha = 1.01,\ \beta = 1.01,\ a = 2,\ b = 1,\ R = 1,\ \lambda = 2\sqrt{2},\ \mu = 4\sqrt{2}$

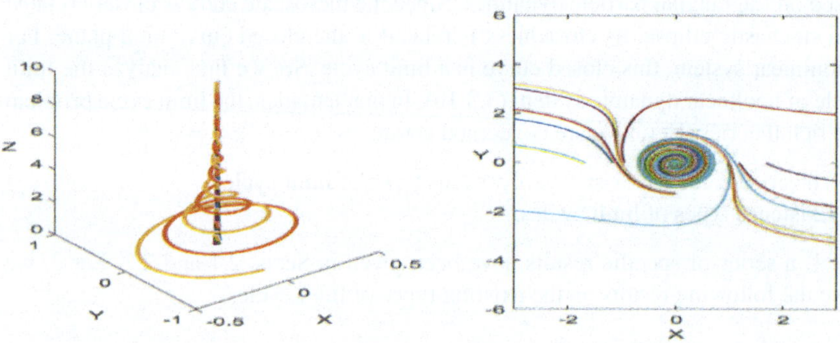

(b) $\alpha = 1.2,\ \beta = 0.1,\ a = 2,\ b = 1,\ R = 1,\ \lambda = 2\sqrt{2},\ \mu = 4\sqrt{2}$

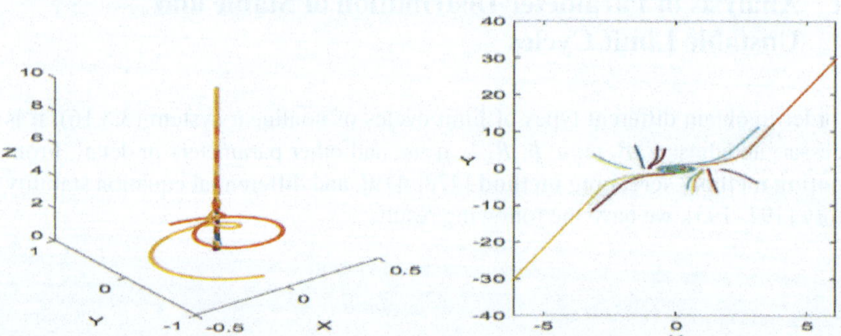

(c) $\alpha = 0.8,\ \beta = 2.1,\ a = 2,\ b = 1,\ R = 1,\ \lambda = 2\sqrt{2},\ \mu = 4\sqrt{2}$

Fig. 4.1 $m = 3$

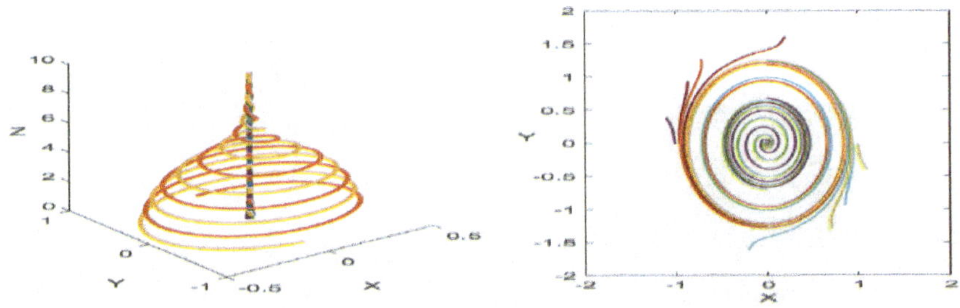

(a) $\alpha = 0.8,\ \beta = 1.1,\ a = 2,\ b = 1,\ R = 1,\ m = 5,\ \lambda = 2\sqrt{2},\ \mu = 4\sqrt{2}$

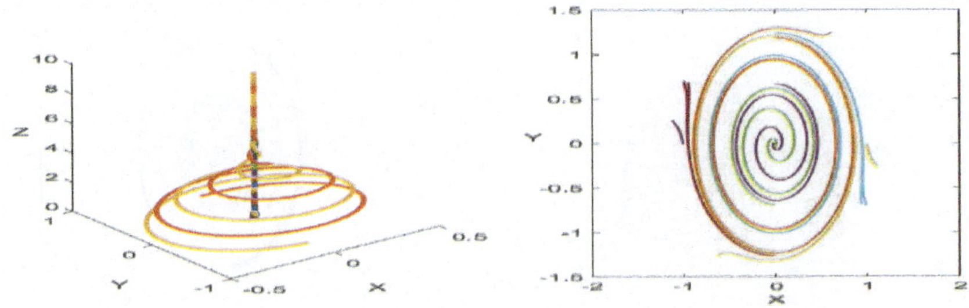

(b) $\alpha = 1.2,\ \beta = 1.9,\ a = 2,\ b = 1,\ R = 1,\ m = 5,\ \lambda = 2\sqrt{2},\ \mu = 4\sqrt{2}$

Fig. 4.2 $m = 5$

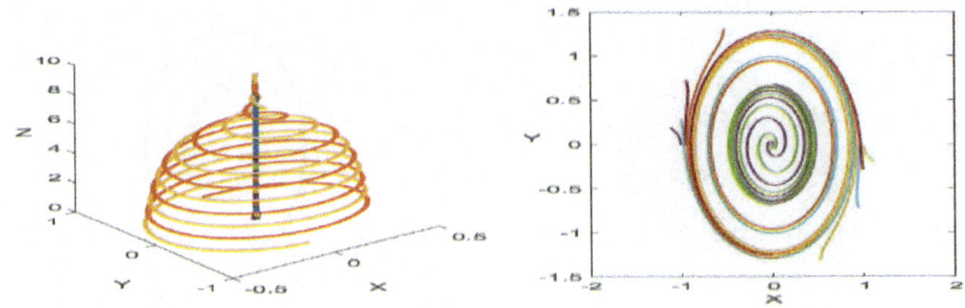

(a) $\alpha = 1.2,\ \beta = 1.9,\ a = 2,\ b = 1,\ R = 1,\ m = 7,\ \lambda = 2\sqrt{2},\ \mu = 4\sqrt{2}$

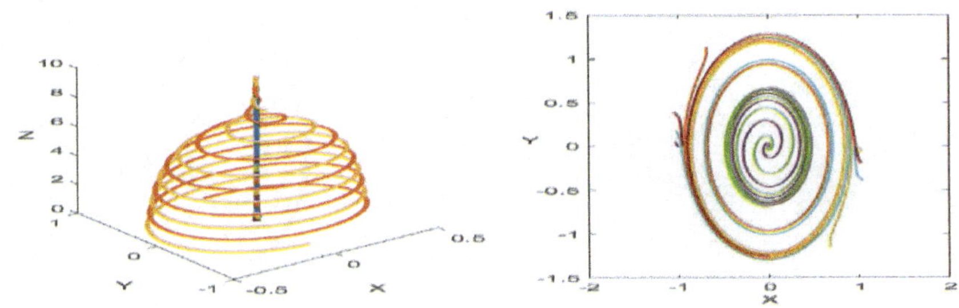

(b) $\alpha = 2.2,\ \beta = 0.9,\ a = 2,\ b = 1,\ R = 1,\ m = 7,\ \lambda = 2\sqrt{2},\ \mu = 4\sqrt{2}$

Fig. 4.3 $m = 7$

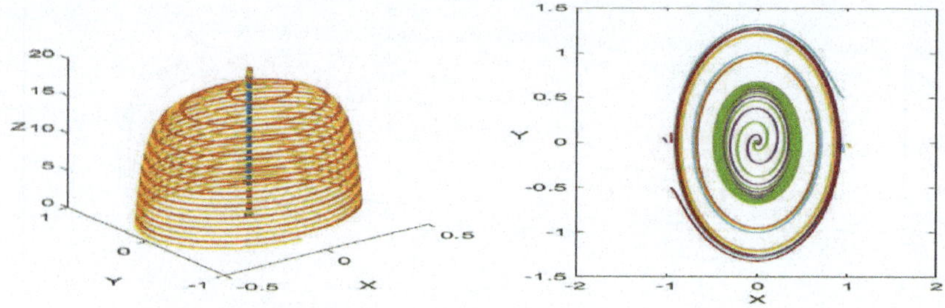

(a) $\alpha = 2.2,\ \beta = 0.9,\ a = 2,\ b = 1,\ R = 1,\ m = 9,\ \lambda = 2\sqrt{2},\ \mu = 4\sqrt{2}$

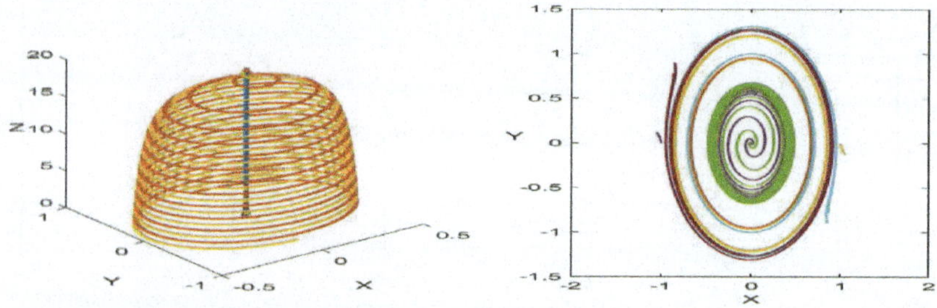

(b) $\alpha = 1.2,\ \beta = 1.7,\ a = 2,\ b = 1,\ R = 1,\ m = 9,\ \lambda = 2\sqrt{2},\ \mu = 4\sqrt{2}$

Fig. 4.4 $m = 9$

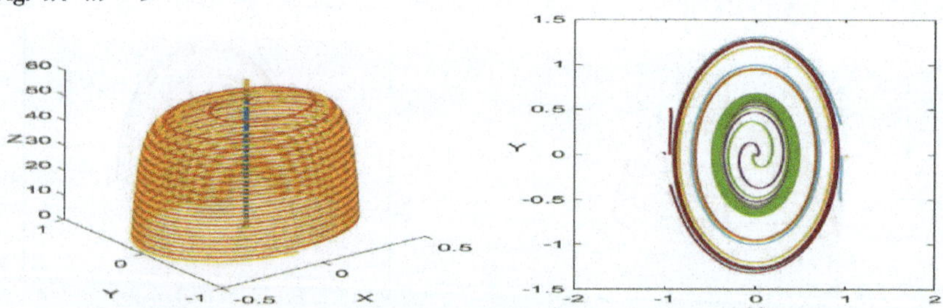

(a) $\alpha = 1.2,\ \beta = 1.7,\ a = 2,\ b = 1,\ R = 1,\ m = 11,\ \lambda = \sqrt{2},\ \mu = \sqrt{2}$

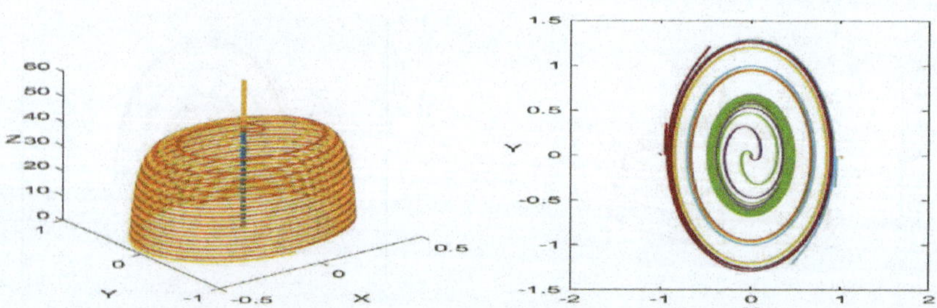

(b) $\alpha = 1.1,\ \beta = 2.7,\ a = 2,\ b = 1,\ R = 1,\ m = 11,\ \lambda = \sqrt{2},\ \mu = 2\sqrt{2}$

Fig. 4.5 $m = 11$

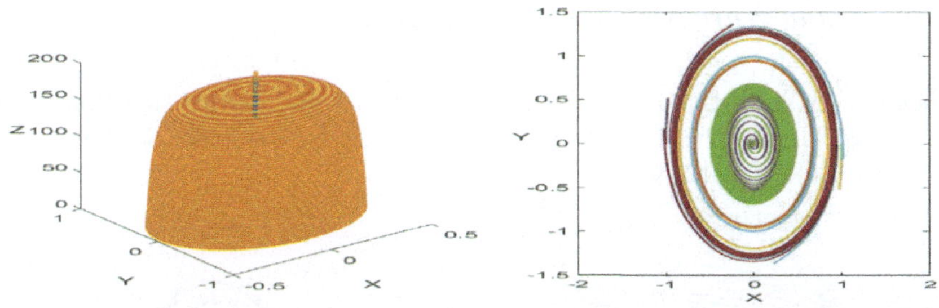

(c) $\alpha=0.1,\ \beta=0.7,\ a=2,\ b=1,\ R=1,\ m=11,\ \lambda=\sqrt{2},\ \mu=2\sqrt{2}$

Fig. 4.5 (continued)

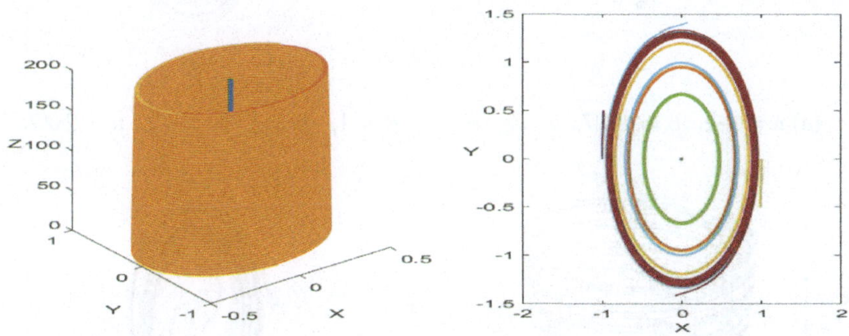

(a) $\alpha=0.1,\ \beta=0.7,\ a=2,\ b=1,\ R=1,\ m=13,\ \lambda=\sqrt{2},\ \mu=2\sqrt{2}$

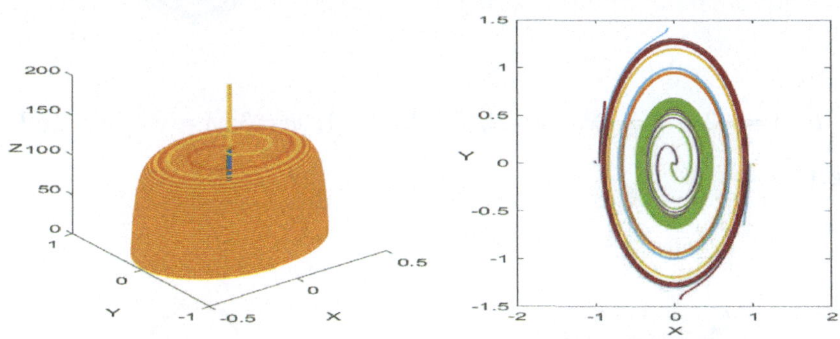

(b) $\alpha=1.1,\ \beta=2.7,\ a=2,\ b=1,\ R=1,\ m=13,\ \lambda=\sqrt{2},\ \mu=2\sqrt{2}$

Fig. 4.6 $m = 13$

It can be seen from the above that m is an odd number, α and β are positive, the limit cycle is internally unstable converging toward the center and externally unstable diverging outward. Therefore, the system has an unstable limit cycle, where m is less than 13.

4.1.2 α and β are Negative and m is Odd

When $m = 3, 9, 11, 15, 17$ and $m \in (18, 21)$, internal and external streamlines of the limit cycle tend to be closed-loop lines, so the limit cycle is stable. See Figs. 4.7, 4.8, 4.9, 4.10, 4.11, 4.12, 4.13.

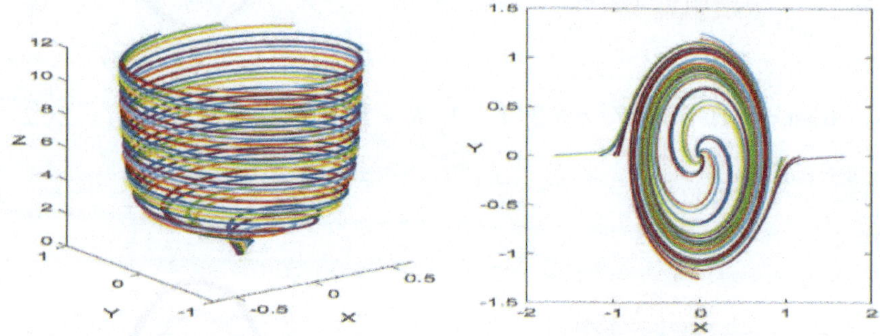

(a) $\alpha = -2,\ \beta = -1.7,\ a = 2,\ b = 1,\ R = 1,\ m = 3,\ \lambda = \sqrt{2},\ \mu = 2\sqrt{2}$

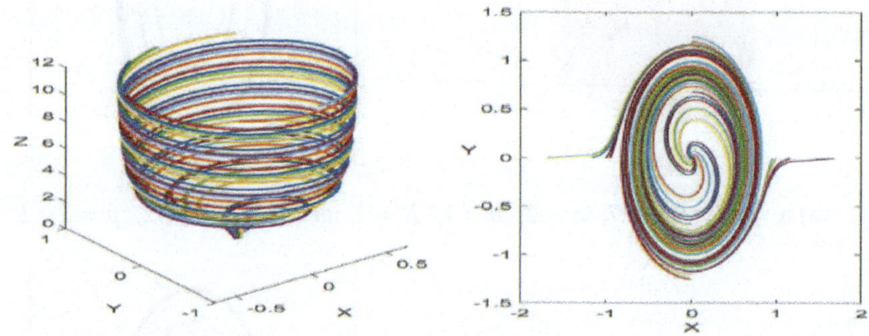

(b) $\alpha = -2.4,\ \beta = -0.7,\ a = 2,\ b = 1,\ R = 1,\ m = 3,\ \lambda = \sqrt{2},\ \mu = 2\sqrt{2}$

Fig. 4.7 $m = 3$

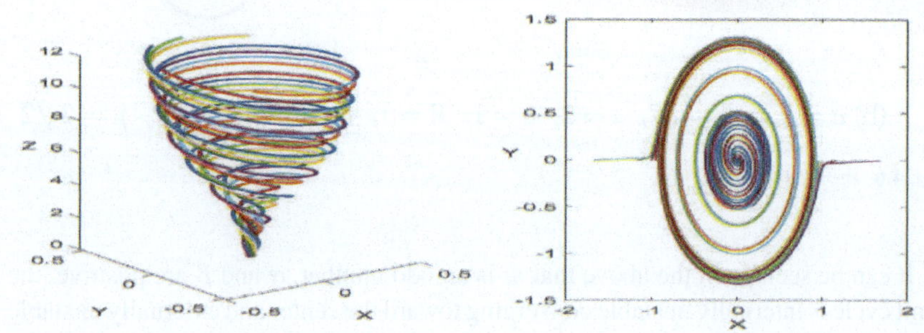

Fig. 4.8 $m = 9$

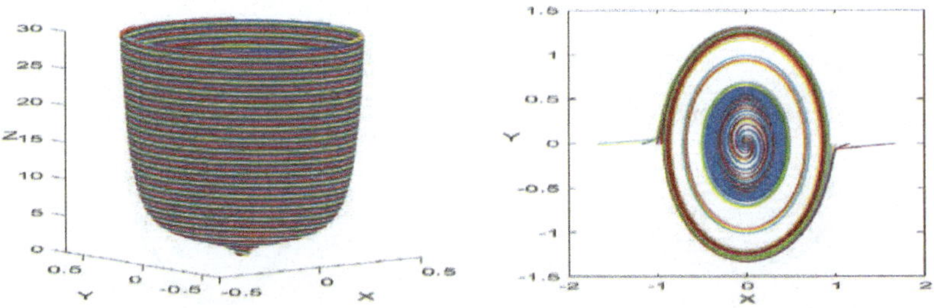

(a) $\alpha=-1.8,\ \beta=-0.8,\ a=2,\ b=1,\ R=1,\ m=11,\ \lambda=2\sqrt{2},\ \mu=4\sqrt{2}$

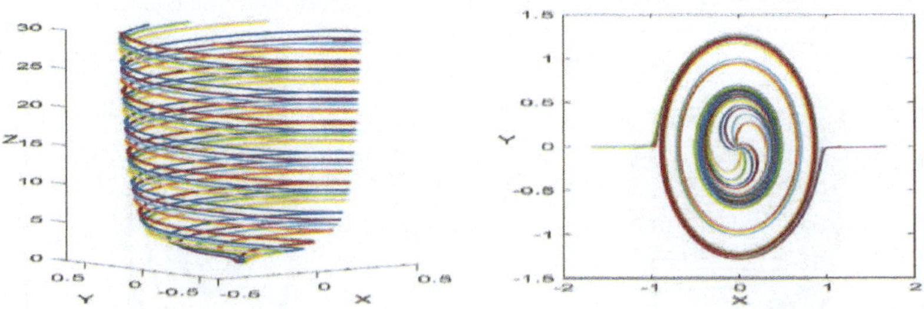

(b) $\alpha=-2.1,\ \beta=-1.7,\ a=2,\ b=1,\ R=1,\ m=11,\ \lambda=0.4\sqrt{2},\ \mu=0.8\sqrt{2}$

Fig. 4.9 $m=11$

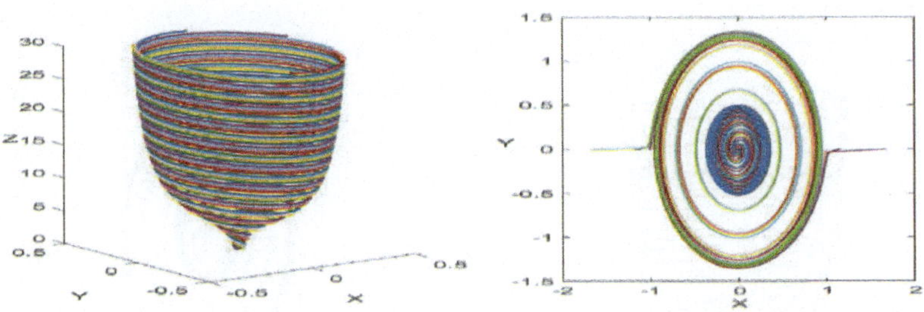

(a) $\alpha=-0.4,\ \beta=-0.7,\ a=2,\ b=1,\ R=1,\ m=15,\ \lambda=\sqrt{2},\ \mu=2\sqrt{2}$

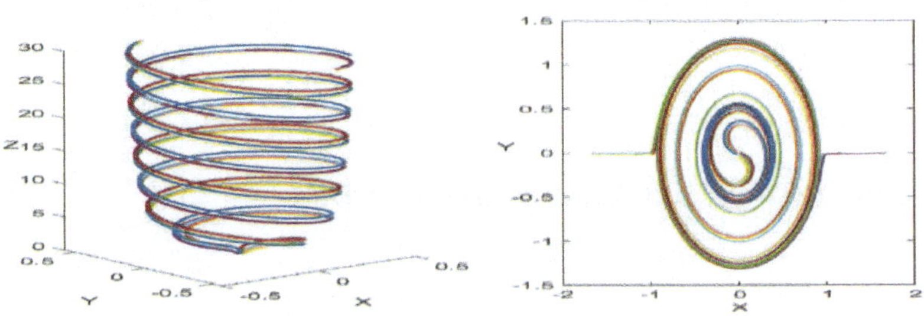

(b) $\alpha=-2.4,\ \beta=-0.7,\ a=2,\ b=1,\ R=1,\ m=15,\ \lambda=0.4\sqrt{2},\ \mu=0.8\sqrt{2}$

Fig. 4.10 $m=15$

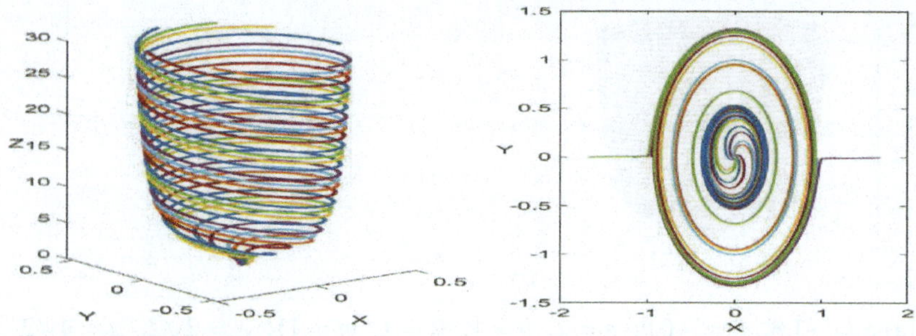

Fig. 4.11 $m = 17$

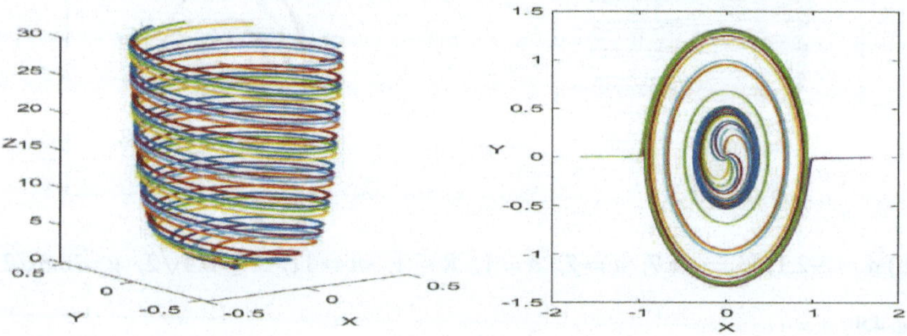

Fig. 4.12 $m = 19$

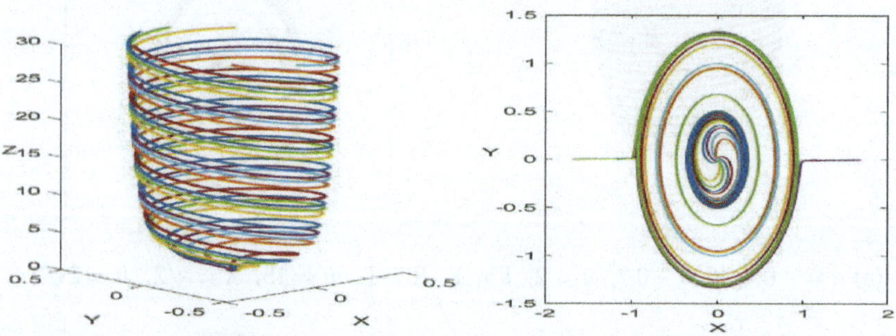

Fig. 4.13 $m = 21$

4.1.3 α is Positive, β is Negative and m is Odd

4.1.3.1 $m = 1$

1. When $|\beta| > \alpha$, the limit cycle is semi-stable. See Fig. 4.14.

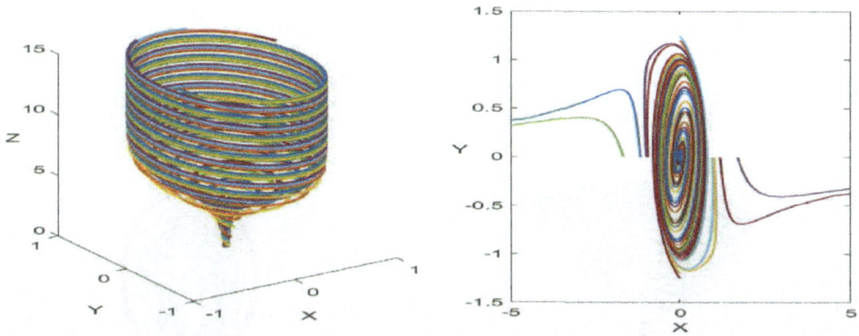

Fig. 4.14 $m = 1, |\beta| > \alpha$

2. When $|\beta| \approx \alpha$, the limit cycle is unstable. See Fig. 4.15.

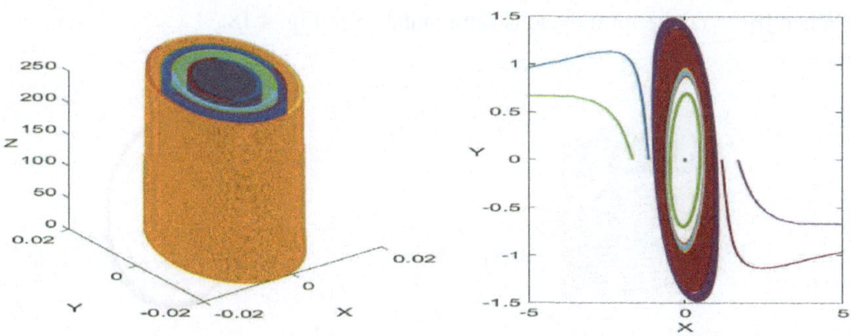

Fig. 4.15 $m = 1, |\beta| \approx \alpha$

3. When $|\beta| < \alpha$, the limit cycle is unstable. See Fig. 4.16.

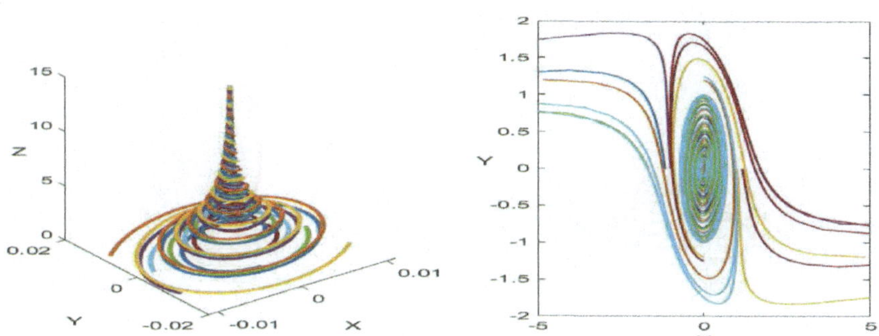

Fig. 4.16 $m = 1, |\beta| < \alpha$

4.1.3.2 $m = 3$

1. When $|\beta| > \alpha$, the limit cycle is semi-stable. See Fig. 4.17.

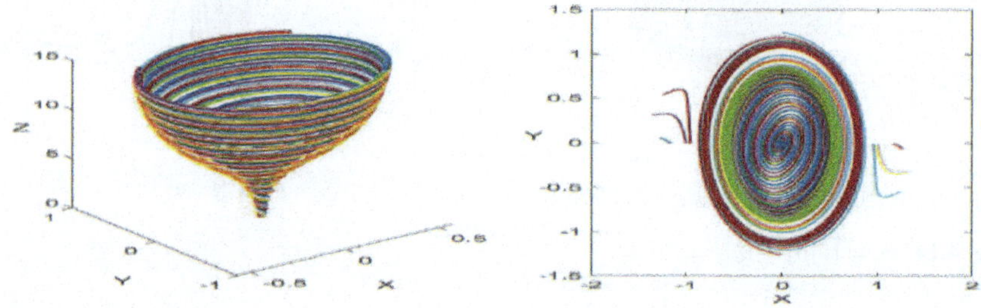

Fig. 4.17 $m = 3, |\beta| > \alpha$

2. When $|\beta| \approx \alpha$, the limit cycle is semi-stable. See Fig. 4.18.

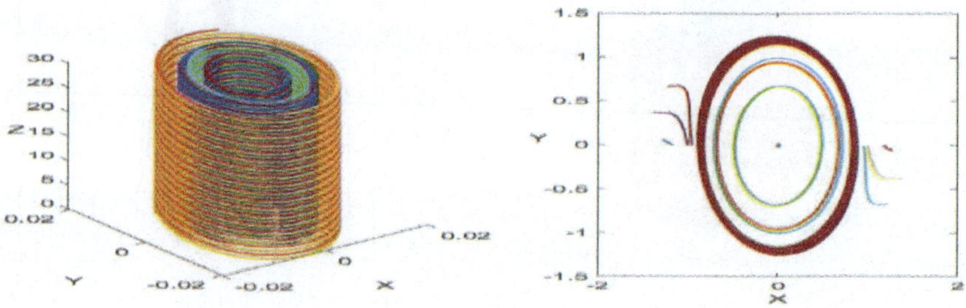

Fig. 4.18 $m = 3, |\beta| \approx \alpha$

3. When $|\beta| < \alpha$, the limit cycle is unstable. See Fig. 4.19.

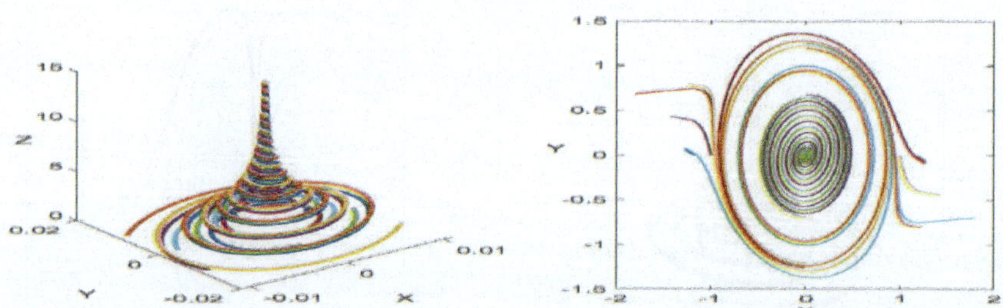

Fig. 4.19 $m = 3, |\beta| < \alpha$

4.1.3.3 $m = 5$

1. When $|\beta| > \alpha$, the limit cycle is semi-stable. See Fig. 4.20.

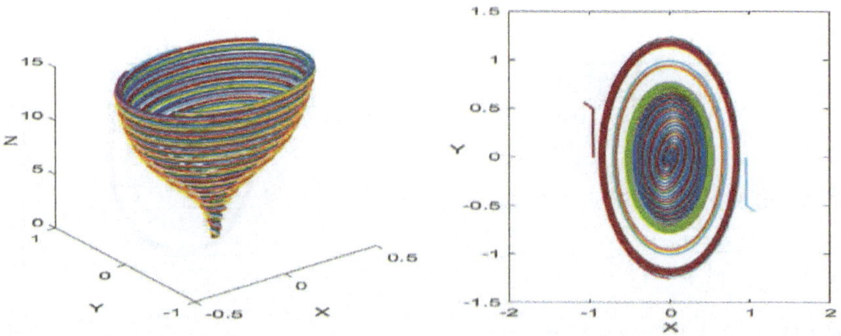

Fig. 4.20 $m = 5, |\beta| > \alpha$

2. When $|\beta| \approx \alpha$, the limit cycle is semi-stable. See Fig. 4.21.

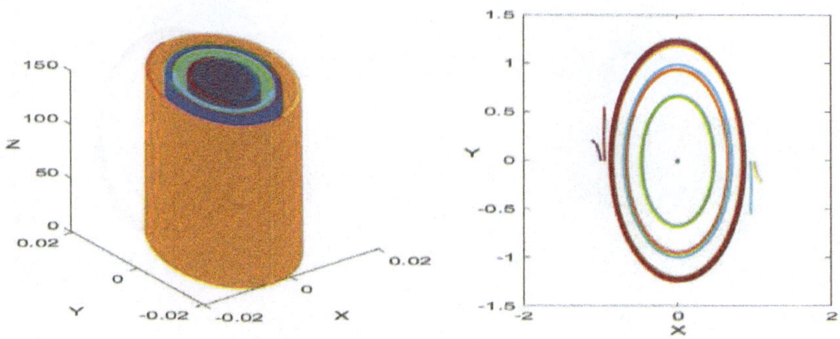

Fig. 4.21 $m = 5, |\beta| \approx \alpha$

3. When $|\beta| < \alpha$, the limit cycle is unstable. See Fig. 4.22.

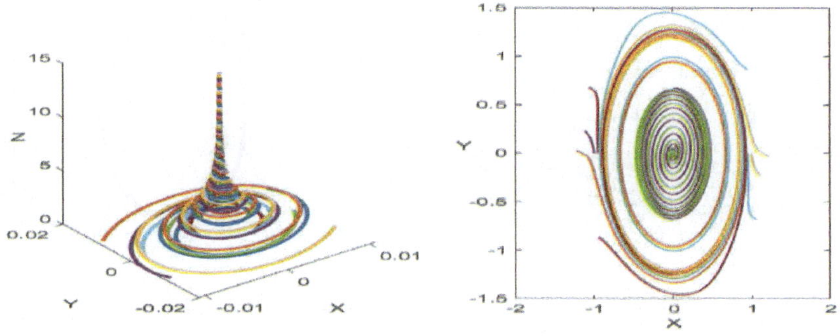

Fig. 4.22 $m = 5, |\beta| < \alpha$

4.1.3.4 $m = 11$

1. $|\beta| > \alpha$, the limit cycle is semi-stable. See Fig. 4.23.

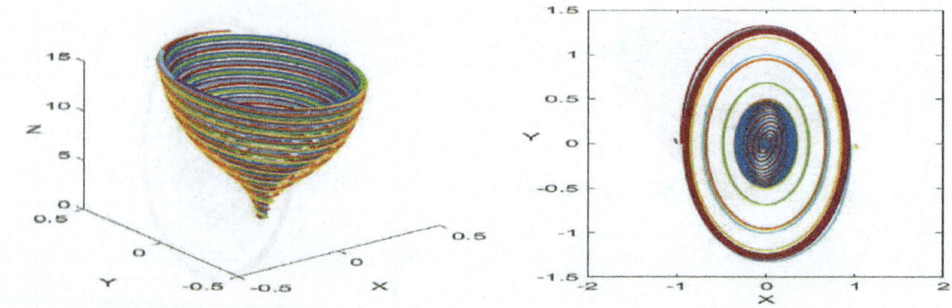

Fig. 4.23 $m = 11, |\beta| > \alpha$

2. $|\beta| \approx \alpha$, the limit cycle is semi-stable. See Fig. 4.24.

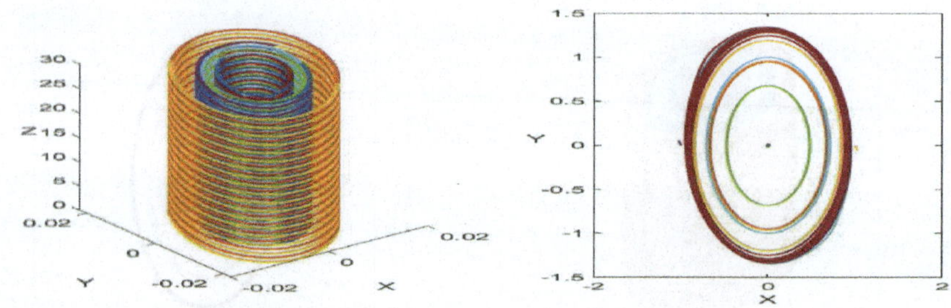

Fig. 4.24 $m = 11, |\beta| \approx \alpha$

3. $|\beta| < \alpha$, the limit cycle is unstable. See Fig. 4.25.

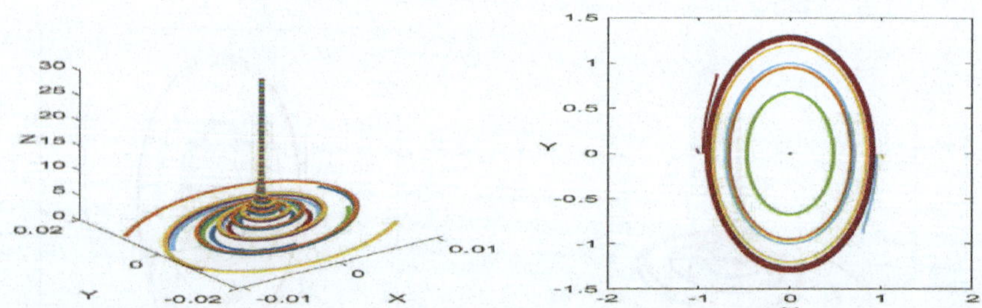

Fig. 4.25 $m = 11, |\beta| < \alpha$

4.1.4 α is Negative, β is Positive and m is Odd

4.1.4.1 $m = 1$

1. When $|\beta| > \alpha$, the limit cycle is stable. See Fig. 4.26.

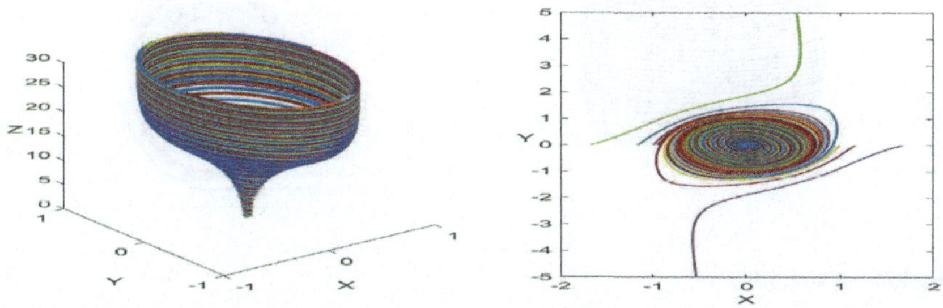

Fig. 4.26 $m = 1, |\beta| > \alpha$

2. $|\beta| \approx \alpha$, the limit cycle is semi-stable. See Fig. 4.27.

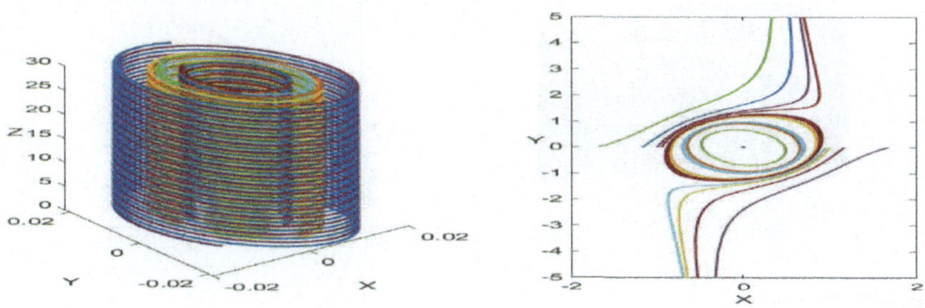

Fig. 4.27 $m = 1, |\beta| \approx \alpha$

3. $|\beta| < \alpha$, the limit cycle is unstable. See Fig. 4.28.

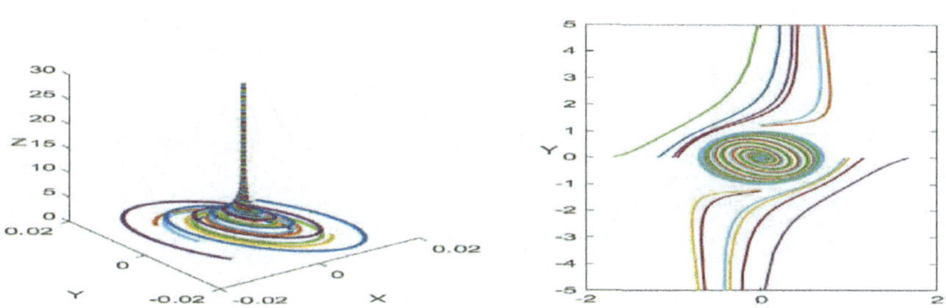

Fig. 4.28 $m = 1, |\beta| < \alpha$

4.1.4.2 $m = 3$

1. $|\beta| > \alpha$, the limit cycle is semi-stable. See Fig. 4.29.

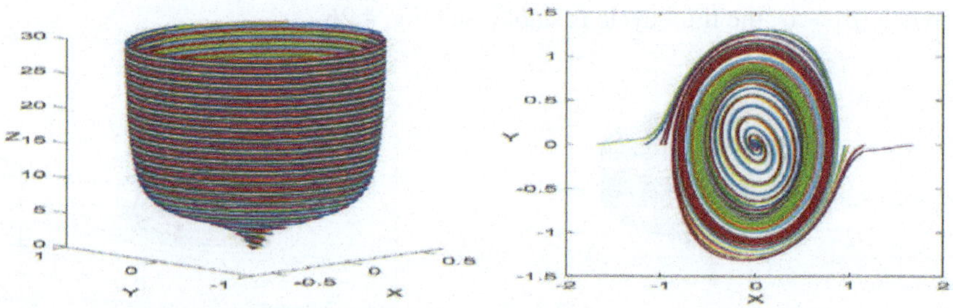

Fig. 4.29 $m = 3$, $|\beta| > \alpha$

2. $|\beta| \approx \alpha$, the limit cycle is semi-stable. See Fig. 4.30.

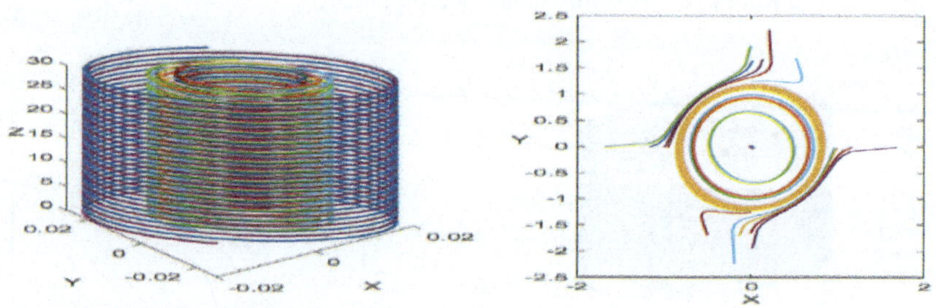

Fig. 4.30 $m = 3$, $|\beta| \approx \alpha$

3. $|\beta| < \alpha$, the limit cycle is unstable. See Fig. 4.31.

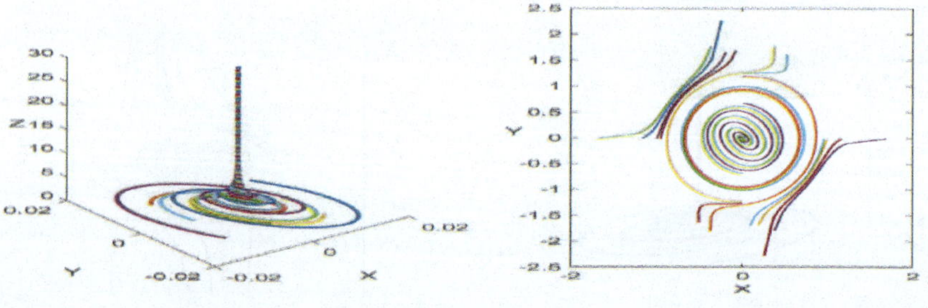

Fig. 4.31 $m = 3$, $|\beta| < \alpha$

4.1.4.3 $m = 5$

1. $|\beta| > \alpha$, the limit cycle is semi-stable. See Fig. 4.32.

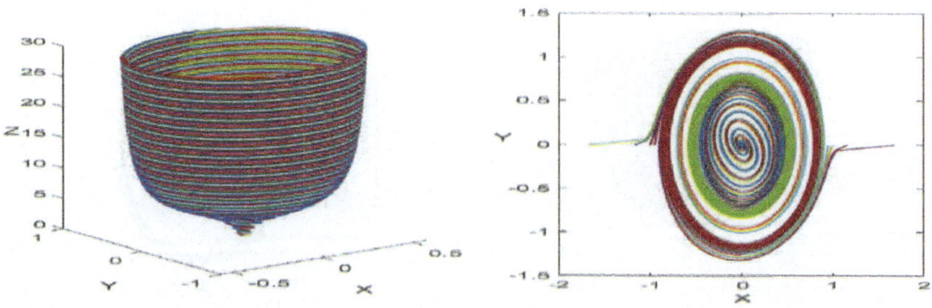

Fig. 4.32 $m = 5, |\beta| > \alpha$

2. $|\beta| \approx \alpha$, the limit cycle is semi-stable. See Fig. 4.33.

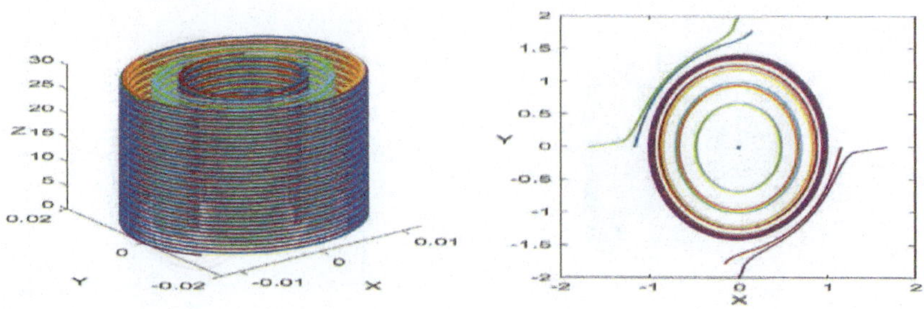

Fig. 4.33 $m = 5, |\beta| \approx \alpha$

3. $|\beta| < \alpha$, the limit cycle is unstable. See Fig. 4.34.

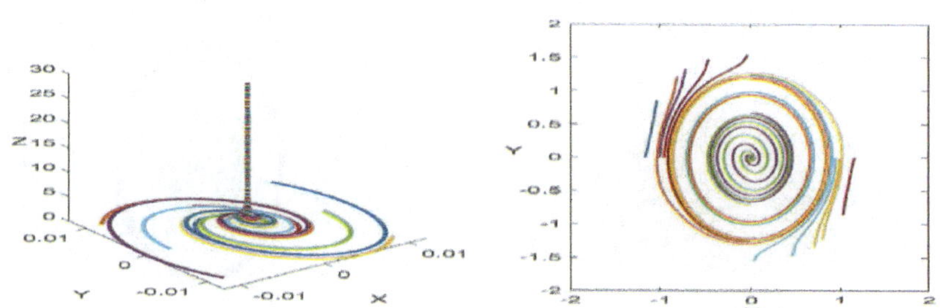

Fig. 4.34 $m = 5, |\beta| < \alpha$

4.1.4.4 $m = 11$

1. $|\beta| > \alpha$, the limit cycle is stable. See Fig. 4.35.

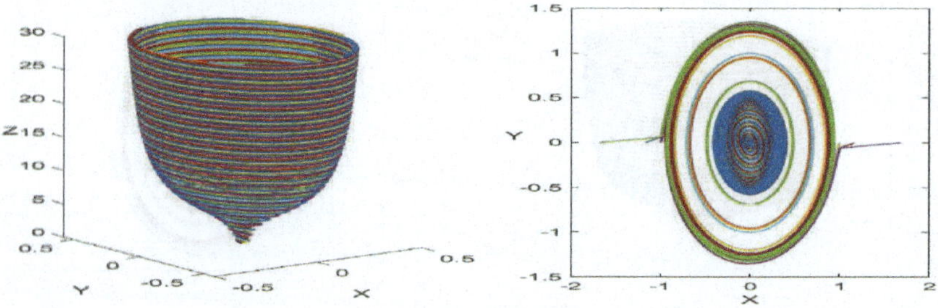

Fig. 4.35 $m = 11, |\beta| > \alpha$

2. $|\beta| \approx \alpha$, the limit cycle is semi-stable. See Fig. 4.36.

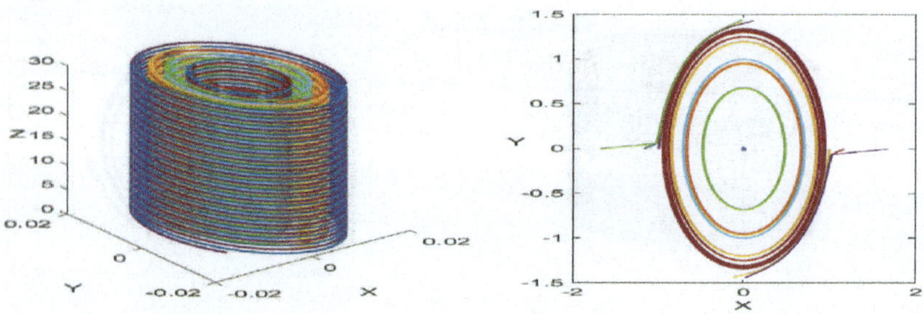

Fig. 4.36 $m = 11, |\beta| \approx \alpha$

3. $|\beta| < \alpha$, the limit cycle is unstable. See Fig. 4.37.

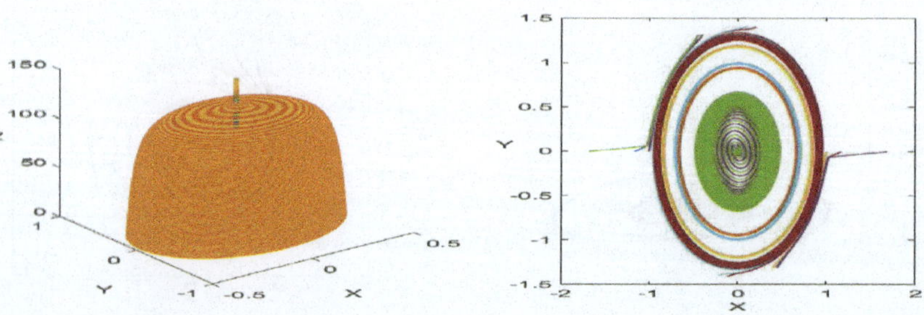

Fig. 4.37 $m = 11, |\beta| < \alpha$

4.1.5 m is a Decimal

4.1.5.1 α and β Are Positive Numbers

1. There is an unstable limit cycle at $m \in (0.5, 1.5)$. See Fig. 4.38.

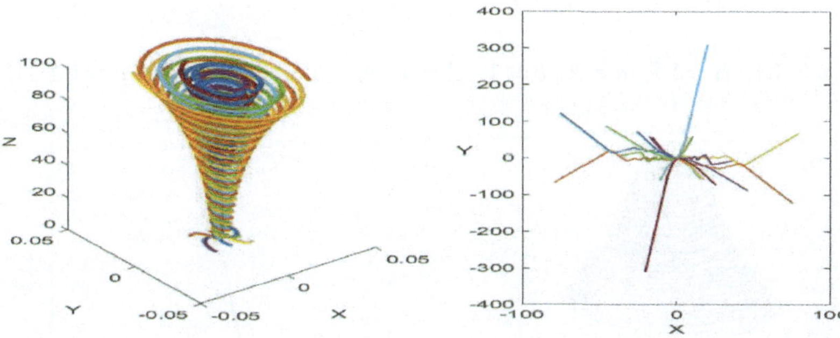

(a) $\alpha = 1.1,\ \beta = 1.7,\ a = 2,\ b = 1,\ R = 1,\ m = 0.5,\ \lambda = 0.4\sqrt{2},\ \mu = 0.8\sqrt{2}$

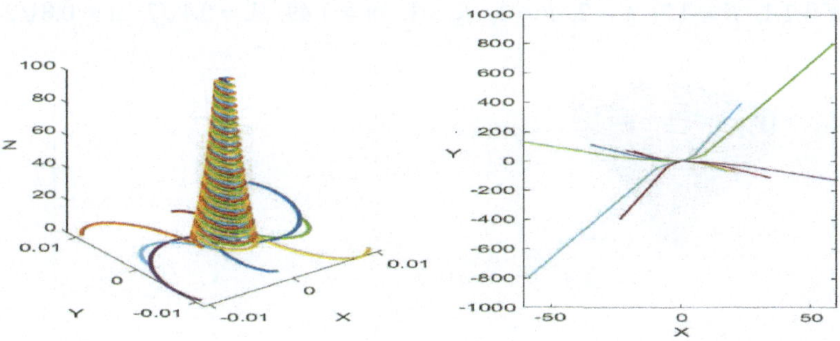

(b) $\alpha = 1.1,\ \beta = 1.7,\ a = 2,\ b = 1,\ R = 1,\ m = 0.51,\ \lambda = 0.4\sqrt{2},\ \mu = 0.8\sqrt{2}$

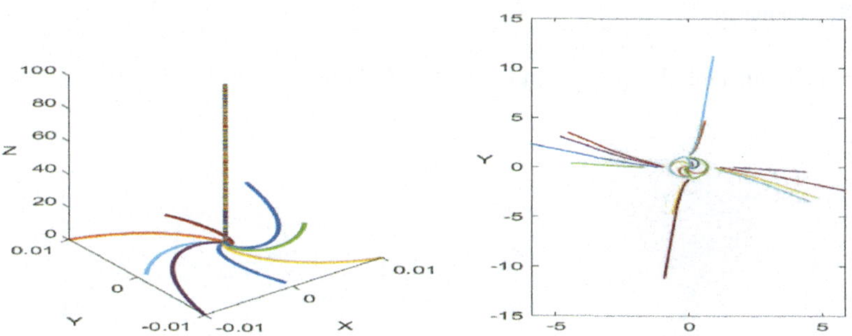

(c) $\alpha = 1.1,\ \beta = 1.7,\ a = 2,\ b = 1,\ R = 1,\ m = 1.11,\ \lambda = 0.4\sqrt{2},\ \mu = 0.8\sqrt{2}$

Fig. 4.38 $m \in (0.5, 1.5]$

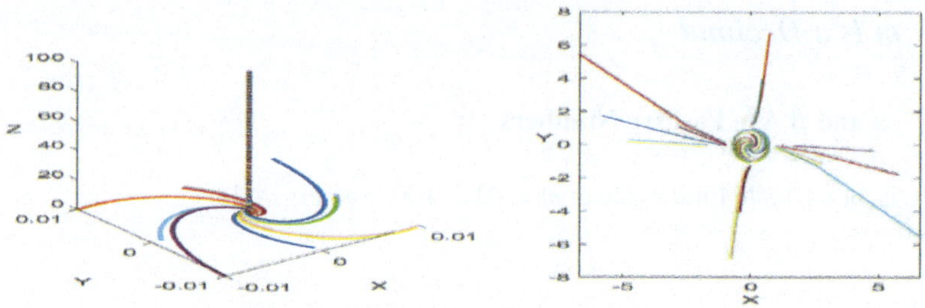

(d) $\alpha = 1.1,\ \ \beta = 1.7,\ \ a = 2,\ \ b = 1,\ \ R = 1,\ \ m = 1.3,\ \ \lambda = 0.4\sqrt{2},\ \ \mu = 0.8\sqrt{2}$

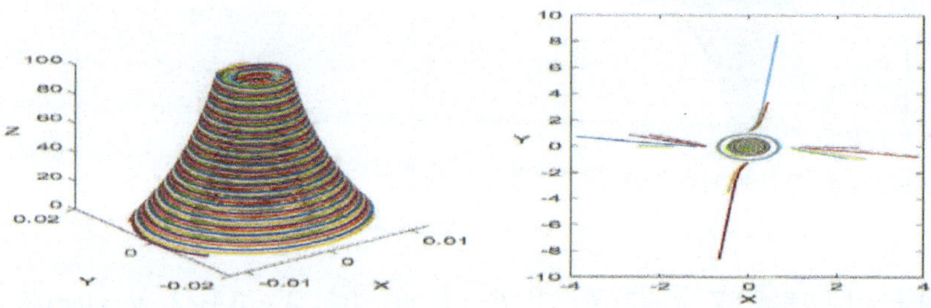

(e) $\alpha = 1.1,\ \ \beta = 1.7,\ \ a = 2,\ \ b = 1,\ \ R = 1,\ \ m = 1.49,\ \ \lambda = 0.4\sqrt{2},\ \ \mu = 0.8\sqrt{2}$

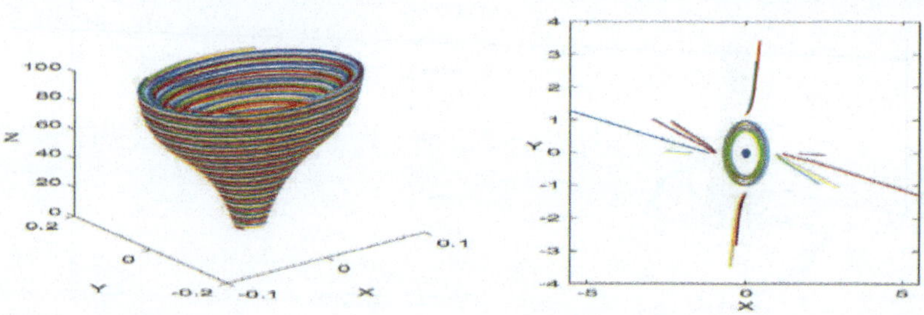

(f) $\alpha = 1.1,\ \ \beta = 1.7,\ \ a = 2,\ \ b = 1,\ \ R = 1,\ \ m = 1.5,\ \ \lambda = 0.4\sqrt{2},\ \ \mu = 0.8\sqrt{2}$

Fig. 4.38 (continued)

2. There is an unstable limit cycle at $m \in (2.5, 3.5)$. See Fig. 4.39.

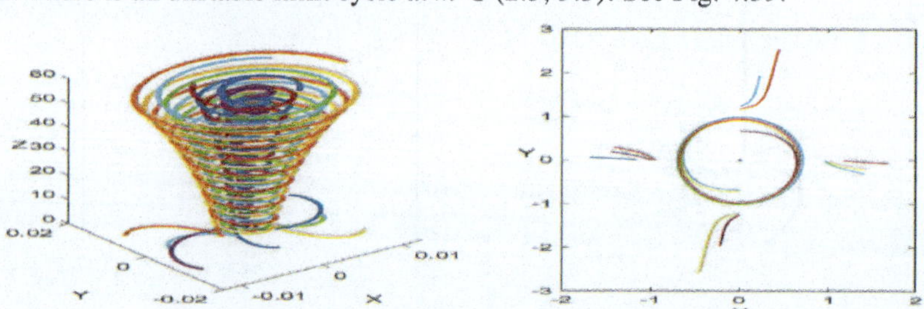

(a) $\alpha = 1.1,\ \ \beta = 1.7,\ \ a = 2,\ \ b = 1,\ \ R = 1,\ \ m = 2.5,\ \ \lambda = 0.4\sqrt{2},\ \ \mu = 0.8\sqrt{2}$

Fig. 4.39 $m \in (2.5, 3.5]$

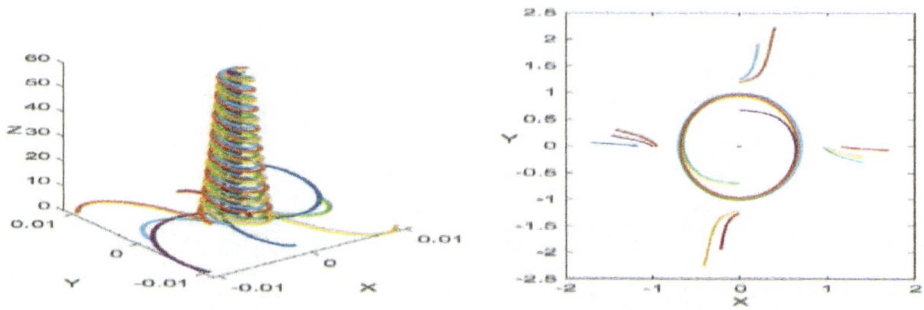

(b) $\alpha = 1.1,\ \beta = 1.7,\ a = 2,\ b = 1,\ R = 1,\ m = 2.51,\ \lambda = 0.4\sqrt{2},\ \mu = 0.8\sqrt{2}$

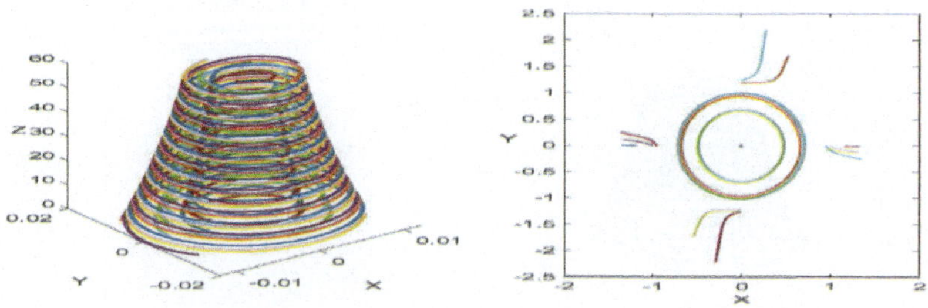

(c) $\alpha = 1.1,\ \beta = 1.7,\ a = 2,\ b = 1,\ R = 1,\ m = 3.49,\ \lambda = 0.4\sqrt{2},\ \mu = 0.8\sqrt{2}$

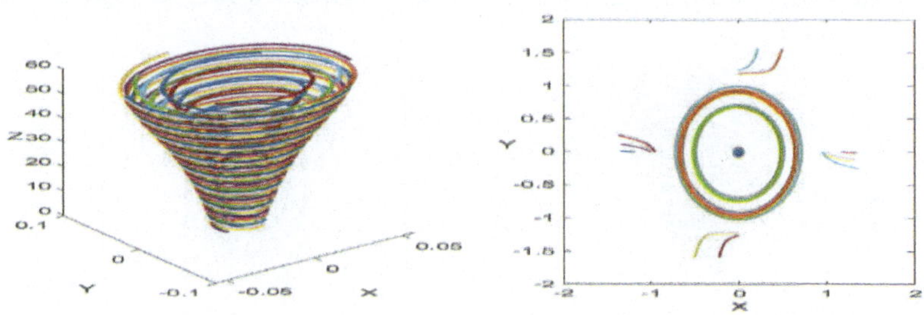

(d) $\alpha = 1.1,\ \beta = 1.7,\ a = 2,\ b = 1,\ R = 1,\ m = 3.5,\ \lambda = 0.4\sqrt{2},\ \mu = 0.8\sqrt{2}$

Fig. 4.39 (continued)

3. There is an unstable limit cycle at $m \in (4.5, 5.5)$. See Fig. 4.40.

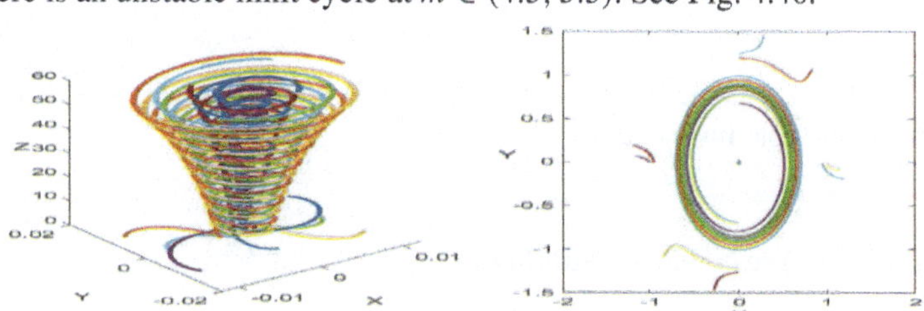

(a) $\alpha = 1.1,\ \beta = 1.7,\ a = 2,\ b = 1,\ R = 1,\ m = 4.5,\ \lambda = 0.4\sqrt{2},\ \mu = 0.8\sqrt{2}$

Fig. 4.40 $m \in (4.5, 5.5]$

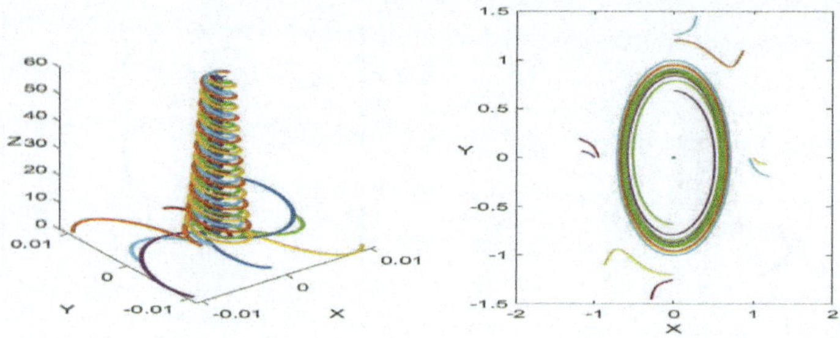

(b) $\alpha = 1.1,\ \beta = 1.7,\ a = 2,\ b = 1,\ R = 1,\ m = 4.51,\ \lambda = 0.4\sqrt{2},\ \mu = 0.8\sqrt{2}$

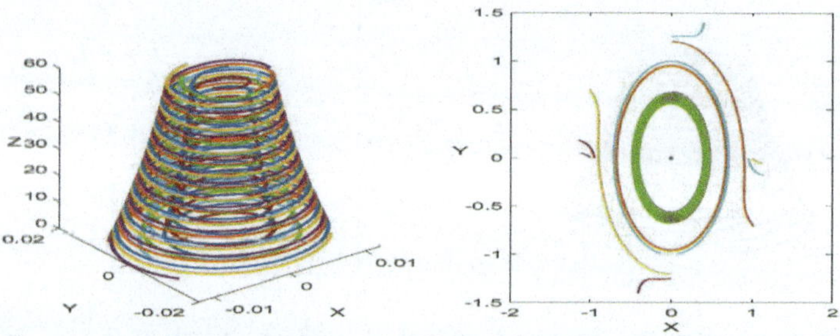

(c) $\alpha = 1.1,\ \beta = 1.7,\ a = 2,\ b = 1,\ R = 1,\ m = 5.49,\ \lambda = 0.4\sqrt{2},\ \mu = 0.8\sqrt{2}$

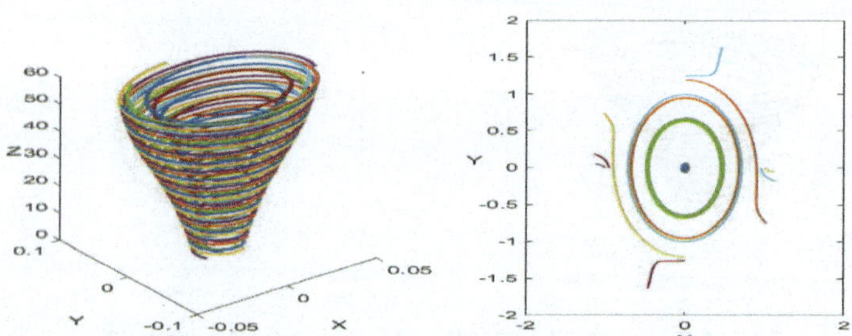

(d) $\alpha = 1.1,\ \beta = 1.7,\ a = 2,\ b = 1,\ R = 1,\ m = 5.5,\ \lambda = 0.4\sqrt{2},\ \mu = 0.8\sqrt{2}$

Fig. 4.40 (continued)

From the above, when m is $(N - 0.5, N + 0.5)$ and $N = 1, 3, 5, 7, 9, 11$, the system has an unstable limit cycle.

4.1.5.2 α and β Are Negative Numbers

1. There is a stable limit cycle at $m \in (0.5, 1.13)$. See Fig. 4.41.

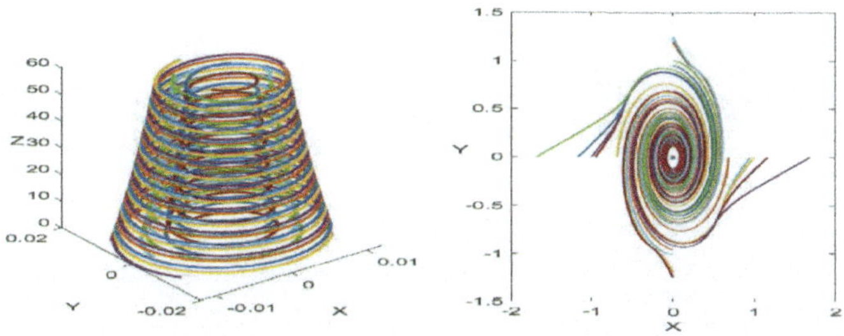

(a) $\alpha=-1.1,\ \beta=-1.3,\ a=2,\ b=1,\ R=1,\ m=0.49,\ \lambda=0.4\sqrt{2},\ \mu=0.8\sqrt{2}$

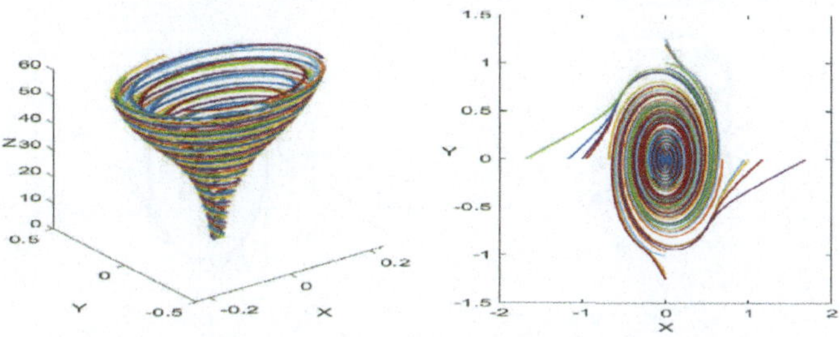

(b) $\alpha=-1.1,\ \beta=-1.3,\ a=2,\ b=1,\ R=1,\ m=0.501,\ \lambda=0.4\sqrt{2},\ \mu=0.8\sqrt{2}$

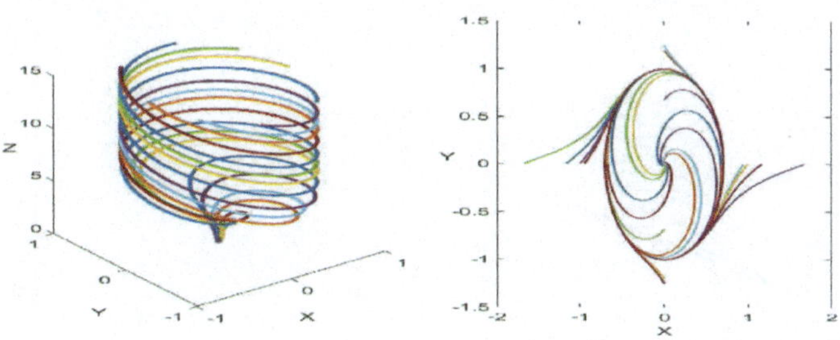

(c) $\alpha=-1.1,\ \beta=-1.3,\ a=2,\ b=1,\ R=1,\ m=0.8,\ \lambda=0.4\sqrt{2},\ \mu=0.8\sqrt{2}$

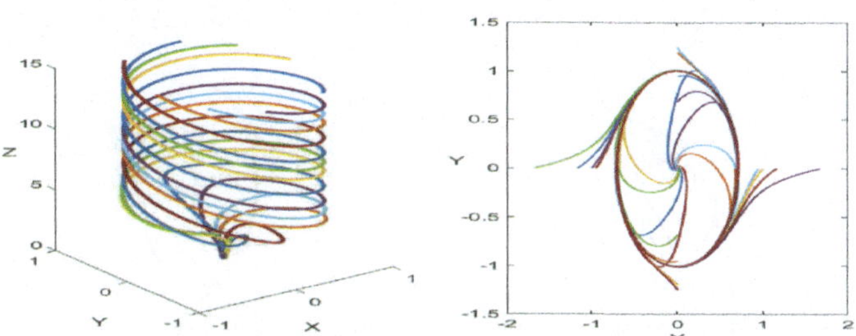

(d) $\alpha=-1.1,\ \beta=-1.3,\ a=2,\ b=1,\ R=1,\ m=1.07,\ \lambda=0.4\sqrt{2},\ \mu=0.8\sqrt{2}$

Fig. 4.41 $m \in (0.5, 1.13]$

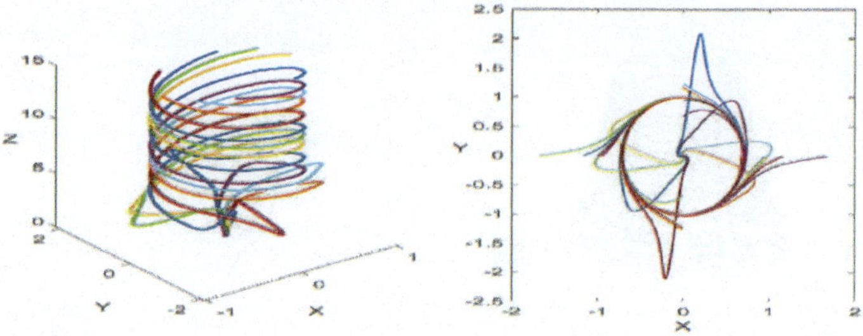

(e) $\alpha=-1.1,\ \beta=-1.3,\ a=2,\ b=1,\ R=1,\ m=1.1,\ \lambda=0.4\sqrt{2},\ \mu=0.8\sqrt{2}$

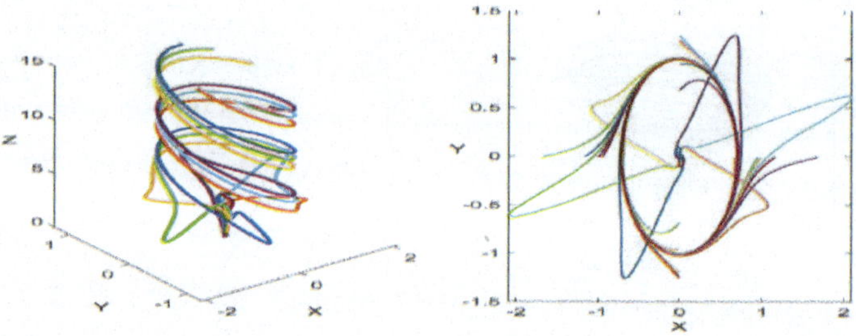

(f) $\alpha=-1.1,\ \beta=-1.3,\ a=2,\ b=1,\ R=1,\ m=1.11,\ \lambda=0.4\sqrt{2},\ \mu=0.8\sqrt{2}$

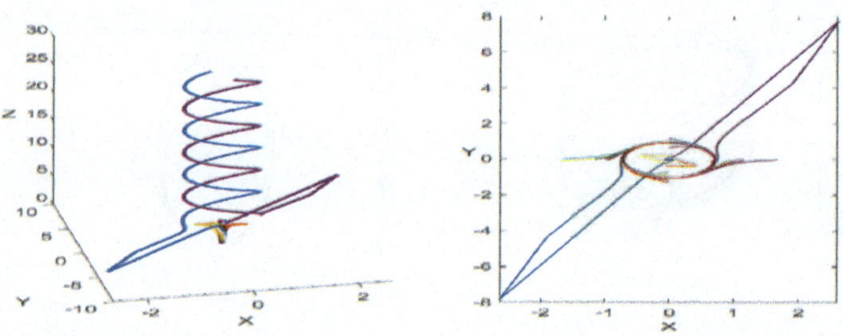

(g) $\alpha=-1.1,\ \beta=-1.3,\ a=2,\ b=1,\ R=1,\ m=1.12,\ \lambda=0.4\sqrt{2},\ \mu=0.8\sqrt{2}$

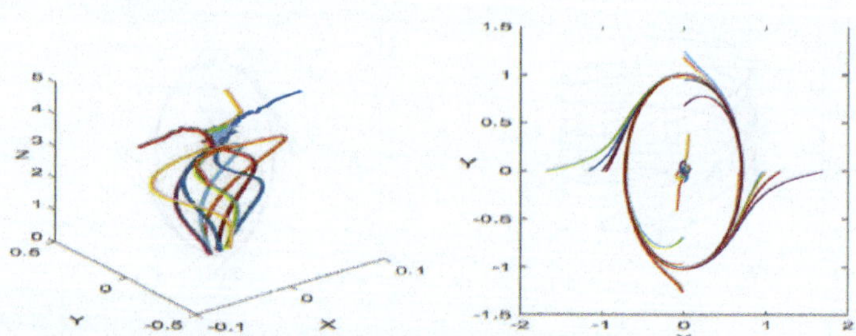

(h) $\alpha=-1.1,\ \beta=-1.3,\ a=2,\ b=1,\ R=1,\ m=1.13,\ \lambda=0.4\sqrt{2},\ \mu=0.8\sqrt{2}$

Fig. 4.41 (continued)

2. There is a stable limit cycle at $m \in (2.5, 3.15)$. See Fig. 4.42.

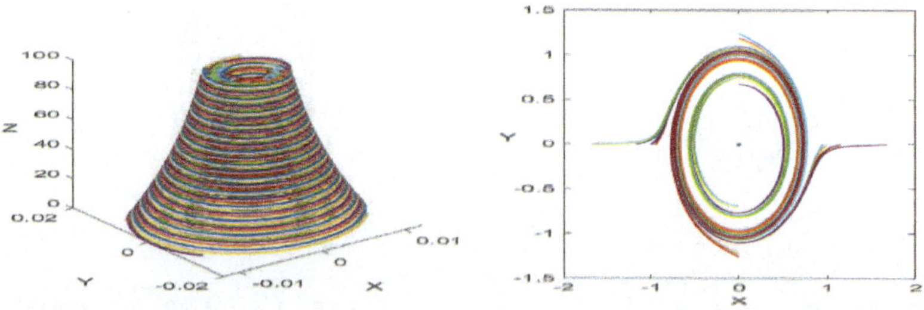

(a) $\alpha = -1.6,\ \beta = -1.4,\ a = 2,\ b = 1,\ R = 1,\ m = 2.49,\ \lambda = 0.4\sqrt{2},\ \mu = 0.8\sqrt{2}$

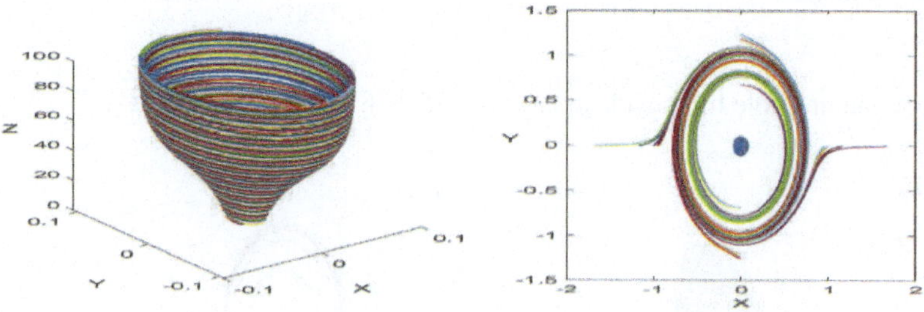

(b) $\alpha = -1.6,\ \beta = -1.4,\ a = 2,\ b = 1,\ R = 1,\ m = 2.5,\ \lambda = 0.4\sqrt{2},\ \mu = 0.8\sqrt{2}$

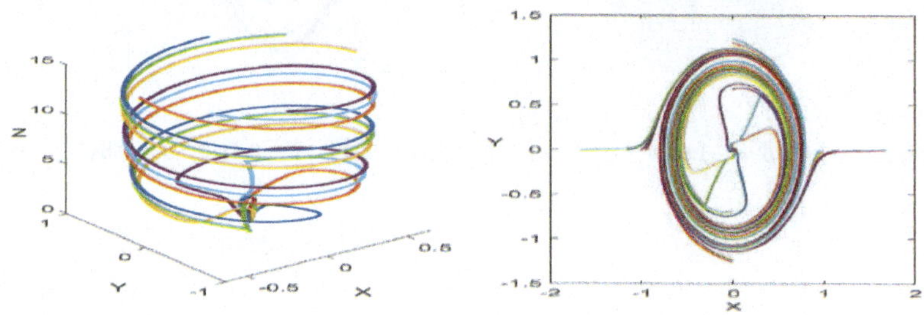

(c) $\alpha = -1.6,\ \beta = -1.4,\ a = 2,\ b = 1,\ R = 1,\ m = 3.13,\ \lambda = 0.4\sqrt{2},\ \mu = 0.8\sqrt{2}$

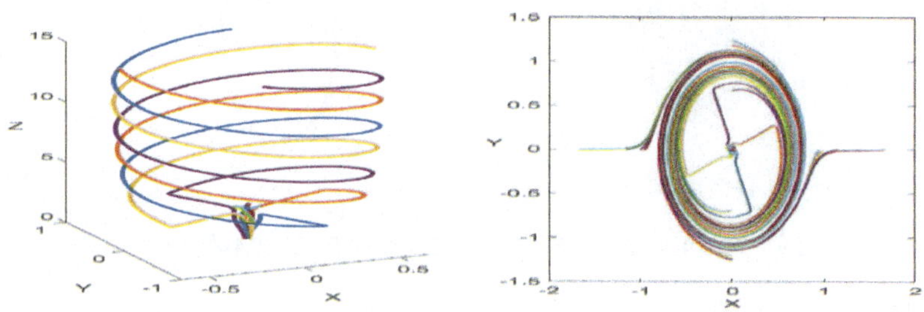

(d) $\alpha = -1.6,\ \beta = -1.4,\ a = 2,\ b = 1,\ R = 1,\ m = 3.14,\ \lambda = 0.4\sqrt{2},\ \mu = 0.8\sqrt{2}$

Fig. 4.42 $m \in (2.5, 3.15]$

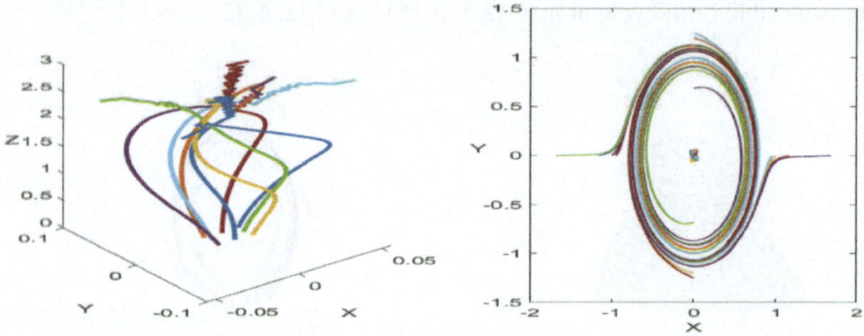

(e) $\alpha = -1.6,\ \ \beta = -1.4,\ \ a = 2,\ \ b = 1,\ \ R = 1,\ \ m = 3.15,\ \ \lambda = 0.4\sqrt{2},\ \ \mu = 0.8\sqrt{2}$

Fig. 4.42 (continued)

3. There is an unstable limit cycle at $m \in (4.5, 5.16)$. See Fig. 4.43.

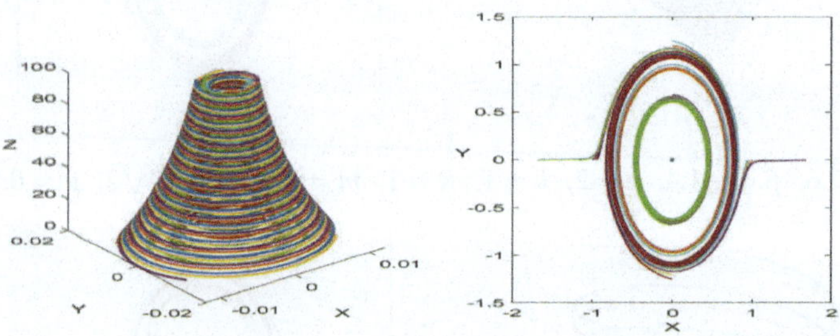

(a) $\alpha = -1.9,\ \ \beta = -1.6,\ \ a = 2,\ \ b = 1,\ \ R = 1,\ \ m = 4.49,\ \ \lambda = 0.4\sqrt{2},\ \ \mu = 0.8\sqrt{2}$

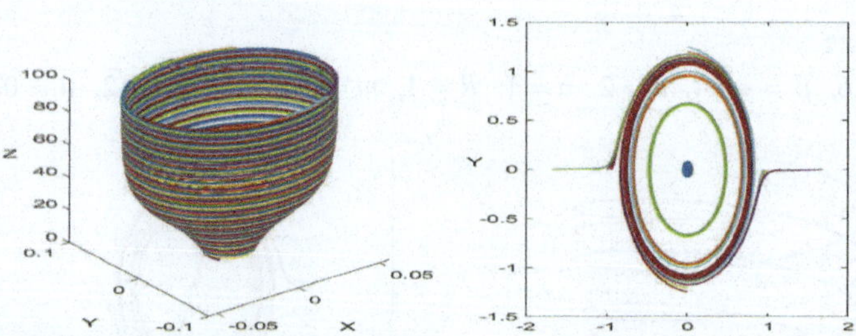

(b) $\alpha = -1.9,\ \ \beta = -1.6,\ \ a = 2,\ \ b = 1,\ \ R = 1,\ \ m = 4.5,\ \ \lambda = 0.4\sqrt{2},\ \ \mu = 0.8\sqrt{2}$

Fig. 4.43 $m \in (4.5, 5.16]$

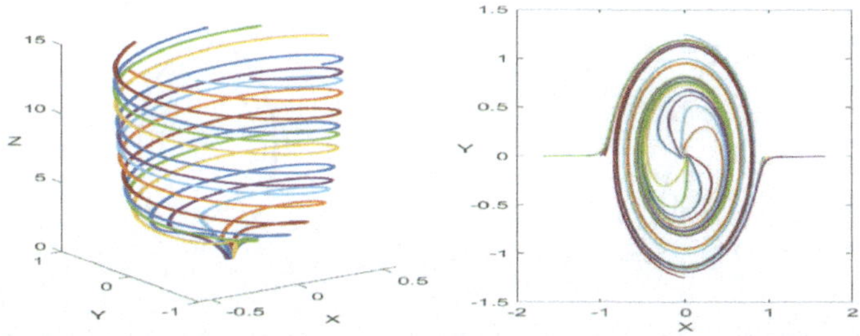

(c) $\alpha = -1.9,\ \beta = -1.6,\ a = 2,\ b = 1,\ R = 1,\ m = 5.11,\ \lambda = 0.4\sqrt{2},\ \mu = 0.8\sqrt{2}$

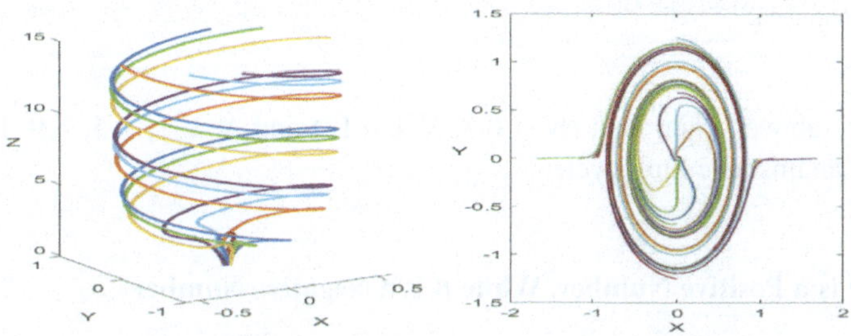

(d) $\alpha = -1.9,\ \beta = -1.6,\ a = 2,\ b = 1,\ R = 1,\ m = 5.13,\ \lambda = 0.4\sqrt{2},\ \mu = 0.8\sqrt{2}$

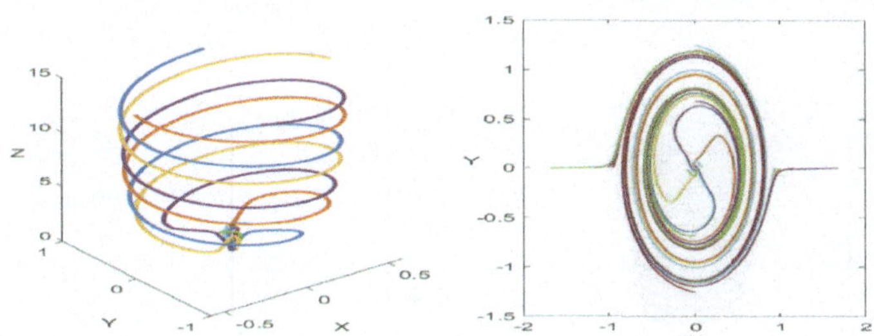

(e) $\alpha = -1.9,\ \beta = -1.6,\ a = 2,\ b = 1,\ R = 1,\ m = 5.14,\ \lambda = 0.4\sqrt{2},\ \mu = 0.8\sqrt{2}$

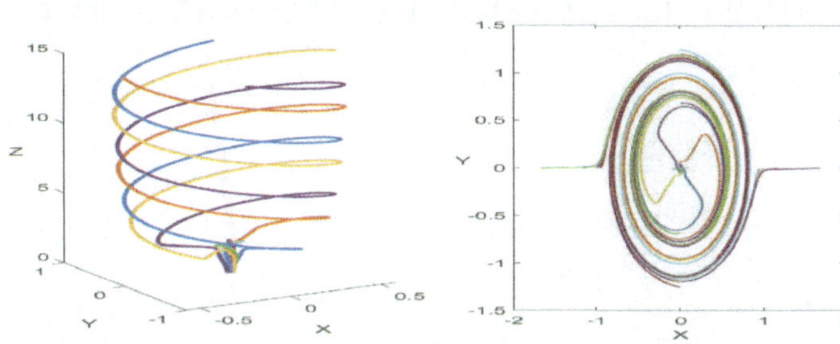

(f) $\alpha = -1.9,\ \beta = -1.6,\ a = 2,\ b = 1,\ R = 1,\ m = 5.15,\ \lambda = 0.4\sqrt{2},\ \mu = 0.8\sqrt{2}$

Fig. 4.43 (continued)

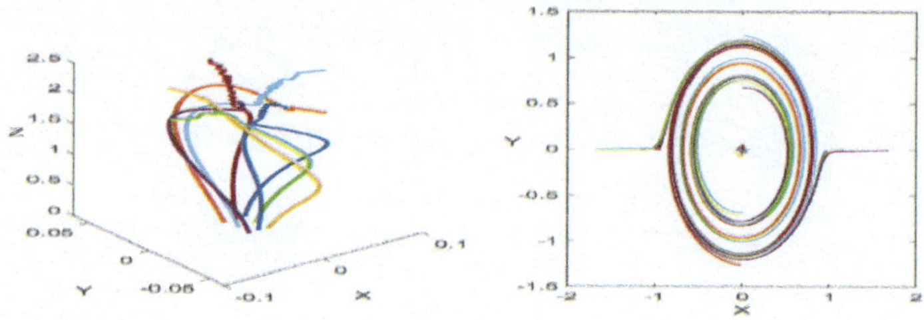

(g) $\alpha=-1.9,\ \beta=-1.6,\ a=2,\ b=1,\ R=1,\ m=5.16,\ \lambda=0.4\sqrt{2},\ \mu=0.8\sqrt{2}$

Fig. 4.43 (continued)

From the above, when $m \in (N-0.5, N+0.13)$ and $N = 1, 3, 5, 7, 9, 11$, the system has an unstable limit cycle.

4.1.5.3 α is a Positive Number, While β is a Negative Number

1. When $|\beta| > \alpha$ and m is about 1, the limit cycle does not change regularly. See Fig. 4.44.

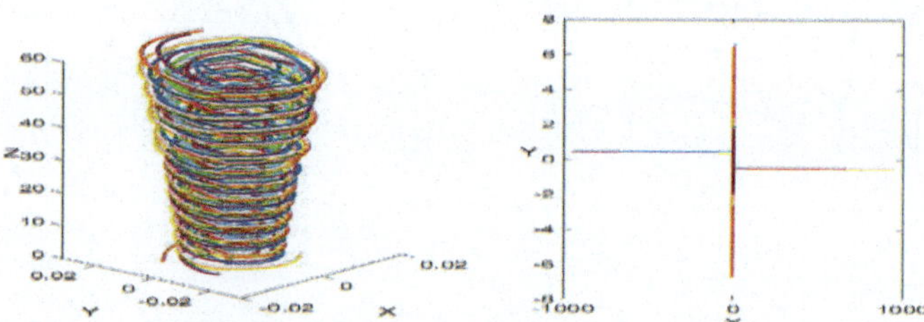

(a) $\alpha=1.1,\ \beta=-1.7,\ a=2,\ b=1,\ R=1,\ m=0.5,\ \lambda=0.4\sqrt{2},\ \mu=0.8\sqrt{2}$

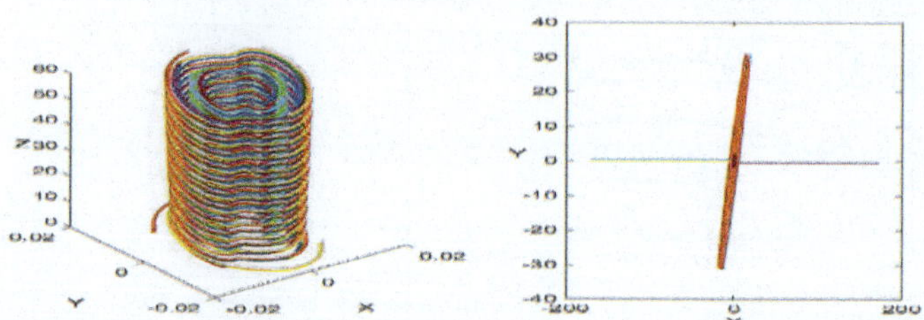

(b) $\alpha=1.1,\ \beta=-1.7,\ a=2,\ b=1,\ R=1,\ m=0.51,\ \lambda=0.4\sqrt{2},\ \mu=0.8\sqrt{2}$

Fig. 4.44 $|\beta| > \alpha$

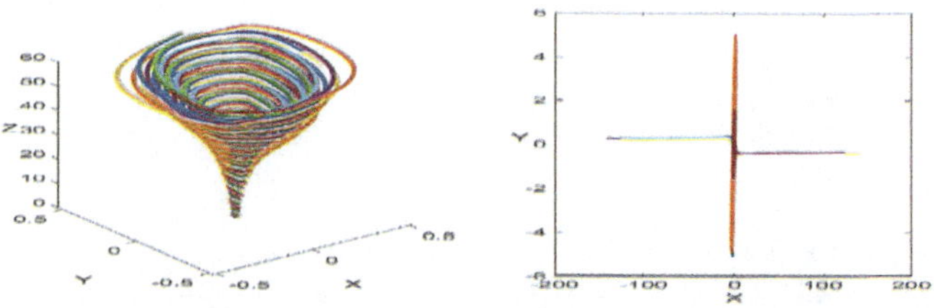

(c) $\alpha = 1.1,\ \beta = -1.7,\ a = 2,\ b = 1,\ R = 1,\ m = 0.55,\ \lambda = 0.4\sqrt{2},\ \mu = 0.8\sqrt{2}$

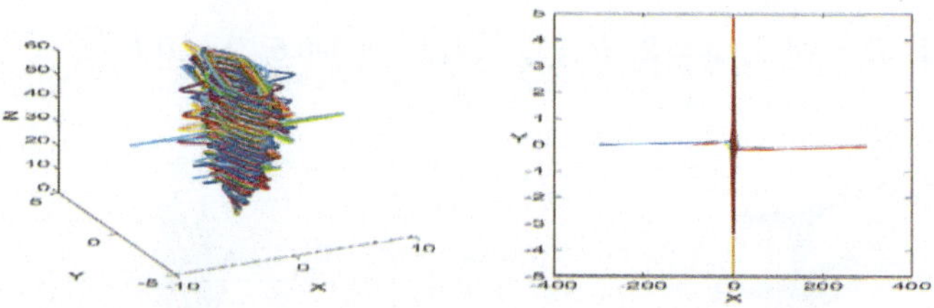

(d) $\alpha = 1.1,\ \beta = -1.7,\ a = 2,\ b = 1,\ R = 1,\ m = 0.8,\ \lambda = 0.4\sqrt{2},\ \mu = 0.8\sqrt{2}$

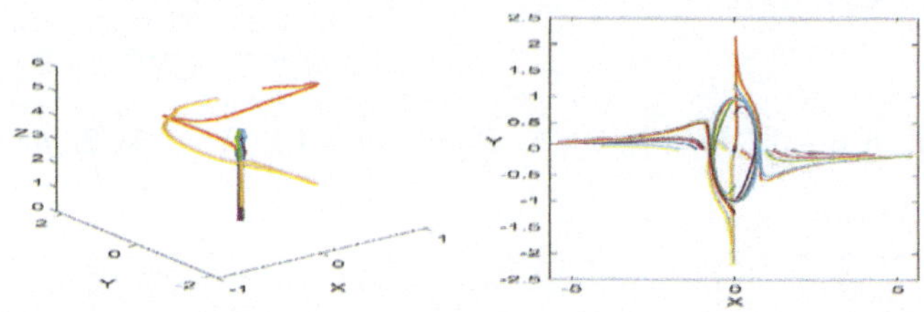

(e) $\alpha = 1.1,\ \beta = -1.7,\ a = 2,\ b = 1,\ R = 1,\ m = 1.1,\ \lambda = 0.4\sqrt{2},\ \mu = 0.8\sqrt{2}$

Fig. 4.44 (continued)

2. When $|\beta| < \alpha$ and m is about 1, the limit cycle does not change regularly. See Fig. 4.45.

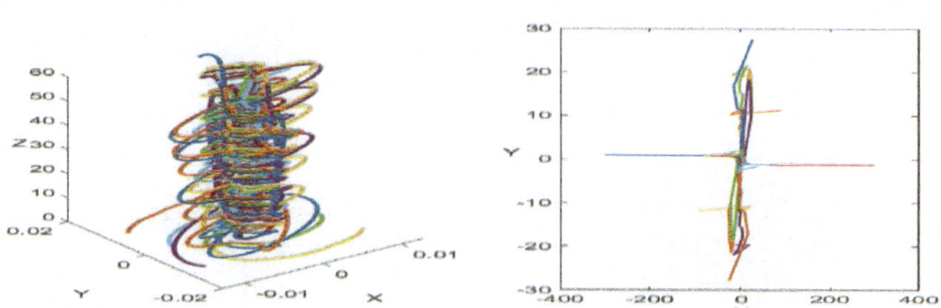

(a) $\alpha = 1.1,\ \beta = -0.7,\ a = 2,\ b = 1,\ R = 1,\ m = 0.5,\ \lambda = 0.4\sqrt{2},\ \mu = 0.8\sqrt{2}$

Fig. 4.45 $|\beta| < \alpha$

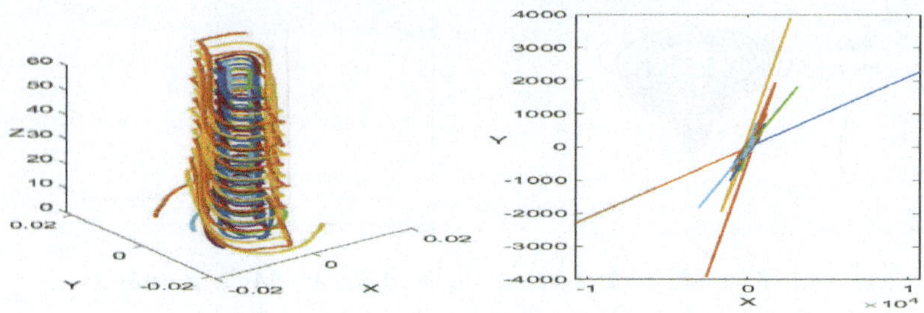

(b) $\alpha = 1.1,\ \ \beta = -0.7,\ \ a = 2,\ \ b = 1,\ \ R = 1,\ \ m = 0.6,\ \ \lambda = 0.4\sqrt{2},\ \ \mu = 0.8\sqrt{2}$

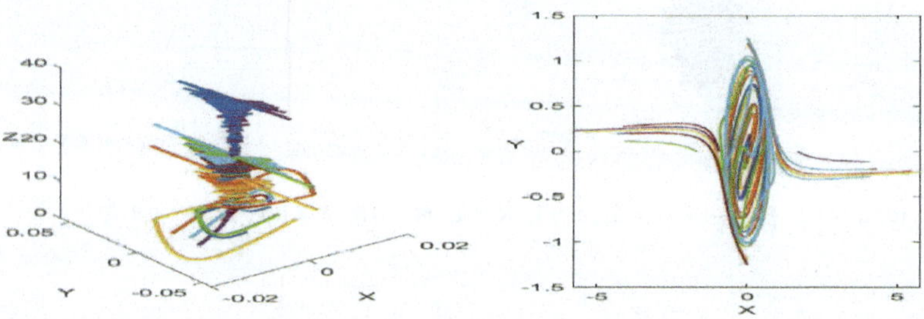

(c) $\alpha = 1.1,\ \ \beta = -0.7,\ \ a = 2,\ \ b = 1,\ \ R = 1,\ \ m = 1.1,\ \ \lambda = 0.4\sqrt{2},\ \ \mu = 0.8\sqrt{2}$

Fig. 4.45 (continued)

From the above, when α is a positive number, while β is a negative number, the limit cycle does not change regularly.

4.1.5.4 α is a Negative Number, While β is a Positive Number

1. When $|\beta| > \alpha$ and m is about 1, the limit cycle does not change regularly. See Fig. 4.46.

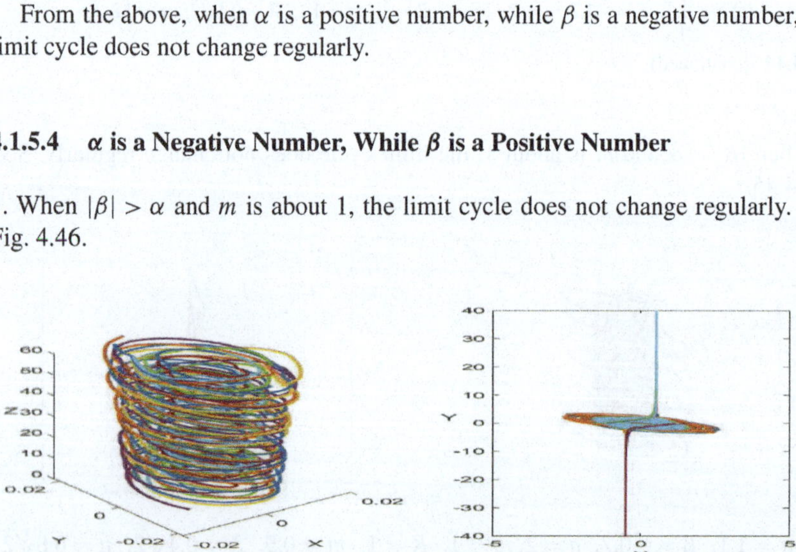

(a) $\alpha = -1.1,\ \ \beta = 0.7,\ \ a = 2,\ \ b = 1,\ \ R = 1,\ \ m = 0.5,\ \ \lambda = 0.4\sqrt{2},\ \ \mu = 0.8\sqrt{2}$

Fig. 4.46 $|\beta| > \alpha$

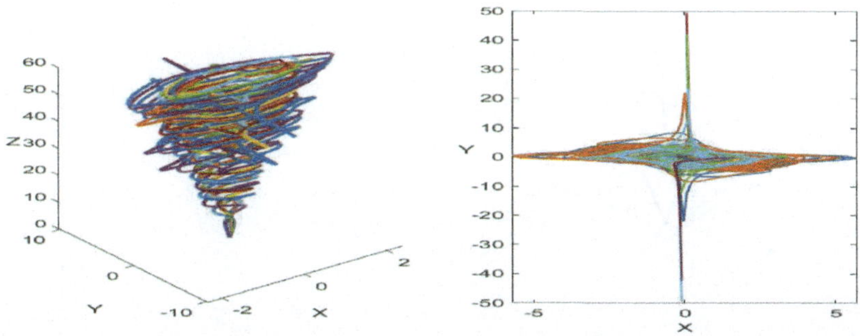

(b) $\alpha = -1.1,\ \beta = 0.7,\ a = 2,\ b = 1,\ R = 1,\ m = 0.8,\ \lambda = 0.4\sqrt{2},\ \mu = 0.8\sqrt{2}$

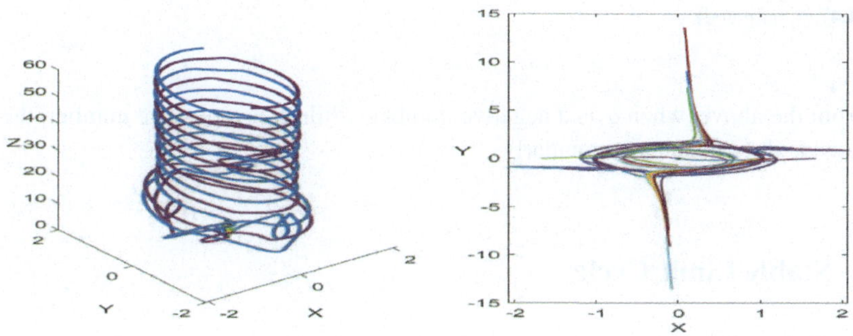

(c) $\alpha = -1.1,\ \beta = 0.7,\ a = 2,\ b = 1,\ R = 1,\ m = 1.07,\ \lambda = 0.4\sqrt{2},\ \mu = 0.8\sqrt{2}$

Fig. 4.46 (continued)

2. When $|\beta| < \alpha$ and m is about 1, the limit cycle does not change regularly. See Fig. 4.47.

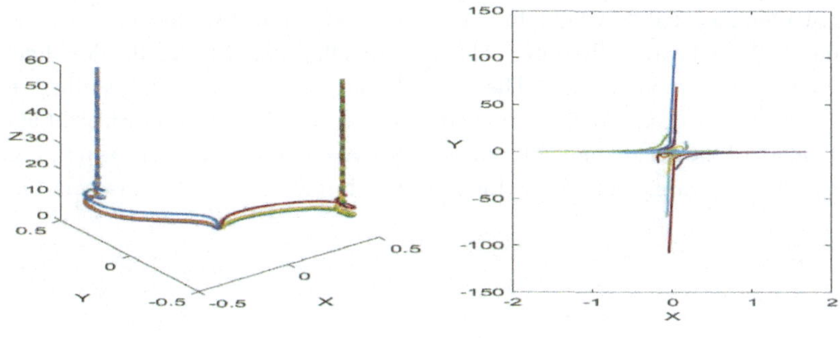

(a) $\alpha = -1.1,\ \beta = 2.3,\ a = 2,\ b = 1,\ R = 1,\ m = 0.8,\ \lambda = 0.4\sqrt{2},\ \mu = 0.8\sqrt{2}$

Fig. 4.47 $|\beta| < \alpha$

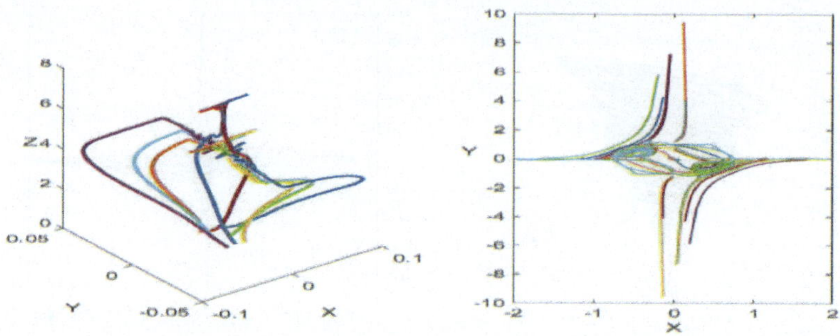

(b) $\alpha=-1.1,\ \beta=2.3,\ a=2,\ b=1,\ R=1,\ m=1.1,\ \lambda=0.4\sqrt{2},\ \mu=0.8\sqrt{2}$

Fig. 4.47 (continued)

From the above, when α is a negative number, while β is a positive number, the limit cycle does not change regularly.

4.2 Stable Limit Cycle

Based on the conclusion of Sect. 4.1, we get that when parameters $\alpha, \beta, m, a, b, R, \lambda, \mu$, and ω satisfy the range in Table 4.1.

Then system (3.3.16) has a stable limit cycle C. For example, take $a = b = \omega = m = \lambda = \mu = 1$ and $\alpha = \beta = -1$. Then system (3.3.16) becomes

$$\begin{cases} \dot{x} = y - x(x^2 + y^2 - 1), \\ \dot{y} = -x - y(x^2 + y^2 - 1). \end{cases} \tag{4.2.1}$$

According to the limit cycle theory [193], there is a stable limit cycle C. That is, in the two-dimensional flow field, limit cycle C is within the two divided sea areas, both internally and externally stable. That is to say, the area of C and the flow lines of the area tend to limit cycle C. There is only one stable closed orbit, so there is no mesoscale eddy. And take the vertical direction $\dot{z} = \varepsilon(f)t$. On sea surface or at a certain depth of the ocean, only one stable limit cycle is generated as shown in Fig. 4.48. As time increases, only one cylinder can be formed in space as shown in Fig. 4.49.

Table 4.1 Stable limit cycle parameter range

α	β	a	b	m	R	ω	$\lambda\omega$	$\mu\omega$
[0, 3]	[0, 3]			$(N-0.5$ $N+0.13)$				
$\alpha+\beta<0$	$\alpha+\beta<0$	(0, 400]	(0, 400]	$N=$ $1, 3, \ldots, 11$	[0.8, 1.1]	$\omega \geq 0.8$	$\lambda\omega \geq 0.8\sqrt{\frac{b}{a}}$	$\mu\omega \geq 0.8\sqrt{\frac{a}{b}}$
[0, 3]	[−3, 0]	(0, 400]	(0, 400]	$N=$ $1, 3, 5, 7, 9$	[0.8, 1.1]	$\omega \geq 0.8$	$\lambda\omega \geq 0.8\sqrt{\frac{b}{a}}$	$\mu\omega \geq 0.8\sqrt{\frac{a}{b}}$
$\vert\beta\vert<\alpha$	$\vert\beta\vert<\alpha$							

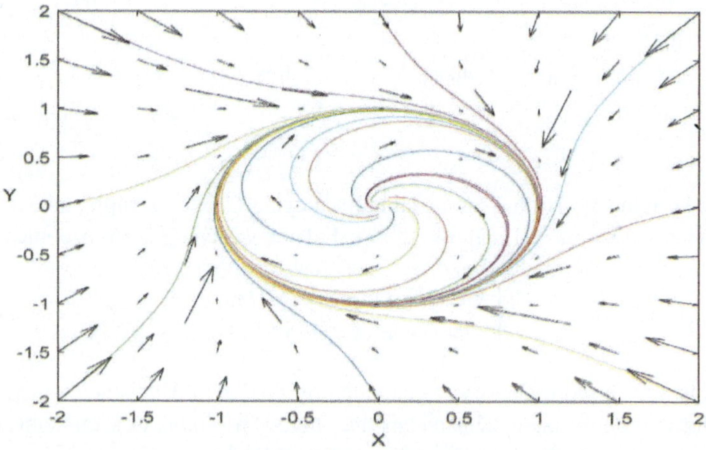

Fig. 4.48 The surface flow field of a stable limit cycle on a plane

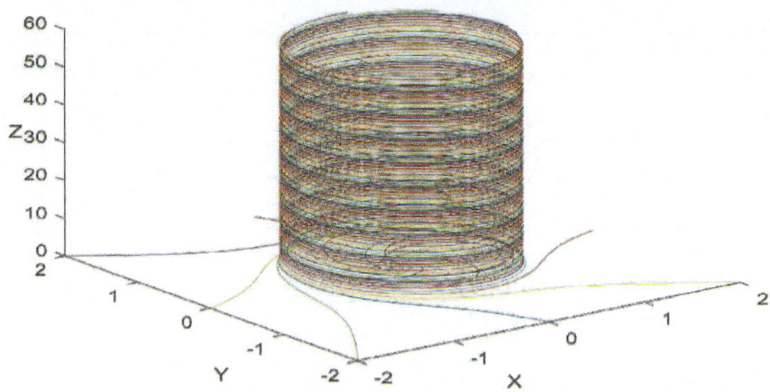

Fig. 4.49 Space structure of stable limit cycle

4.3 Unstable Limit Cycle

Based on the conclusion of Sect. 4.1, we get that when parameters α, β, m, a, b, R, λ, μ, and ω satisfy the range in Table 4.2.

Table 4.2 Unstable limit cycle parameter range

α	β	a	b	m	R	ω	$\lambda\omega$	$\mu\omega$
$[0, 3]$	$[0, 3]$			$(N - 0.5N + 0.5)$				
$\alpha + \beta > 0$	$\alpha + \beta > 0$	$(0, 400]$	$(0, 400]$	$N = 1, 3, \ldots, 11$	$[0.8, 1.1]$	$\omega \geq 0.8$	$\lambda\omega \geq 0.8\sqrt{\frac{b}{a}}$	$\mu\omega \geq 0.8\sqrt{\frac{a}{b}}$
$[0, 3]$	$[-3, 0]$							
$\lvert\beta\rvert < \alpha$	$\lvert\beta\rvert < \alpha$	$(0, 400]$	$(0, 400]$	$N = 1, 3, 5, 7, 9$	$[0.8, 1.1]$	$\omega \geq 0.8$	$\lambda\omega \geq 0.8\sqrt{\frac{b}{a}}$	$\mu\omega \geq 0.8\sqrt{\frac{a}{b}}$
$[-3, 0]$	$[0, 3]$							
$\lvert\alpha\rvert < \beta$	$\lvert\alpha\rvert < \beta$	$(0, 400]$	$(0, 400]$	$N = 1, 3, 5, 7, 9$	$[0.8, 1.1]$	$\omega \geq 0.8$	$\lambda\omega \geq 0.8\sqrt{\frac{b}{a}}$	$\mu\omega \geq 0.8\sqrt{\frac{a}{b}}$

Then system (3.3.16) has an unstable limit cycle C. For example, take $a = b = R = \omega = m = \lambda = \mu = 1$ and $\alpha = \beta = -1$. Then system (3.3.16) becomes

$$\begin{cases} \dot{x} = y + x(x^2 + y^2 - 1), \\ \dot{y} = -x + y(x^2 + y^2 - 1). \end{cases} \tag{4.3.1}$$

That is, the closed trajectory C generated by system (4.3.1) is neither externally stable nor internally stable. This means that there is no stable or semi-stable closed loop. And take the vertical direction $\dot{z} = \varepsilon(f)t$. When $t \to \infty(-\infty)$, C is unstable, as shown in Fig. 4.50, so that there is no mesoscale eddy, as shown in Fig. 4.51.

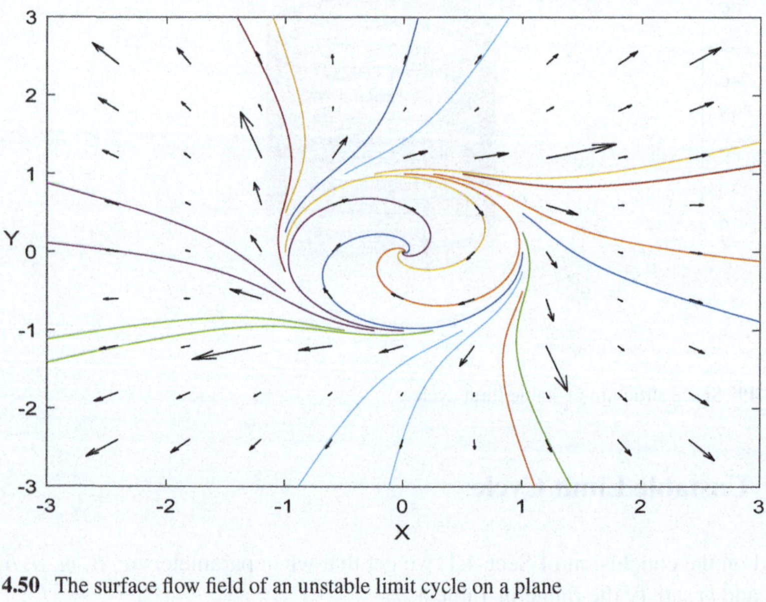

Fig. 4.50 The surface flow field of an unstable limit cycle on a plane

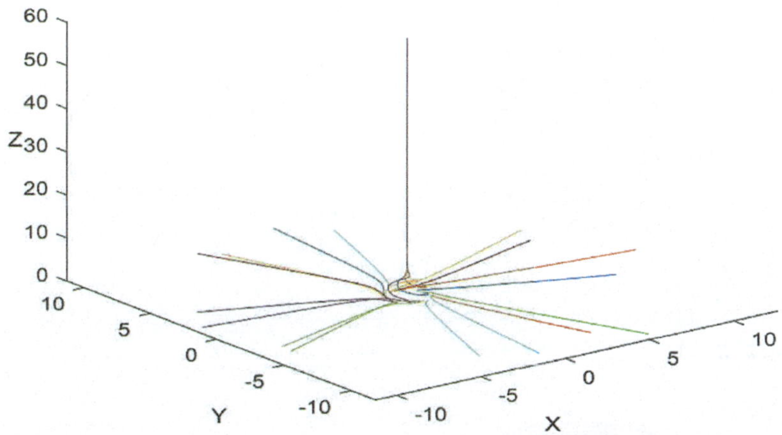

Fig. 4.51 Space structure of unstable limit cycle

4.4 Parameter Distribution Analysis of Semi-stable Limit Cycle

For the semi-stable limit cycle, we analyze the situation separately.

4.4.1 Special System

In order to obtain a general conclusion, first, pay attention to a special dynamic system

$$\begin{cases} \frac{dx}{dt} = \alpha x(x^2 + y^2 - R^2) + f\omega y, \\ \frac{dy}{dt} = \beta y(x^2 + y^2 - R^2) - f\omega x, \end{cases} \tag{4.4.1}$$

where α, β, m are three parameters, R is the radius, ω is the perturbation coefficient, and f is the Coriolis-force. Next, we fix one or several ranges of these parameters to examine parameters one by one. The shooting method and the screening method [478, 479] are still used.

4.4.1.1 About Radius *R*

When $\alpha = 0$, $\beta = 3$, $m = 6$, $f\omega = 0.8$, and $R \in (0.6, 1.1)$, system (4.4.1) has a semi-stable limit cycle. See Figs. 4.52, 4.53, 4.54.

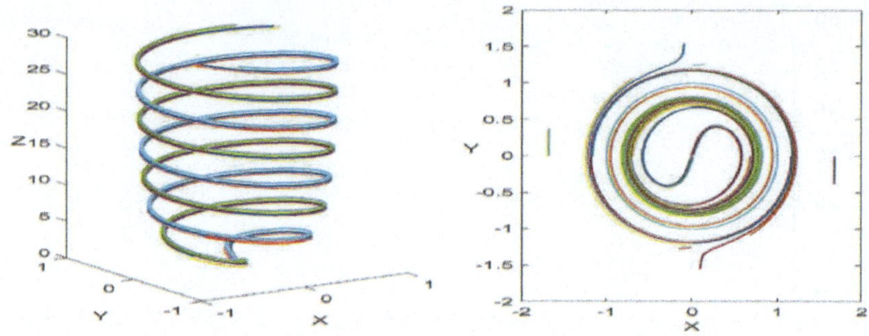

Fig. 4.52 $R = 1$

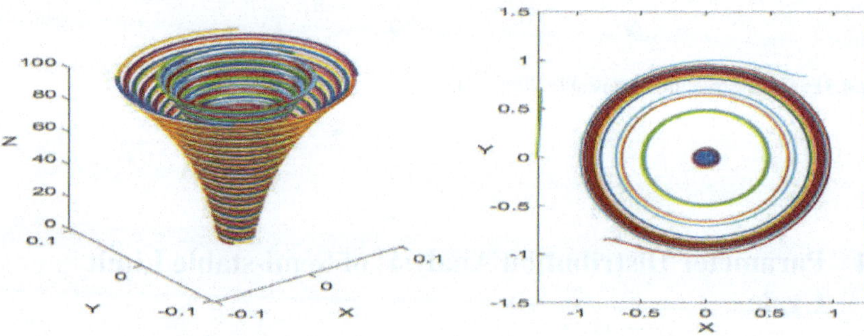

Fig. 4.53 $R = 0.7$

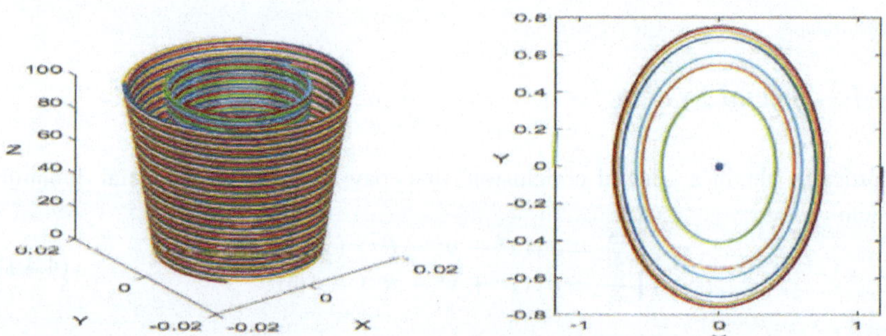

Fig. 4.54 $R = 0.8$

4.4.1.2 About Parameter *m*

Assuming that α, β, R, and f are fixed, examine the value of m that determines the existence of the semi-stable limit cycle of the system.

When $m = 8$, it can be seen from Fig. 4.55 that although the inside of the limit cycle converges to itself, the trend is very slow. So, $m < 8$.

When $m = 7$, although the inside of the limit cycle converges to itself, the trend is still very slow. See Fig. 4.56. Therefore, $m < 7$.

When $m = 6$, the inside of the limit cycle converges to itself from different initial points. See Fig. 4.57.

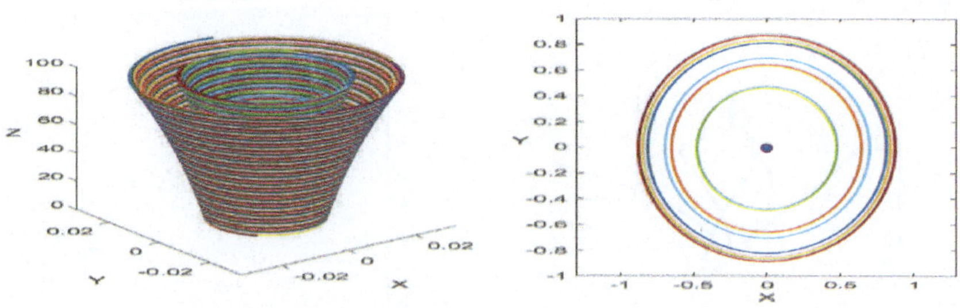

Fig. 4.55 $m = 8$

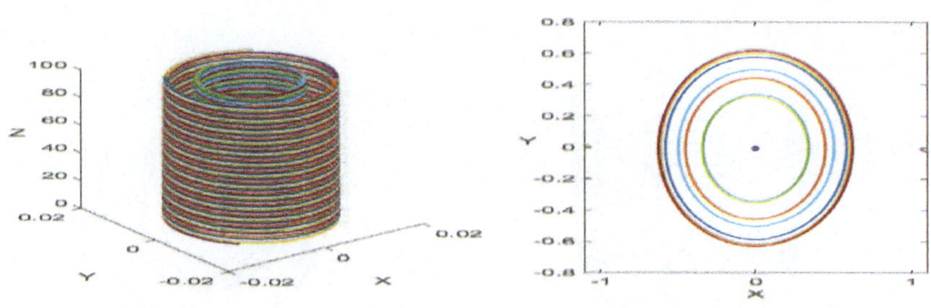

Fig. 4.56 $m = 7$

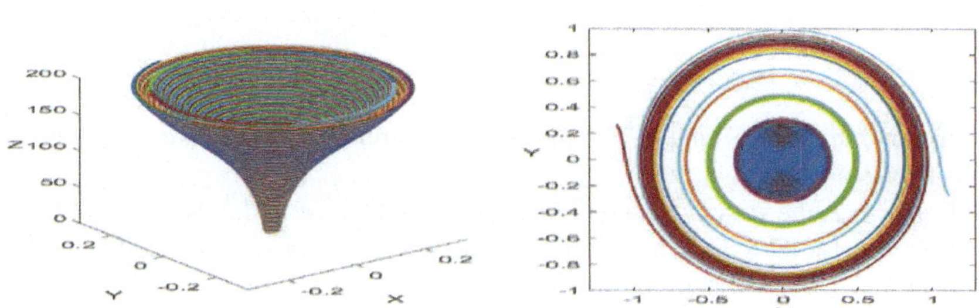

Fig. 4.57 $m = 6$

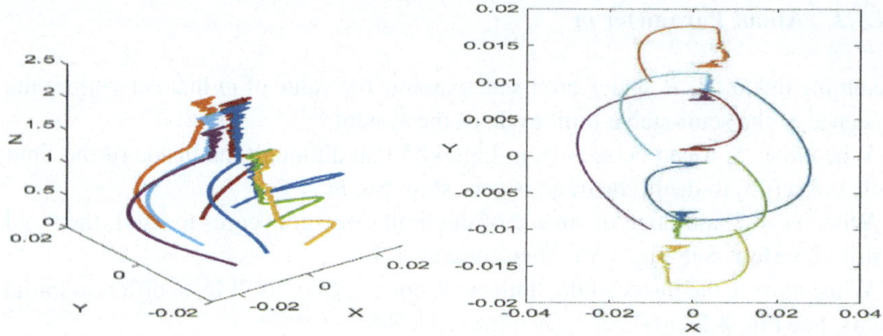

Fig. 4.58 $m = 6.2$

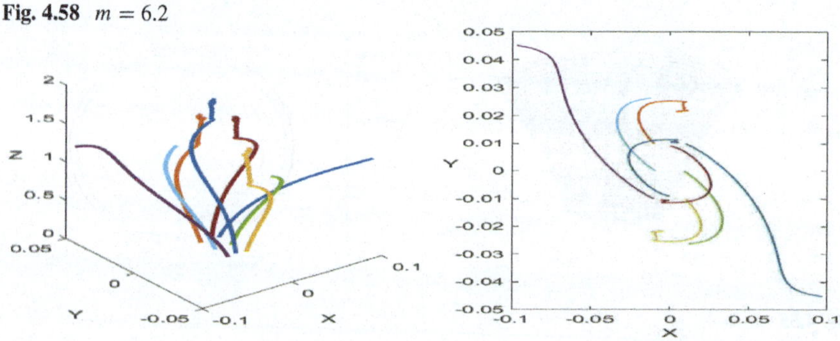

Fig. 4.59 $m = 6.14$

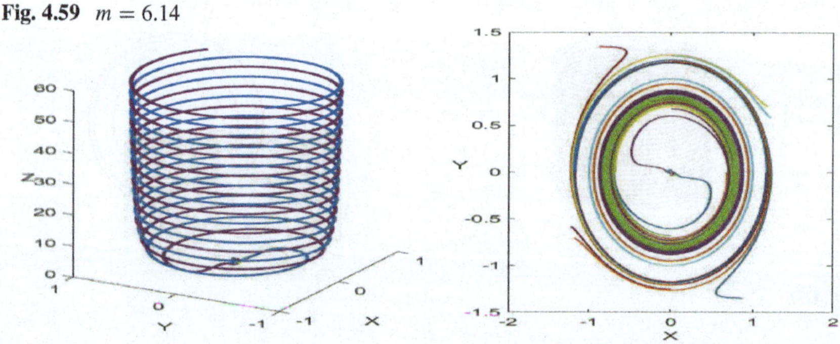

Fig. 4.60 $m = 6.13$

In summary, $m < 7$ and $m = 6$ meets the requirements. Therefore, the range of m is between 6 and 7.

If $m = 6.2$, the system has no limit cycle. Therefore, $m < 6.2$. See Fig. 4.58.

If $m = 6.14$, the system has no limit cycle. Therefore, $m < 6.14$. See Fig. 4.59.

If $m = 6.13$, the system has just one limit cycle. Therefore, $m \leq 6.13$. See Fig. 4.60.

In summary, the upper limit of m range is 6.13.

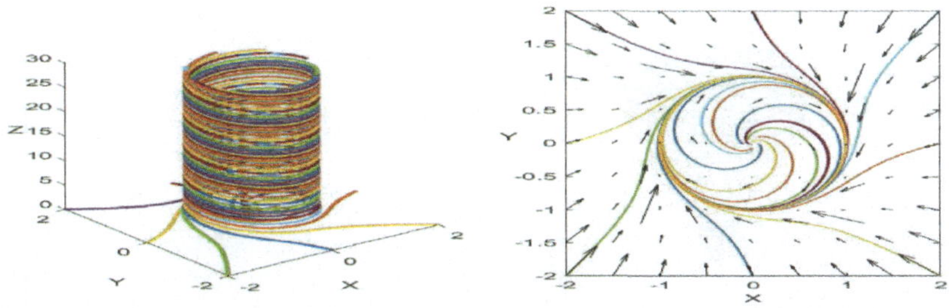

Fig. 4.61 $m = 6.2$

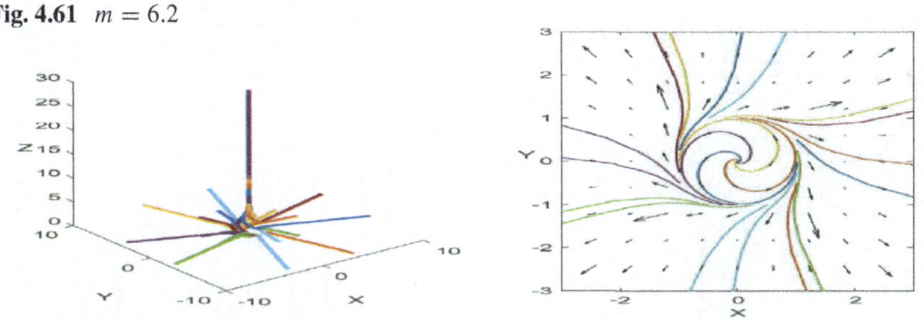

Fig. 4.62 $m = 1$

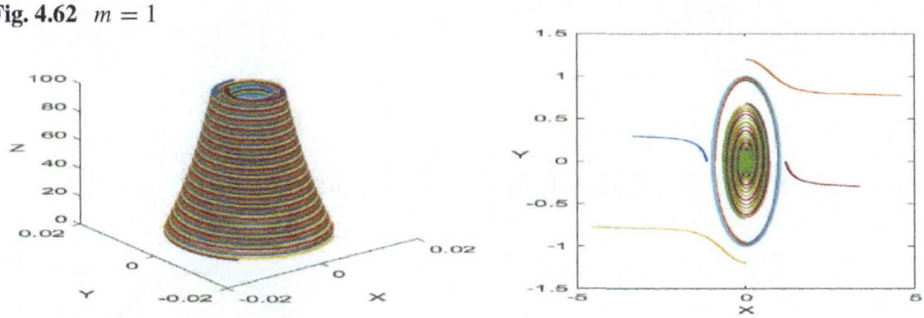

Fig. 4.63 $m = 1.49$

When $m = 6.2$, if $\alpha = \beta = -1$, $R = 1$ and $f\omega = 1$, the system has a stable limit cycle. See Fig. 4.61.

If $m = 1$, $\alpha = \beta = 1$, $R = 1$, $f\omega = 1$, the system has unstable limit cycles. See Fig. 4.62.

It cannot produce a semi-stable limit cycle when $m = 1$. Therefore, $m > 1$. When $m = 1.49$, the system has a semi-stable limit cycle, that is, $m > 1.49$. See Fig. 4.63.

When $m = 1.5$, the system has a semi-stable limit cycle. Therefore, $m \geq 1.5$. See Fig. 4.64.

From the above, we get the approximate range of m is $[1.5, 6.13]$. However, when $m = 3$ or 5, there is no semi-stable limit cycle. Therefore, the range of m needs to be further refined. See Fig. 4.65.

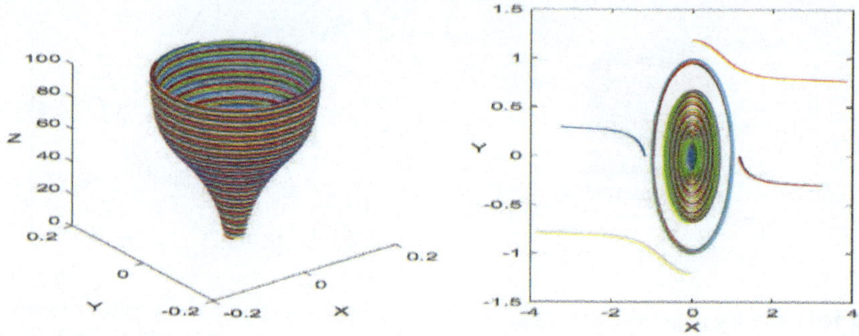

Fig. 4.64 $m = 1.5$

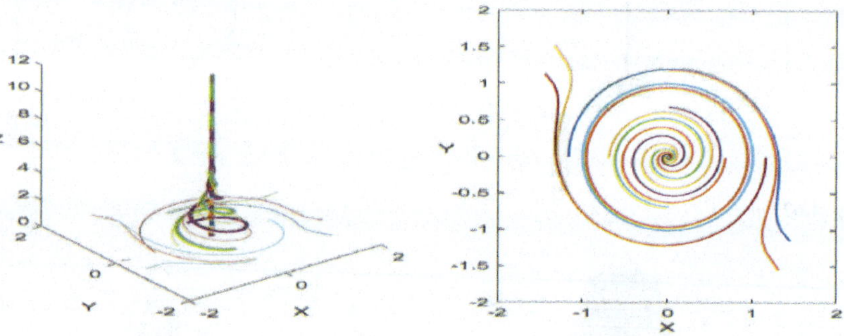

(a) $\alpha = 1,\ \beta = 1,\ a = 1,\ b = 1,\ R = 1,\ m = 3,\ \lambda = 3,\ \mu = 3$

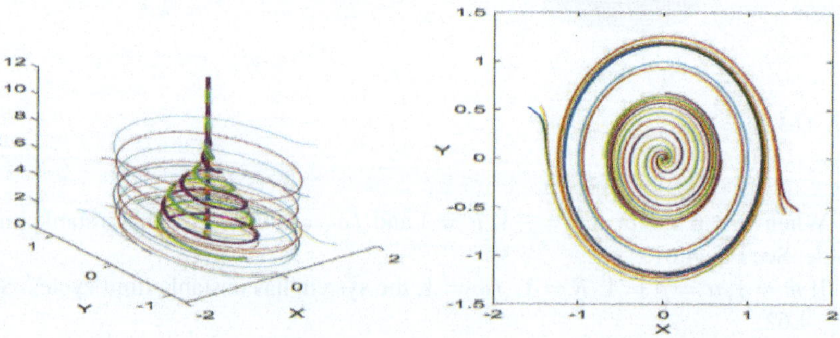

(b) $\alpha = 1,\ \beta = 1,\ a = 1,\ b = 1,\ R = 1,\ m = 5,\ \lambda = 1,\ \mu = 1$

Fig. 4.65 $m = 3$ or 5

The following reanalysis about $m \in [1.5, 6.13]$. Since $m = 2, 3, 5$ does not meet the requirements but $m = 4$ is compliant, the approximate range is divided into three segments [1.5, 3), [3, 5), and [5, 6.13] for analysis.

1. $m \in [1.5, 3)$.

   (a) When $m = 2.11$, the system has a semi-stable limit cycle. See Fig. 4.66.
   (b) When $m = 2.12$, the system has no semi-stable limit cycle. Therefore, $m < 2.12$. See Fig. 4.67.

   Thus, $m \in [1.5, 2.11)$.
2. $m \in [3, 5)$.

   (a) When $m = 3.49$, the system has an unstable limit cycle. See Fig. 4.68. Therefore, $m > 3.49$.
   (b) When $m = 3.5$, the system has a semi-stable limit cycle. See Fig. 4.69. Therefore, $m \geq 3.5$.
   (c) When $m = 4.11$, the system has a semi-stable limit cycle. See Fig. 4.70. Therefore, $m \geq 4.11$.
   (d) When When $m = 4.12$, the system has no semi-stable limit cycle. See Fig. 4.71. Therefore, $m < 4.12$.

   Finally, $m \in [3.5, 4.11]$.
3. $m \in [5, 7)$.

   (a) When $m = 5.49$, the system has an unstable limit cycle. See Fig. 4.72. Therefore, $m > 5.49$.
   (b) When $m = 5.5$, the system has a semi-stable limit cycle. See Fig. 4.73. Therefore, $m \geq 5.5$.

In summary, $m \in [5.5, 6.13]$. We use the shooting method and the screening method to obtain the three ranges of parameter m, that is, $[1.5, 2.11]$, $[3.5, 4.11]$, and $[5.5, 6.13]$, which make the system has a semi-stable limit cycle.

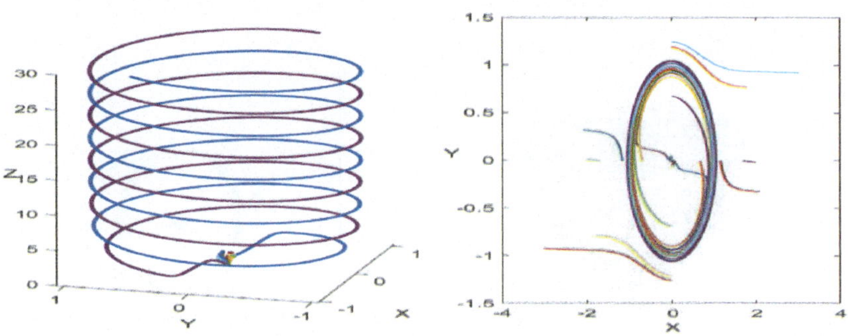

Fig. 4.66 $m = 2.11$

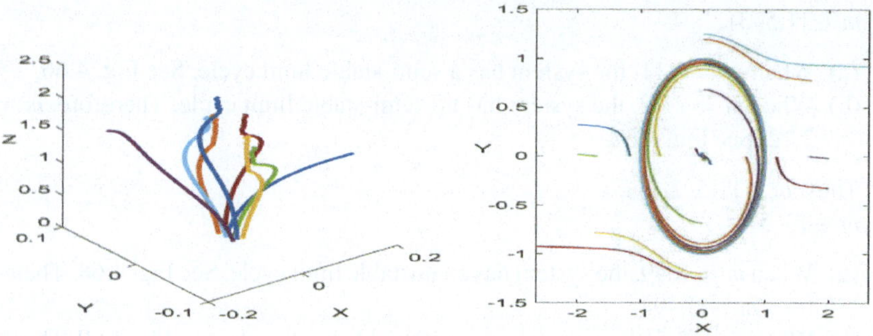

Fig. 4.67 $m = 2.12$

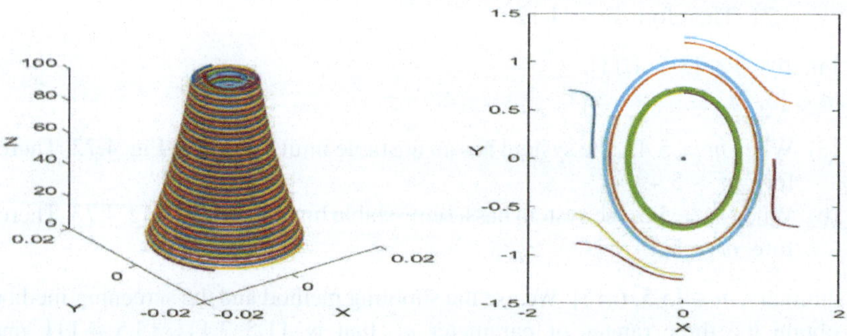

Fig. 4.68 $m = 3.49$

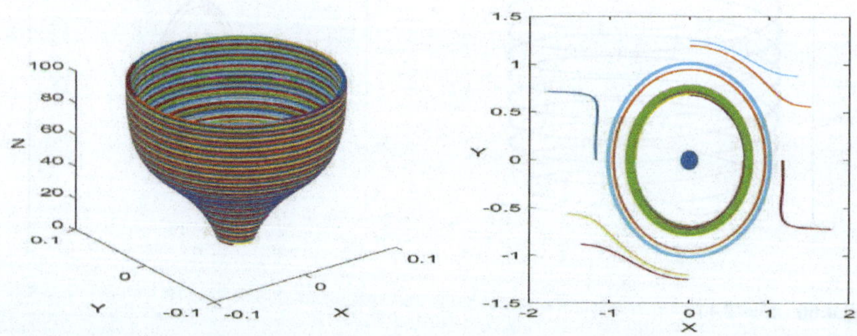

Fig. 4.69 $m = 3.5$

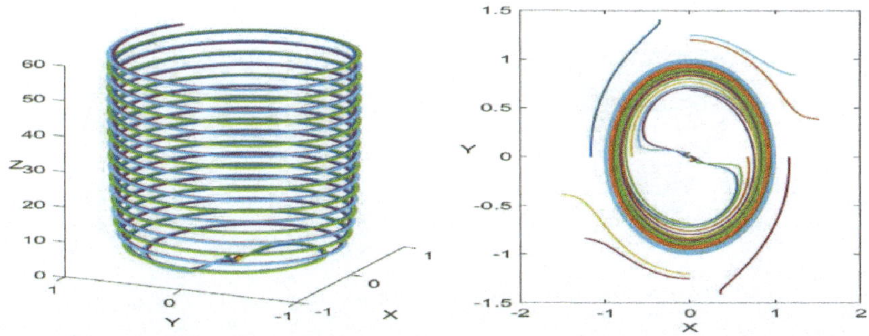

Fig. 4.70 $m = 4.11$

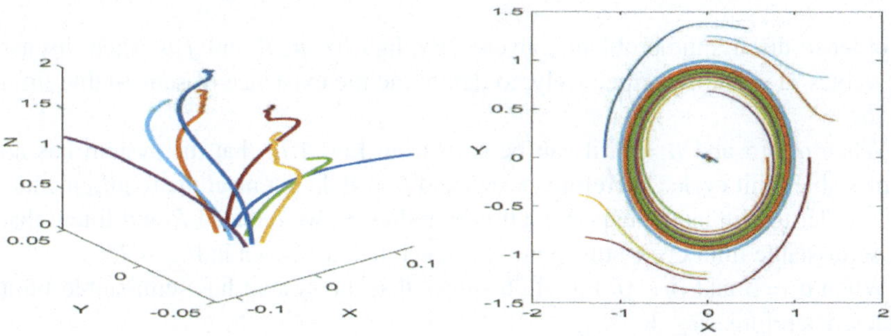

Fig. 4.71 $m = 4.12$

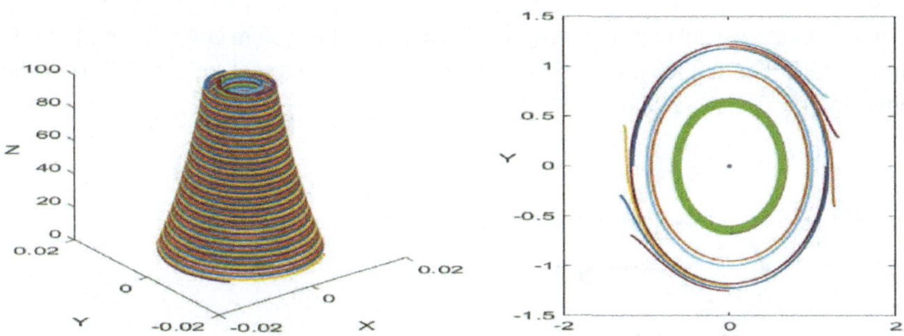

Fig. 4.72 $m = 5.49$

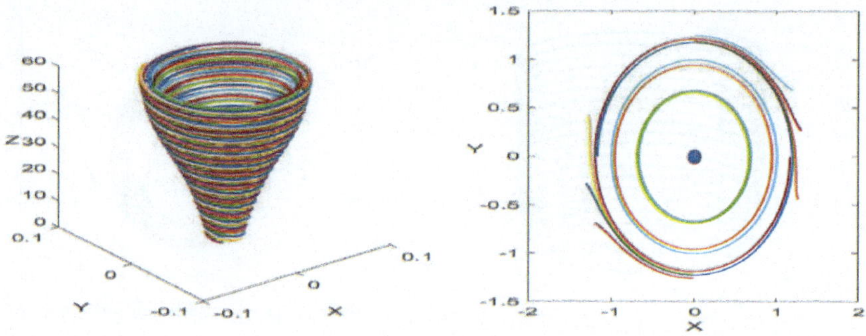

Fig. 4.73 $m = 5.5$

4.4.1.3 About Parameters α and β

In order to discuss the problem conveniently, first fix m, R and $f\omega$. Then discuss the values of α and β, respectively, to determine the existence of semi-stable limit cycles.

When $\alpha = 5$ and $\beta = 0$, it can be seen from Fig. 4.74 that the system has no semi-stable limit cycle. Therefore, $\alpha = 5$ and $\beta = 0$ do not meet the requirements.

In order to eliminate the perturbation of radius R, we reduced R and found that the semi-stable limit cycle still cannot be generated, as shown in Fig. 4.75.

When $\alpha = 3$ and $\beta = 0$, Fig. 4.76 shows that the system has semi-stable limit cycles. Therefore, $\alpha \leq 3$.

When $\alpha = 0$ and $\beta = 3$, the system has semi-stable limit cycles. See Fig. 4.77. Therefore, $\beta \leq 3$.

When $\alpha = 3$ and $\beta = 3$, the system has semi-stable limit cycles. See Fig. 4.78.

When $\alpha = -3$ and $\beta = -3$, the system has semi-stable limit cycles. See Fig. 4.79.

In summary, we get parameters α and β that make the system generate semi-stable limit cycles. The ranges are $[-3, 3]$, $\alpha\beta \geq 0$, α and β cannot be equal to 0 at the same time.

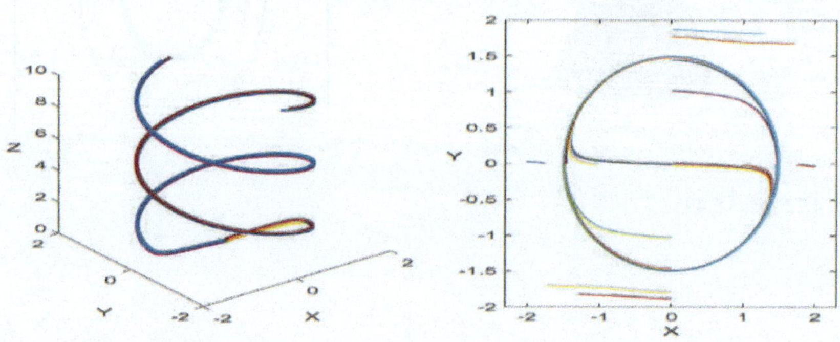

Fig. 4.74 $\alpha = 5$ and $\beta = 0$

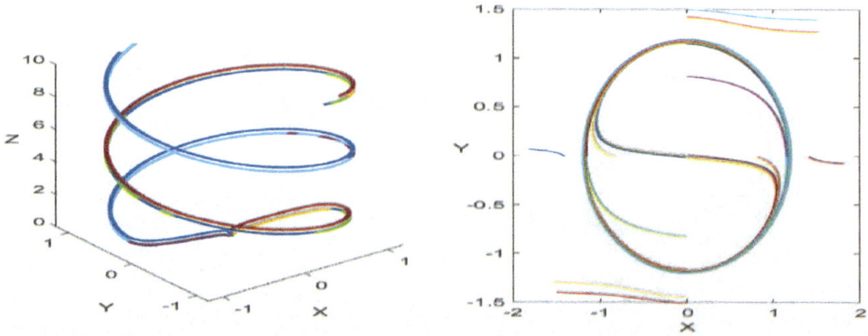

Fig. 4.75 Perturbation of radius R

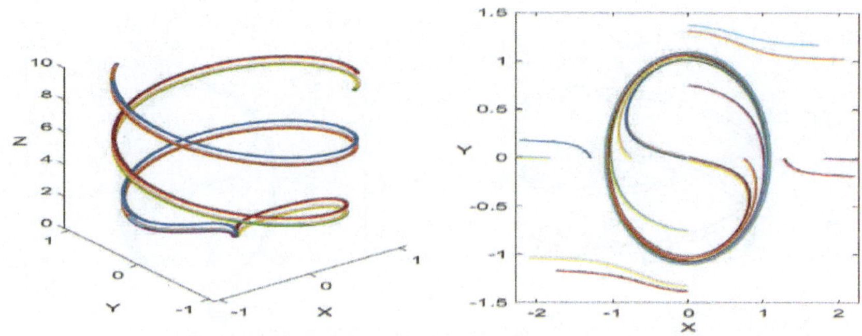

(a) $\alpha=3,\ \beta=0,\ a=1,\ b=1,\ R=1.1,\ m=2,\ \lambda=0.8,\ \mu=0.8$

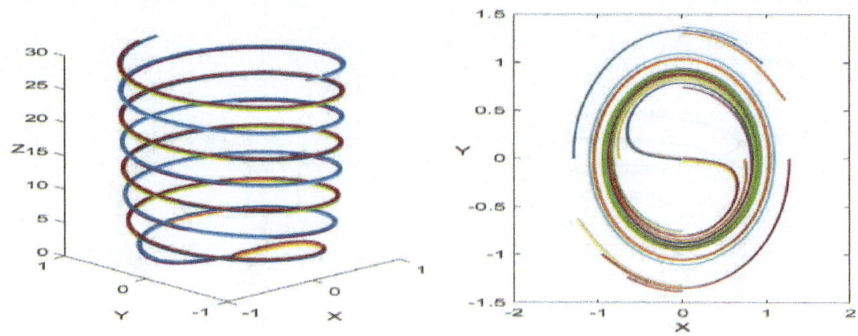

(b) $\alpha=3,\ \beta=0,\ a=1,\ b=1,\ R=1.1,\ m=6,\ \lambda=0.8,\ \mu=0.8$

Fig. 4.76 $\alpha=3$ and $\beta=0$

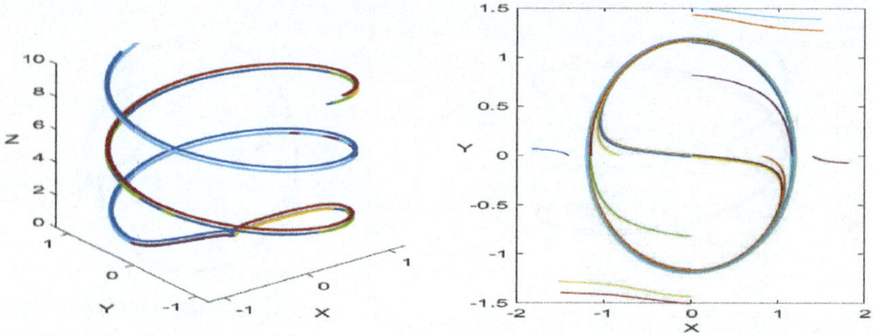

Fig. 4.77 $\alpha = 0, \beta = 3$

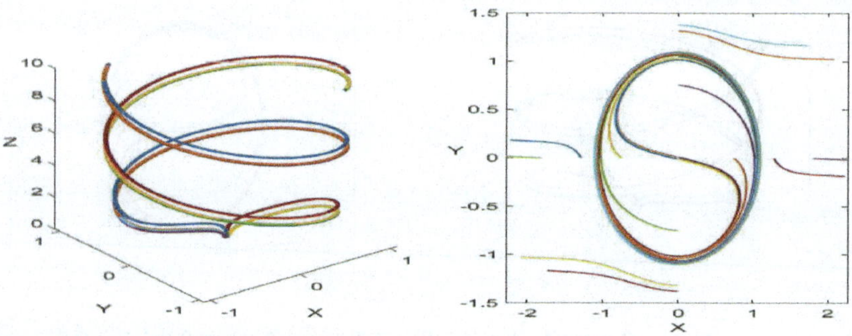

Fig. 4.78 $\alpha = 3, \beta = 3$

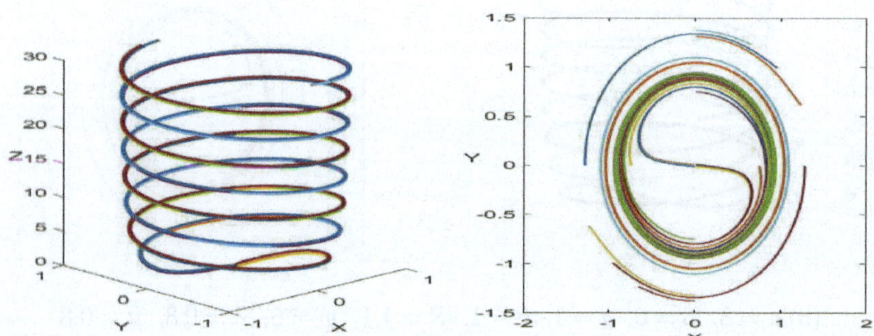

Fig. 4.79 $\alpha = -3, \beta = -3$

4.4.1.4 About Parameter f

Fix α, β, m, and R. For the convenience of discussion, take the case of $f_0 = f\omega$ to study the existence of the semi-stable limit cycle of the system.

α, β, and R take arbitrary values within their range. In order to show that it can be suitable in extreme cases, we take $\alpha = 3$, $\beta = 0$ and $R = 1.1$.

When $f_0 = f\omega = 0.1$, Fig. 4.80 shows that the system has no semi-stable limit cycles.

When $f = 0.7$, the system has no semi-stable limit cycles. See Fig. 4.81.

When $f = 0.8$, the system has semi-stable limit cycles. See Fig. 4.82.

When $f = 5$, the system has a semi-stable limit cycle. See Fig. 4.83.

Based on the above, we have obtained parameter $f_0 = f\omega \geq 0.8$ which makes the system generate a semi-stable limit cycle. And for the larger f_0, according to [209–211], it can generate semi-stable limit cycles. For parameter ω, similarly system (3.3.16) is semi-stable when $\omega \geq 0.8$.

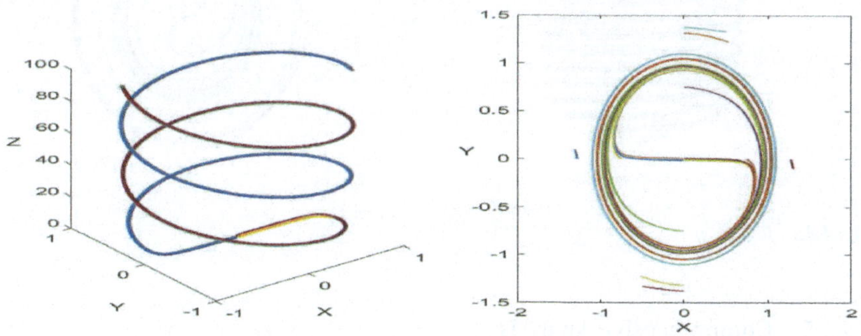

Fig. 4.80 $f = 0.1$

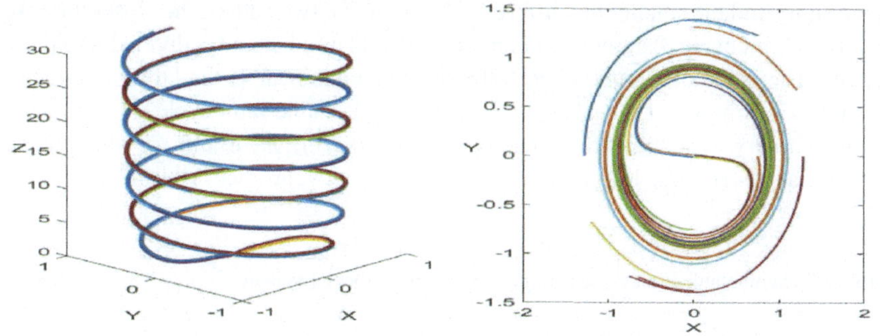

Fig. 4.81 $f = 0.7$

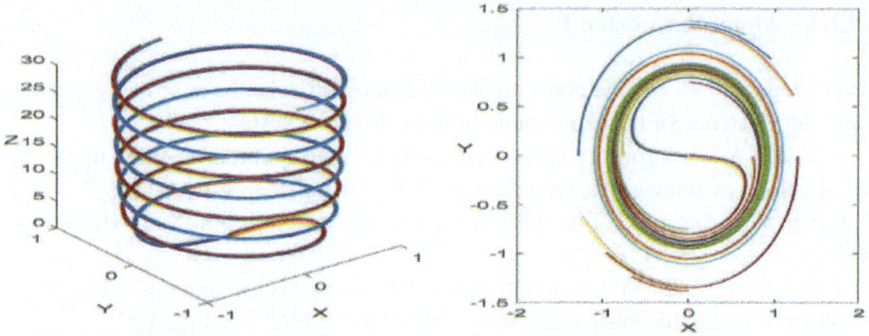

Fig. 4.82 $f = 0.8$

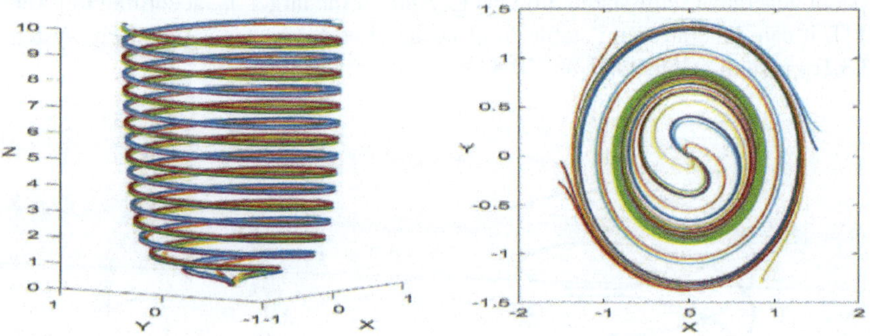

Fig. 4.83 $f = 5$

4.4.1.5 Comprehensive Analysis

Through the above simulation, it can be seen that parameters α and β must have the same sign. And the smaller the difference between the two values, the closer they are to form a limit cycle. Parameter m cannot be an odd number and is divided into three ranges. The value of parameter R is the most suitable near 1; the larger parameter f_0, the more simple for the system can form a semi-stable limit cycle.

In summary, we can get the range of parameters α, β, m, R and f_0, which make the special system (4.4.1) generate semi-stable limit cycles. See Table 4.3.

Table 4.3 Semi-stable limit cycle parameter range for a special system

α	β	m	R	f_0
$[-3, 3]$	$[-3, 3]$	$[1.5, 2.11]$	$(0.6, 1.1]$	$f_0 \geq 0.8$
$\alpha\beta \geq 0$	$\alpha\beta \geq 0$	$[3.5, 4.11]$		
$\alpha \neq \beta \neq 0$	$\alpha \neq \beta \neq 0$	$[5.5, 6.13]$		

4.4.2 General System

In system (3.3.16), R is the radius, ω is the perturbation factor, λ and μ are the parameters related to Coriolis-force. The following fixes the range of one or several of these parameters and uses shooting method and screening method to investigate the range of parameters one by one.

From Sect. 4.4.1, we have obtained the ranges of α, β, m, R, and f_0 as

$$\begin{aligned} &\lambda \in [-3, 3], \\ &\beta \in [-3, 3], \\ &m \in [1.5, 2.11], [3.5, 4.11], [5.5, 6.13], \\ &R \in (0.6, 1.1] \text{ and,} \\ &f_0 \geq 0.8. \end{aligned}$$

Since system (3.3.16) has added a, b, λ, and μ, first select one or more parameters from these four for discussion.

4.4.2.1 About Parameters a, b, λ, and μ

1. If λ and μ have no specific mathematical relationship with $\sqrt{b/a}$ and $\sqrt{a/b}$ (a, b are greater than zero), there is no semi-stable limit cycles.
   (a) When $\alpha = 1$, $\beta = 1$, $m = 2$, $\lambda\omega = 1$, and $\mu\omega = 1$, two parameters a and b change. As shown in Fig. 4.84, there is no semi-stable limit cycle.
   (b) When $\alpha = 1$, $\beta = 1$, $m = 2$, $R = 1.1$, $\lambda\omega = \sqrt{2}$, and $\mu\omega = \sqrt{2}$, two parameters a and b change. As shown in Fig. 4.85, there is no semi-stable limit cycle.

   Therefore, if $\lambda\omega$ and $\mu\omega$ have no mathematical relations with $\sqrt{b/a}$ and $\sqrt{a/b}$, there is no expected semi-stable limit cycle.
2. If λ, μ, and $\sqrt{b/a}$, $\sqrt{a/b}$ have a specific mathematical relationship, that is, $\lambda = \sqrt{b/a}$, $\mu = \sqrt{a/b}$. Then the system has a semi-stable limit cycle.
   (a) When $\alpha = 1$, $\beta = 1$, $m = 2$, and $R = 1.1$, $\lambda\omega$ and $\mu\omega$ changes with a and b, there is a semi-stable limit cycle. See Fig. 4.86.
   (b) When $\alpha = 2$, $\beta = 1$, $m = 6$, and $R = 1.1$, $\lambda\omega$ and $\mu\omega$ change with a and b, there is a semi-stable limit cycle. See Fig. 4.87.

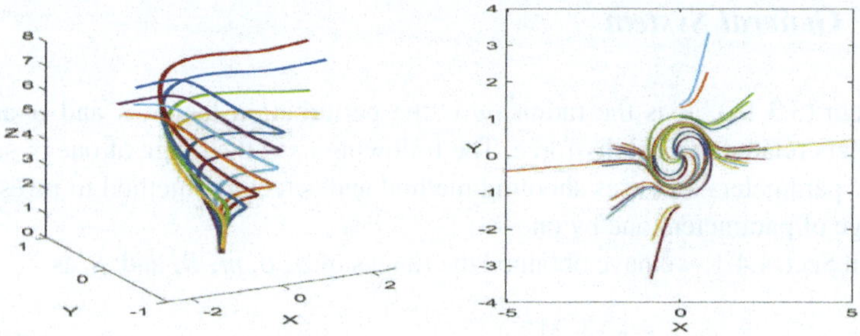

(a) $a=2,\ b=1,\ \lambda=1,\ \mu=1$

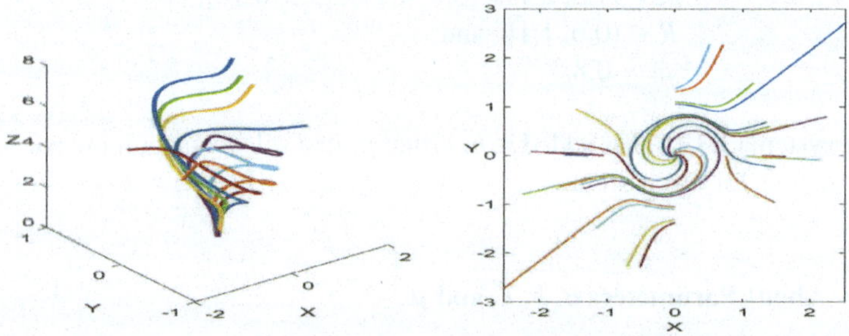

(b) $a=3,\ b=1,\ \lambda=1,\ \mu=1$

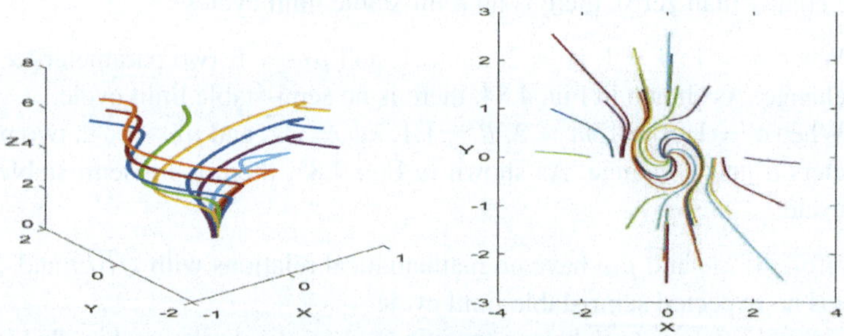

(c) $a=1,\ b=3,\ \lambda=1,\ \mu=1$

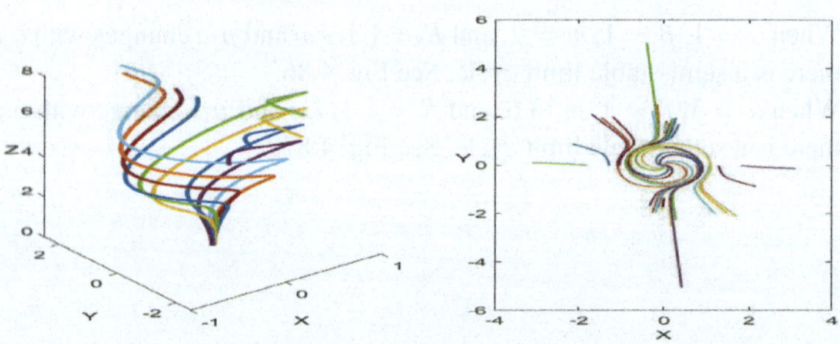

(d) $a=1,\ b=2,\ \lambda=1,\ \mu=1$

Fig. 4.84 $\alpha=1, \beta=1, R=1.1, m=2, \lambda\omega=1$, and $\mu\omega=1$

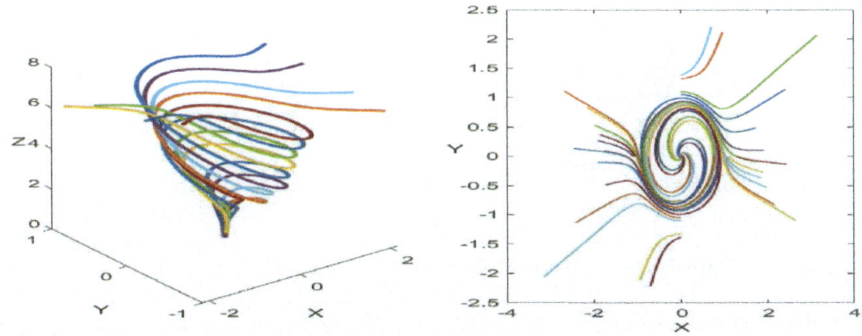

Fig. 4.85 $\alpha = 1, \beta = 1, m = 2, R = 1.1, \lambda\omega = \sqrt{2}$, and $\mu\omega = \sqrt{2}$

It can be seen from the above that a and b (greater than zero) affect the horizontal and vertical size of the limit cycle: a affects x axis direction; b affects y axis direction. The larger the value, the smaller the limit cycle.

3. If a and b are less than zero, there is no limit cycle. See Fig. 4.88.
   From the above, the value of a and b must be greater than zero. If there is less than zero, there are no circles or ellipses.

Next, we will further determine the range of a and b.

When $\alpha = 2$, $\beta = 1$, $m = 6$, and $R = 1$, μ change with a and b, which make the system have a semi-stable limit cycle. See Fig. 4.89. When $a = 400$, the limit cycle is greatly stretched. And a and b are greater than zero, so $a, b \in (0, 400]$. In addition, $\lambda = \sqrt{b/a}$ and $\mu = \sqrt{a/b}$, fix parameters α, β, m, R. Then λ and μ change with a and b. It can be seen from Sect. 4.1.1 that ω is the same as f in system (4.4.1), $f \geq 0.8$, and $\omega \geq 0.8$. See Fig. 4.90.

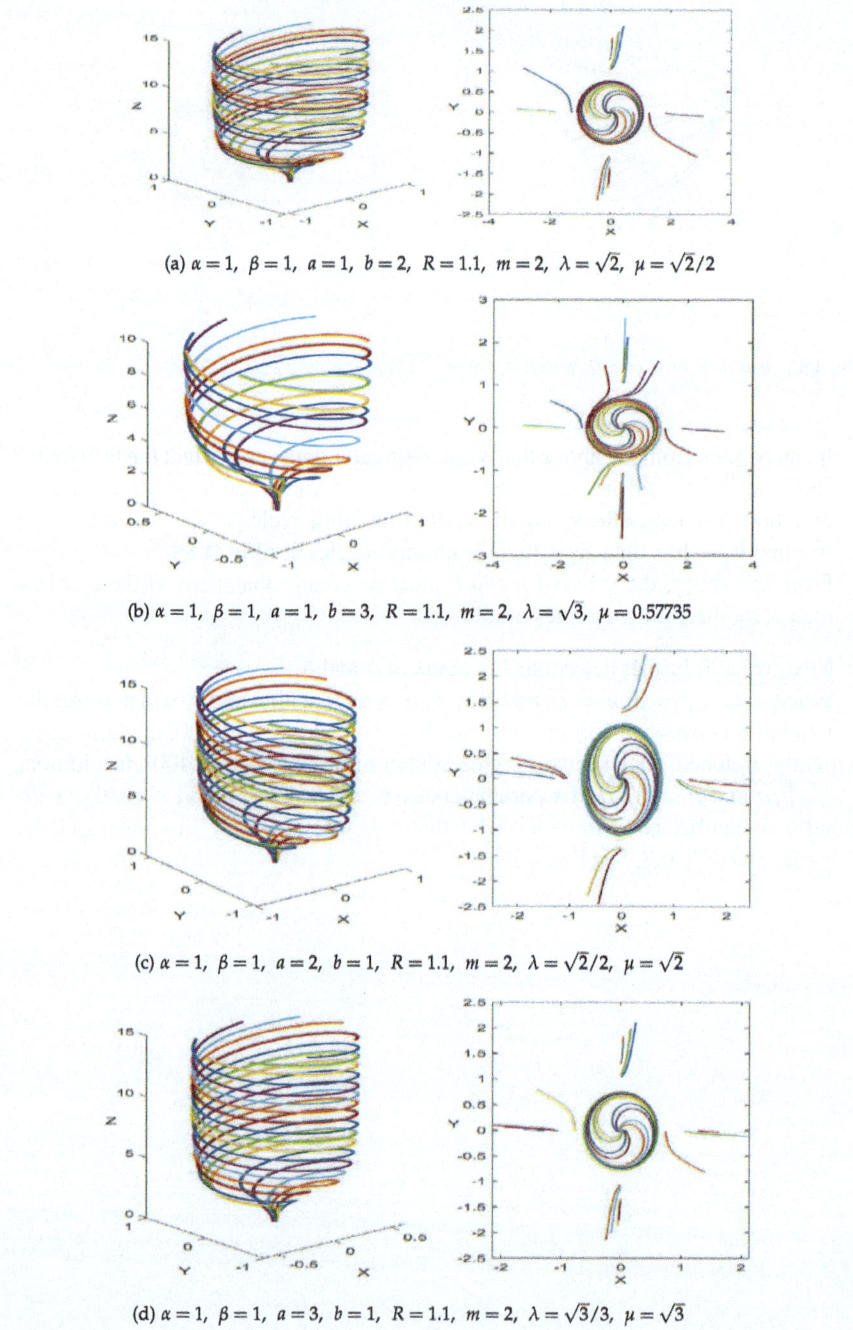

(a) $\alpha = 1,\ \beta = 1,\ a = 1,\ b = 2,\ R = 1.1,\ m = 2,\ \lambda = \sqrt{2},\ \mu = \sqrt{2}/2$

(b) $\alpha = 1,\ \beta = 1,\ a = 1,\ b = 3,\ R = 1.1,\ m = 2,\ \lambda = \sqrt{3},\ \mu = 0.57735$

(c) $\alpha = 1,\ \beta = 1,\ a = 2,\ b = 1,\ R = 1.1,\ m = 2,\ \lambda = \sqrt{2}/2,\ \mu = \sqrt{2}$

(d) $\alpha = 1,\ \beta = 1,\ a = 3,\ b = 1,\ R = 1.1,\ m = 2,\ \lambda = \sqrt{3}/3,\ \mu = \sqrt{3}$

Fig. 4.86 $\alpha = 1, \beta = 1, R = 1.1$, and $m = 2$

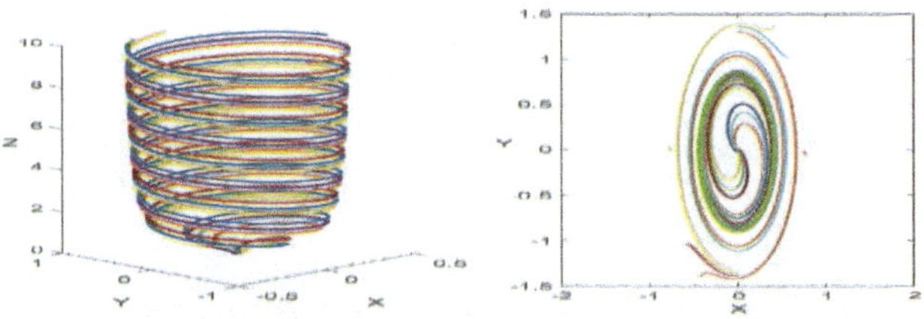

(a) $\alpha=2,\ \beta=1,\ a=4,\ b=1,\ R=1.1,\ m=6,\ \lambda=1.5,\ \mu=3\sqrt{2}$

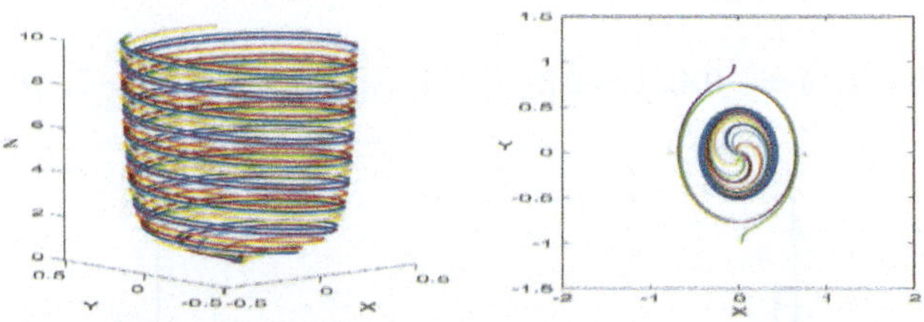

(b) $\alpha=2,\ \beta=1,\ a=4,\ b=3,\ R=1.1,\ m=6,\ \lambda=1.5\sqrt{3},\ \mu=2\sqrt{3}$

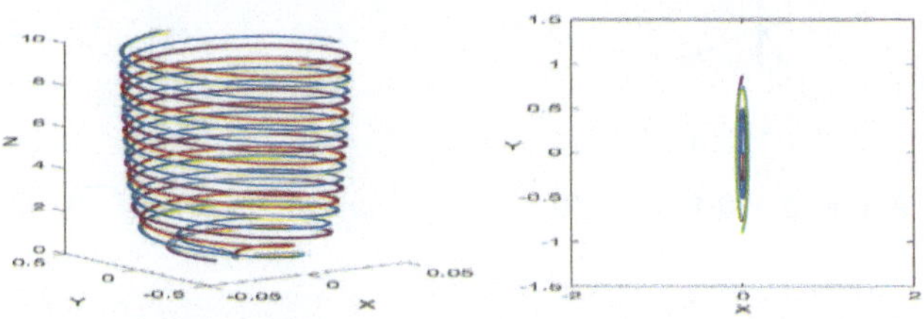

(c) $\alpha=2,\ \beta=1,\ a=400,\ b=3,\ R=1.1,\ m=6,\ \lambda=3\sqrt{3}/20,\ \mu=20\sqrt{3}$

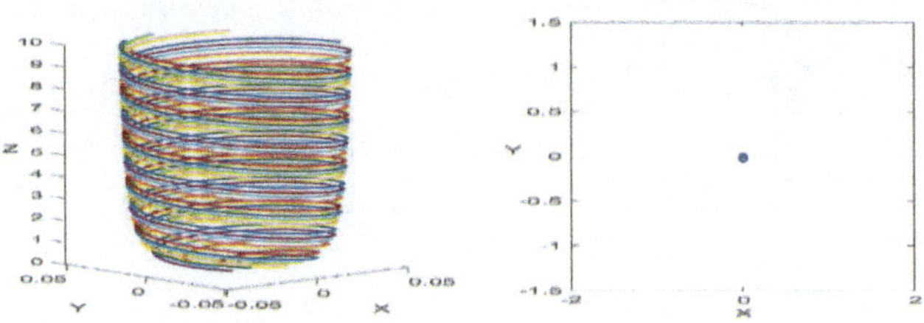

(d) $\alpha=2,\ \beta=1,\ a=400,\ b=300,\ R=1.1,\ m=6,\ \lambda=1.5\sqrt{3},\ \mu=2\sqrt{3}$

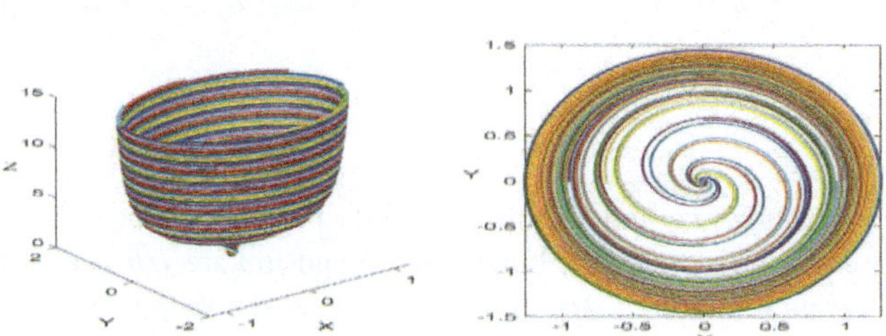

(e) $\alpha=2,\ \beta=1,\ a=0.4,\ b=0.3,\ R=1.1,\ m=6,\ \lambda=1.5\sqrt{3},\ \mu=2\sqrt{3}$

Fig. 4.87 $\alpha = 2, \beta = 1, a = 3, R = 1.1$, and $m = 6$

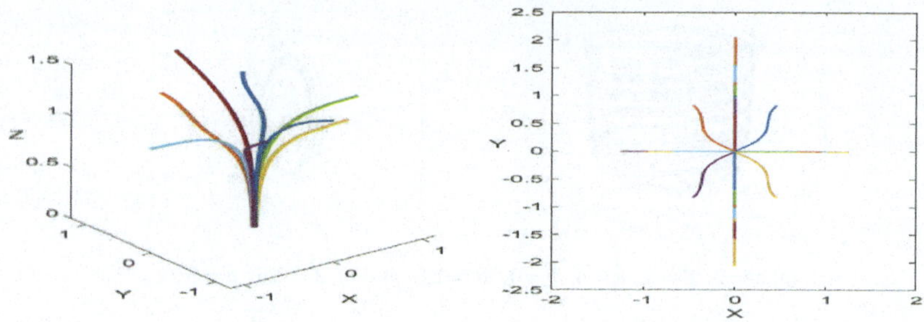

(a) $\alpha = 2,\ \beta = 1,\ a = -0.4,\ b = 0.3,\ R = 1,\ m = 6,\ \lambda = 1.5\sqrt{-3},\ \mu = 2\sqrt{-3}$

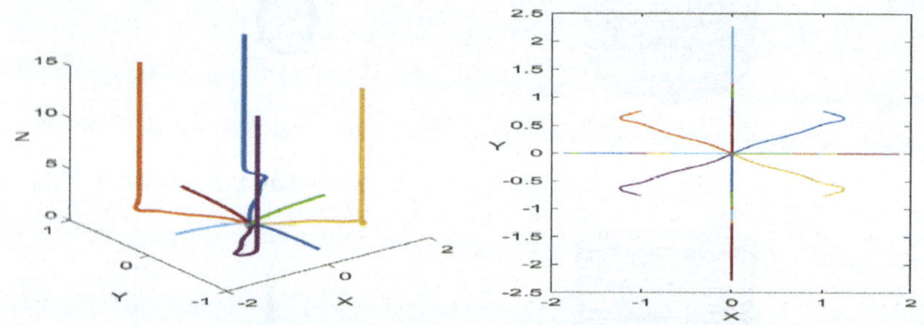

(b) $\alpha = 2,\ \beta = 1,\ a = 0.4,\ b = -0.3,\ R = 1,\ m = 6,\ \lambda = 1.5\sqrt{-3},\ \mu = 2\sqrt{-3}$

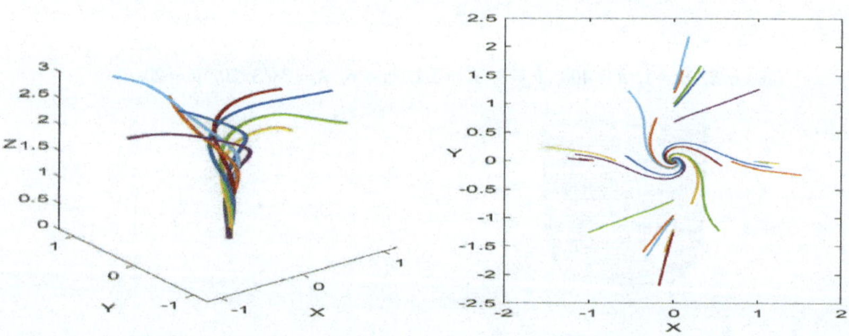

(c) $\alpha = 2,\ \beta = 1,\ a = -0.4,\ b = -0.3,\ R = 1,\ m = 6,\ \lambda = 1.5\sqrt{3},\ \mu = 2\sqrt{3}$

Fig. 4.88 $\alpha = 2$, $\beta = 1$, $R = 1$, and $m = 6$

Therefore, the larger $\lambda\omega$ and $\mu\omega$, the smaller the particle motion trajectory interval, and the denser the trajectory. The ranges of $\lambda\omega$ and $\mu\omega$ are $\sqrt{b/a}$ and $\sqrt{a/b}$ times of f in the original system, respectively. Therefore, the ranges of $\lambda\omega$ and $\mu\omega$ are $\lambda\omega \geq 0.8\sqrt{b/a}$ and $\mu\omega \geq 0.8\sqrt{a/b}$, respectively.

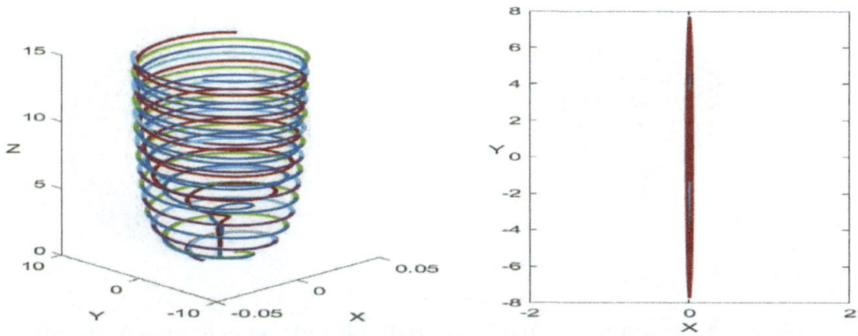

Fig. 4.89 $\alpha = 2, \beta = 1, m = 6$, and $R = 1\alpha = 2, \beta = 1, m = 6$, and $R = 1$

4.4.2.2 About Parameter *R*

Fix α, β, a, b, m, λ, μ, and ω. Then, discuss the value of R in (0.6, 1.1) to determine the existence of the semi-stable limit cycle of the system.

1. When $\alpha = 2$, $\beta = 1$, and $m = 6$, λ and μ change with a and b, the system has a semi-stable limit cycle. It is shown in Fig. 4.91, where $a = 400$, $b = 0.01$, and $R = 0.7$.
2. When $\alpha = 2$, $\beta = 1$, and $m = 6$, λ and μ change with a and b, the system has a semi-stable limit cycle. It is shown in Fig. 4.92, where $a = 0.01$, $b = 0.01$, and $R = 0.7$.
3. When $\alpha = 2$, $\beta = 1$, and $m = 6$, λ and μ change with a and b, the system has a semi-stable limit cycle. It is shown in Fig. 4.93, where $a = 400$, $b = 0.01$, and $R = 1.1$.
4. When $\alpha = 2$, $\beta = 1$, and $m = 6$, λ and μ change with a and b, the system has a semi-stable limit cycle. It is shown in Fig. 4.94, where $a = 400$, $b = 400$, and $R = 1.1$.

From above, we get $R \in [0.8, 1.1]$.

4.4.2.3 About Parameters α and β

From system (4.2.1), it can be seen that the range of α and β is $[-3, 3]$. In the following, we analyze whether the range is suitable for system (3.3.16).

The range of α and β in system (4.4.1) is also suitable for system (3.3.16). So, α, $\beta \in [-3, 3]$, $\alpha\beta \geq 0$, and $\alpha \neq \beta \neq 0$. The simulation is shown in Fig. 4.95.

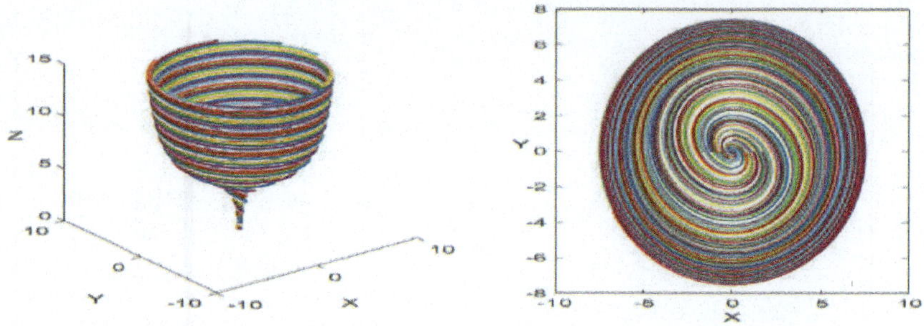

(a) $\alpha = 2,\ \beta = 0.1,\ a = 0.01,\ b = 0.01,\ R = 1,\ m = 6,\ \lambda = 3,\ \mu = 3$

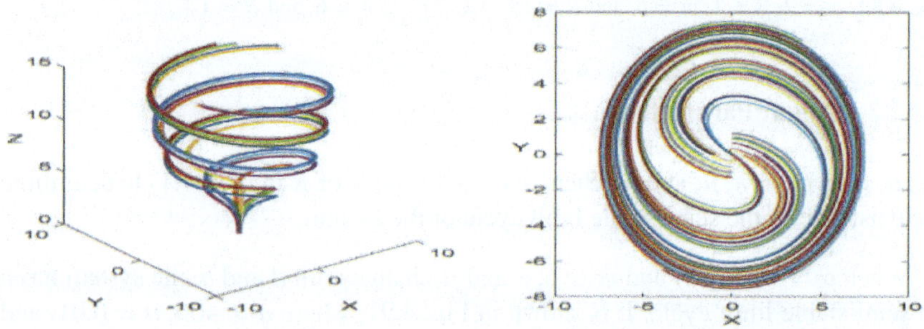

(b) $\alpha = 2,\ \beta = 0.1,\ a = 0.01,\ b = 0.01,\ R = 1,\ m = 6,\ \lambda = 1,\ \mu = 1$

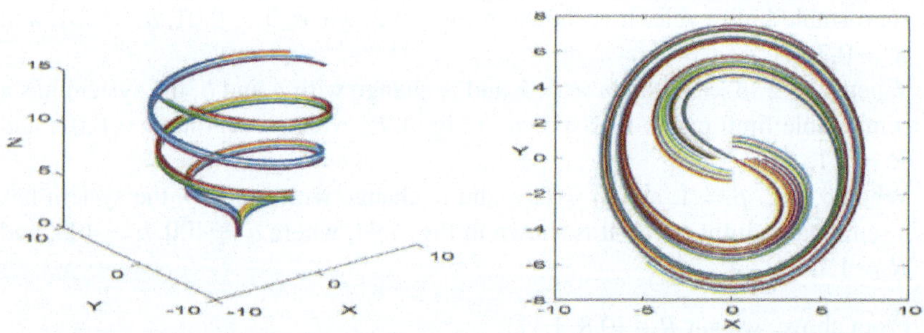

(c) $\alpha = 2,\ \beta = 0.1,\ a = 0.01,\ b = 0.01,\ R = 1,\ m = 6,\ \lambda = 0.8,\ \mu = 0.8$

Fig. 4.90 $\alpha = 2, \beta = 0.1, a = 0.01, b = 0.01, m = 6$, and $R = 1$

4.4.2.4 About Parameter *m*

The ranges of m in system (4.4.1) are $[1.5, 2.11]$, $[3.5, 4.11]$, and $[5.5, 6.13]$. Theses ranges should also suitable for system (3.3.16).

If $\alpha = 2$, $\beta = 1$, $a = 0.4$, and $b = 0.3$, λ and μ change with a and b, the system behavior is shown in Fig. 4.96, where $m = 8$.

Therefore, the range of m is still in $[1.5, 2.11]$, $[3.5, 4.11]$, and $[5.5, 6.13]$.

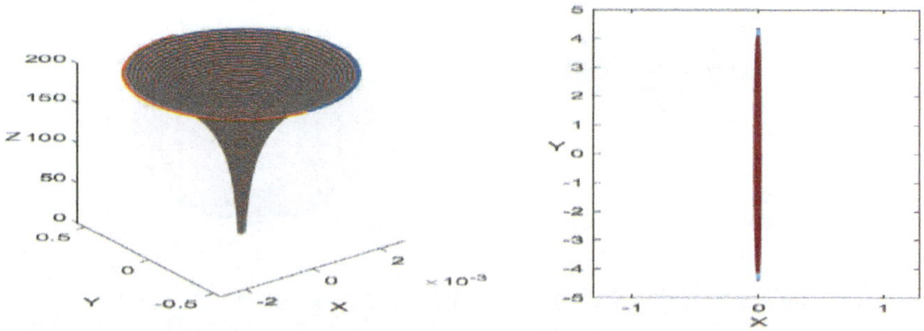

Fig. 4.91 $a = 400, b = 0.01$, and $R = 0.7$

4.4.2.5 Comprehensive Analysis

Parameters α and β must have the same sign, and the smaller the difference between the two values, the more a limit cycle can be formed. Parameter m cannot be an odd number, and it is divided into three ranges. The closer parameter R is to 1, the better the effect is.

1. a and b affect the horizontal and vertical size of the limit cycle. The larger a and b, the smaller the horizontal and vertical direction of the limit cycle.
2. $\lambda\omega$ and $\mu\omega$ affect the density of the particle trajectory interval. The larger $\lambda\omega$ and $\mu\omega$, the smaller the interval, and the denser the trajectory.

In summary, we get ranges of parameters α, β, a, b, m, R, λ, μ, and ω, seeing Table 4.4.

4.4.3 *Different Internal and External Stability*

Based on the above analysis, we have

1. a and b greater than zero, only affect the horizontal and vertical dimensions of the limit cycle;
2. λ and μ affect the density of particle trajectories;

Table 4.4 Semi-stable limit cycle parameter range for a general system

α	β	a	b	m	R	ω	$\lambda\omega$	$\mu\omega$
$[-3, 3]$ $\alpha\beta \geq 0$ $\alpha \neq \beta \neq 0$	$[-3, 3]$ $\alpha\beta \geq 0$ $\alpha \neq \beta \neq 0$	(0, 400]	(0, 400]	[1.5, 2.11] [3.5, 4.11] [5.5, 6.13]	[0.8, 1.1]	$\omega \geq 0.8$	$\lambda\omega \geq 0.8\sqrt{\frac{a}{b}}$	$\mu\omega \geq 0.8\sqrt{\frac{b}{a}}$

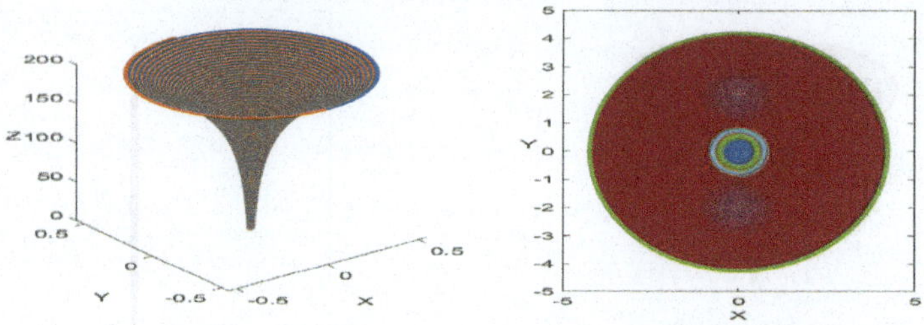

(a) $\alpha=2,\ \beta=1,\ a=0.01,\ b=0.01,\ R=0.7,\ m=6,\ \lambda=3,\ \mu=3$

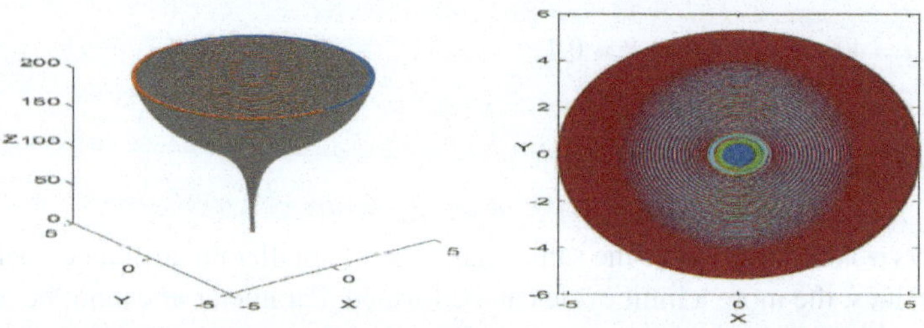

(b) $\alpha=2,\ \beta=1,\ a=0.01,\ b=0.01,\ R=0.75,\ m=6,\ \lambda=3,\ \mu=3$

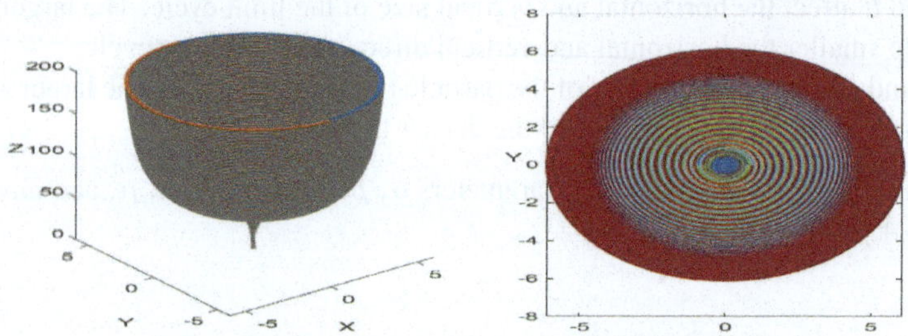

(c) $\alpha=2,\ \beta=1,\ a=0.01,\ b=0.01,\ R=0.8,\ m=6,\ \lambda=3,\ \mu=3$

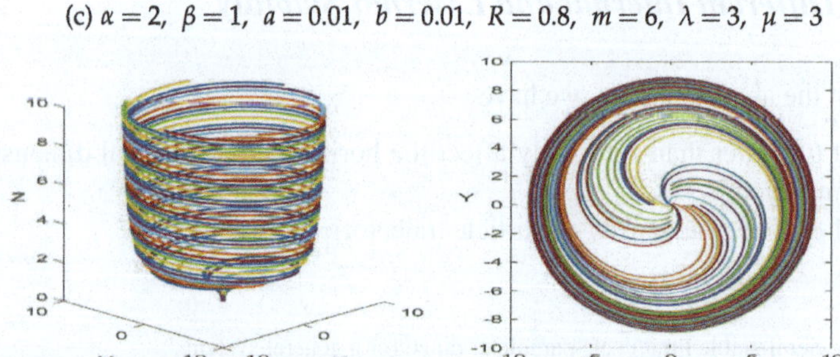

(d) $\alpha=2,\ \beta=1,\ a=0.01,\ b=0.01,\ R=1.1,\ m=6,\ \lambda=3,\ \mu=3$

Fig. 4.92 $\alpha = 2, \beta = 1, a = 0.01, b = 0.01, m = 6, \lambda = 3$, and $\mu = 3$

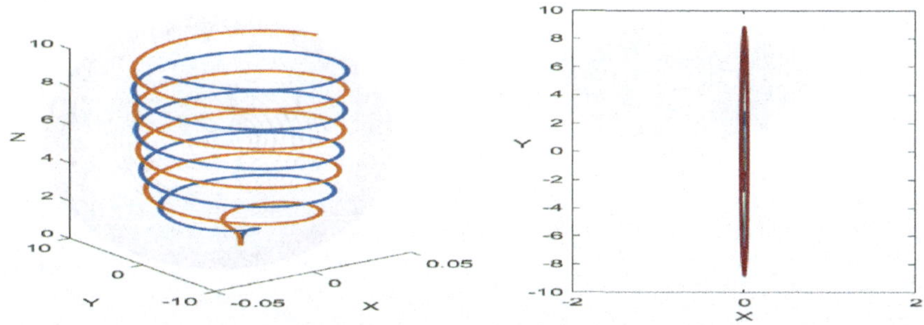

Fig. 4.93 $a = 400, b = 0.01$, and $R = 1.1$

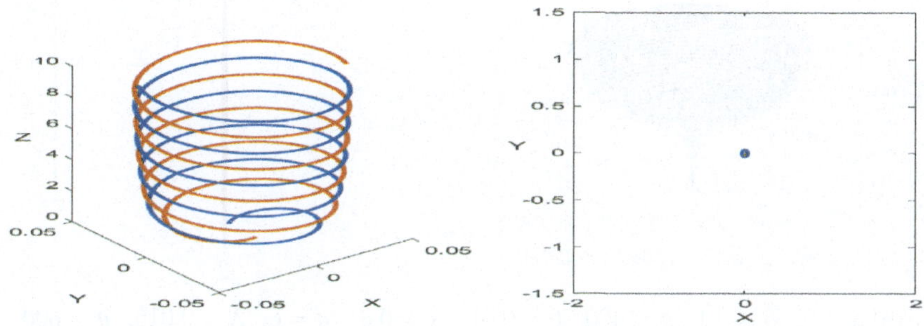

Fig. 4.94 $a = 400, b = 400$, and $R = 1.1$

3. R affects the radius of the limit cycle;
4. no matter how much m changes, there is no limit cycle with the opposite direction of convergence, that is, there are no limit cycles with internal stability and external instability (or internal instability and external stability). Therefore, only α, β may have such an effect on it.

4.4.3.1 α and β are Positive

When parameters are within their respective ranges, the system has a semi-stable limit cycle that is internally stable and externally unstable. See Fig. 4.97.

4.4.3.2 α is Negative, While β is Positive

1. If $|\alpha| < 1.1\beta$ and $|\alpha| \neq \beta$, then a semi-stable limit cycle with internal stability and external instability is obtained. See Fig. 4.98.
2. If $|\alpha| \geq 1.1\beta$, then a semi-stable limit cycle with internal instability and external stability is obtained. See Fig. 4.99.
3. If $|\alpha| = \beta$, then a number of similarly stable limit cycles are obtained. See Fig. 4.100.

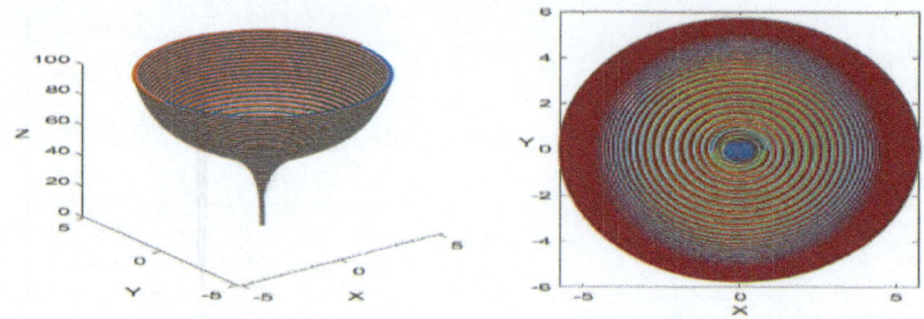

(a) $\alpha = 3,\ \beta = 0.1,\ a = 0.01,\ b = 0.01,\ R = 0.8,\ m = 6,\ \lambda = 3,\ \mu = 3$

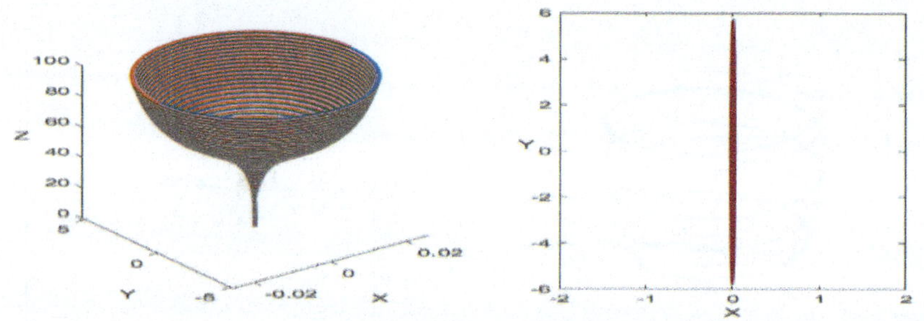

(b) $\alpha = 3,\ \beta = 0.1,\ a = 400,\ b = 0.01,\ R = 0.8,\ m = 6,\ \lambda = 0.015,\ \mu = 600$

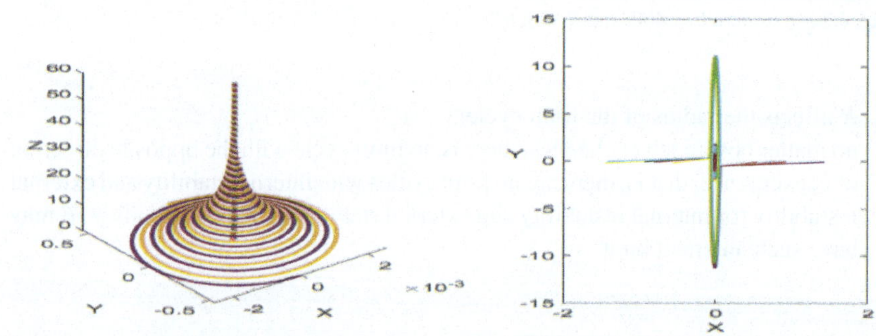

(c) $\alpha = -3,\ \beta = 0.1,\ a = 400,\ b = 0.01,\ R = 0.8,\ m = 6,\ \lambda = 0.015,\ \mu = 600$

Fig. 4.95 $m = 6$ and $R = 0.8$

4.4.3.3 α is Positive While β is Negative

1. If $|\beta| \neq \alpha$, then a semi-stable limit cycle with internal stability and external instability is obtained. See Fig. 4.101.
2. If $|\beta| \approx \alpha$, then a number of similarly stable limit cycles are obtained. See Fig. 4.102.

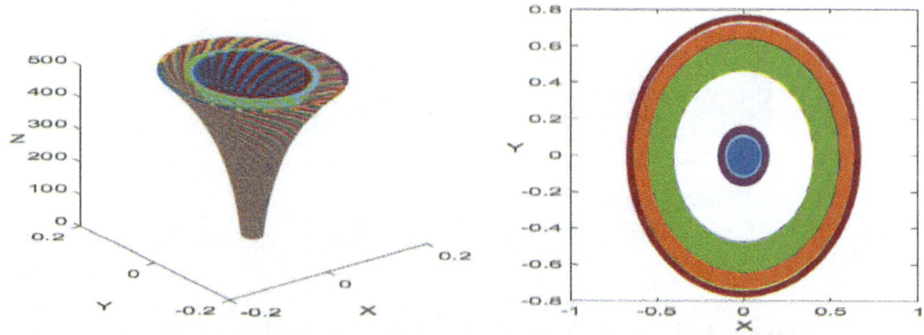

Fig. 4.96 $\alpha = 2, \beta = 1, a = 0.4, b = 0.3$, and $m = 8$

4.4.3.4 α and β are Negative

When α and β are both negative, and other parameters are within ranges obtained, respectively, then there are semi-stable limit cycles with internal instability and external stability. See Fig. 4.103.

4.4.3.5 Comprehensive Analysis

From the above analysis, it can be seen that α and β will affect the internal and external stability of limit cycles.

1. When α and β are positive, a semi-stable limit cycle that is internally stable and externally unstable will surely be produced.
2. When α and β are negative, a semi-stable limit cycle that is internally unstable and externally stable will definitely occur.

When α and β are positive or negative, it needs to be discussed separately. These conclusions are also suitable for general systems. The system has a semi-stable limit cycle, which needs to be satisfied: Parameters α and β must have the same sign, and the smaller the difference between the two, the easier it is to get a limit cycle. Parameter m cannot be an odd number and is divided into three parts. The value of parameter R is the most suitable near 1, while a and b affect the horizontal and vertical size of the limit cycle; $\lambda\omega = \sqrt{b/a}$ and $\mu\omega = \sqrt{a/b}$ affects the density of the particle trajectory interval.

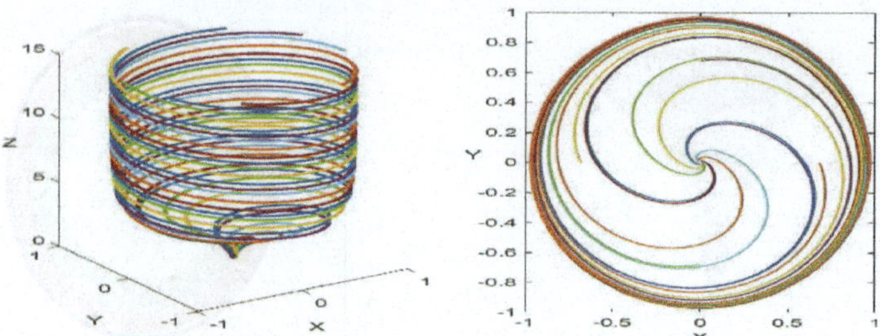

(a) $\alpha = 2,\ \beta = 1,\ a = 1,\ b = 1,\ R = 1,\ m = 2,\ \lambda = 1.5,\ \mu = 1.5$

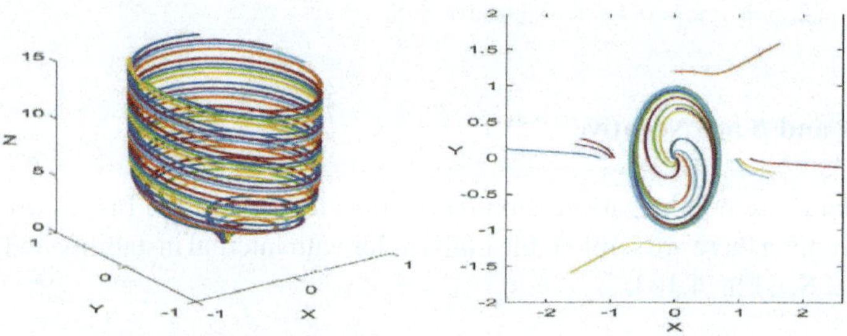

(b) $\alpha = 2,\ \beta = 1,\ a = 2,\ b = 1,\ R = 1,\ m = 2,\ \lambda = 0.75\sqrt{2},\ \mu = 1.5\sqrt{2}$

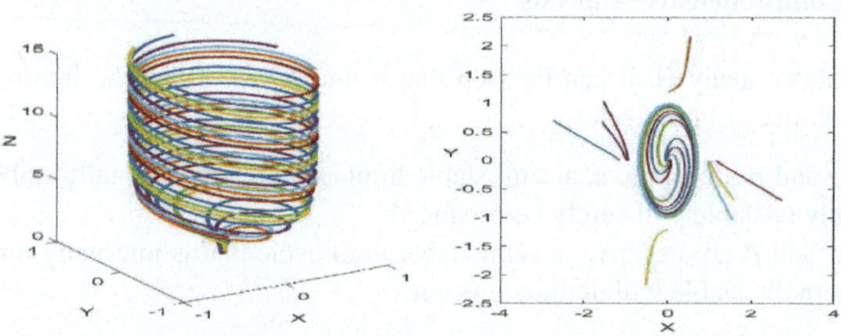

(c) $\alpha = 1,\ \beta = 2,\ a = 2,\ b = 1,\ R = 1,\ m = 2,\ \lambda = 0.75\sqrt{2},\ \mu = 1.5\sqrt{2}$

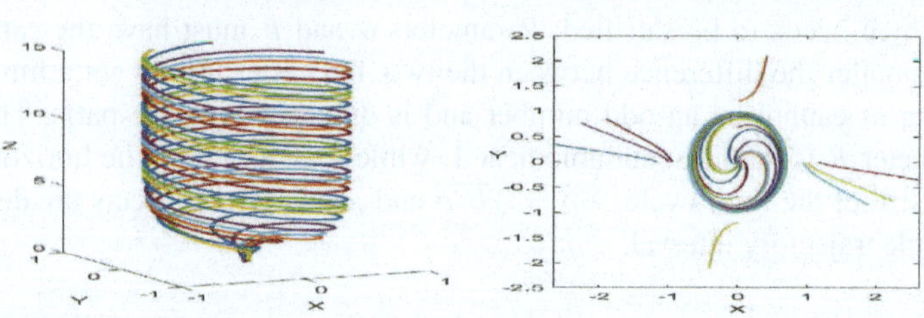

(d) $\alpha = 1.5,\ \beta = 2.5,\ a = 2,\ b = 1,\ R = 1,\ m = 2,\ \lambda = 0.75\sqrt{2},\ \mu = 1.5\sqrt{2}$

Fig. 4.97 $R = 1$ and $m = 2$

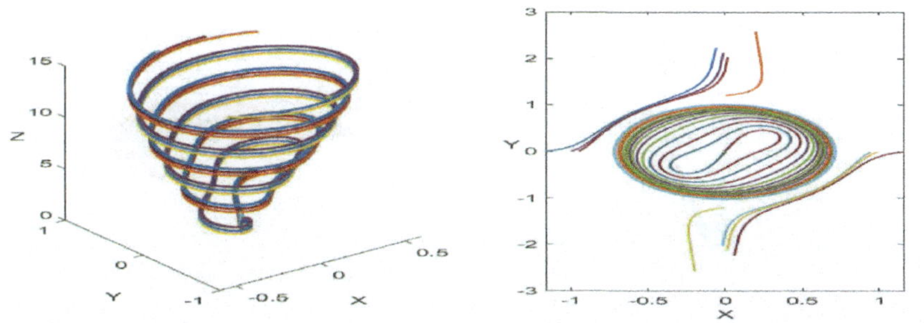

(a) $\alpha = -2,\ \beta = 2.5,\ a = 2,\ b = 1,\ R = 1,\ m = 2,\ \lambda = 0.75\sqrt{2},\ \mu = 1.5\sqrt{2}$

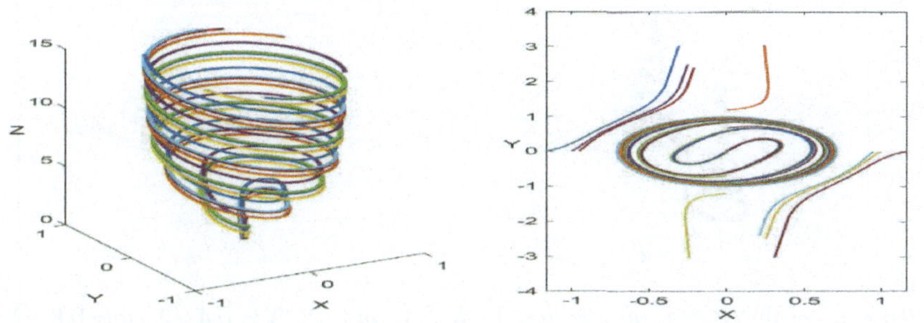

(b) $\alpha = -1,\ \beta = 2.5,\ a = 2,\ b = 1,\ R = 1,\ m = 2,\ \lambda = 0.75\sqrt{2},\ \mu = 1.5\sqrt{2}$

Fig. 4.98 $|\alpha| < 1.1\beta$ and $|\alpha| \neq \beta$

4.5 Externally Unstable and Internally Stable Semi-stable Limit Cycle

In Sect. 4.4, when parameters α, β, a, b, m, R, λ, μ, and ω meet Table 4.3, then the system has a semi-stable limit cycle that is internally stable and externally unstable. For example, in system (3.3.16), take $a = 2$, $b = R = \omega = \alpha = \beta = 1$, $\lambda = \sqrt{1/2}$, $\mu = \sqrt{2}$, and $m = 2$. Then system (3.3.16) becomes

$$\begin{cases} \dot{x} = \frac{y}{\sqrt{2}} + x(2x^2 + y^2 - 1)^2, \\ \dot{y} = -\sqrt{2}x + y(2x^2 + y^2 - 1)^2, \\ \dot{z} = \varepsilon(f). \end{cases} \tag{4.5.1}$$

According to the limit cycle theory [209–211], system (4.5.1) has a semi-stable limit cycle. As shown in Table 4.5, it is internally stable and externally unstable, that is, in the flow field in two dimensions, the internal streamline of the semi-stable limit cycle tends to it while the external streamline is far away from it, as shown in Fig. 4.104.

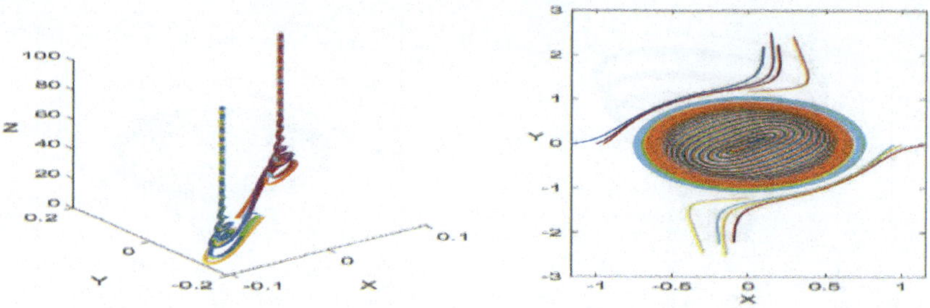

(a) $\alpha=-2.2,\ \beta=2,\ a=2,\ b=1,\ R=1,\ m=2,\ \lambda=\sqrt{2},\ \mu=2\sqrt{2}$

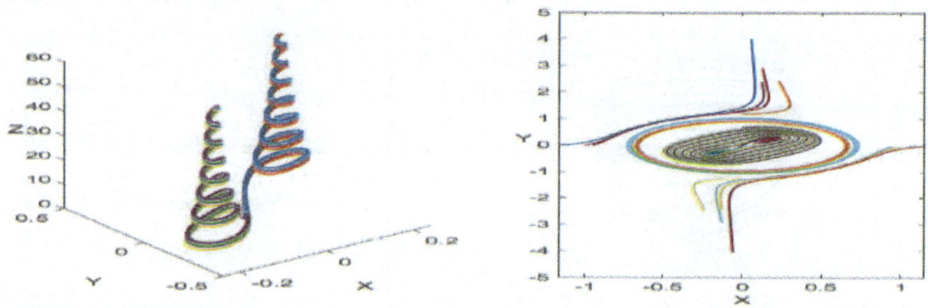

(b) $\alpha=-1.09,\ \beta=1,\ a=2,\ b=1,\ R=1,\ m=2,\ \lambda=0.4\sqrt{2},\ \mu=0.8\sqrt{2}$

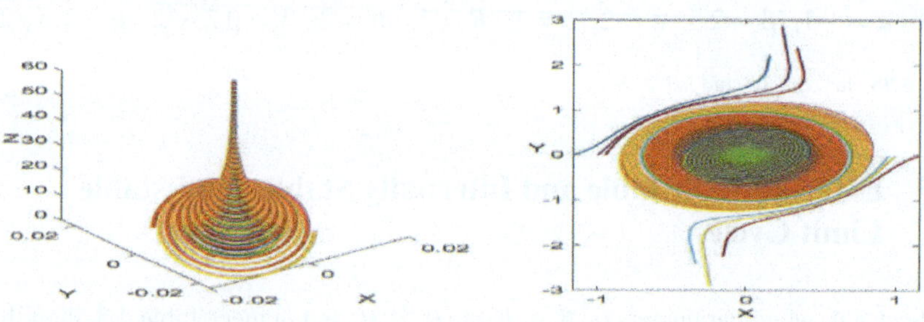

(c) $\alpha=-2.2,\ \beta=2,\ a=2,\ b=1,\ R=1,\ m=2,\ \lambda=2\sqrt{2},\ \mu=4\sqrt{2}$

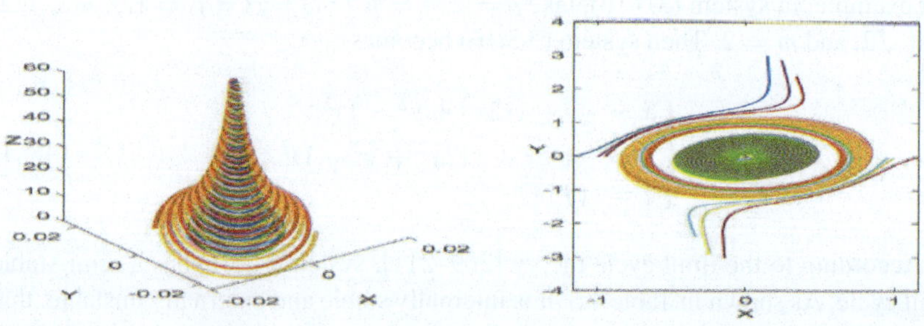

(d) $\alpha=-1.1,\ \beta=1,\ a=2,\ b=1,\ R=1,\ m=2,\ \lambda=\sqrt{2},\ \mu=2\sqrt{2}$

Fig. 4.99 $|\alpha| \geq 1.1\beta$

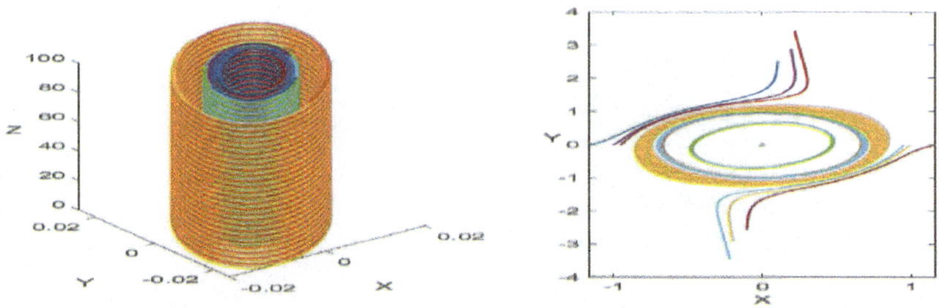

(a) $\alpha = -1.1,\ \beta = 1.1,\ a = 2,\ b = 1,\ R = 1,\ m = 2,\ \lambda = \sqrt{2},\ \mu = 2\sqrt{2}$

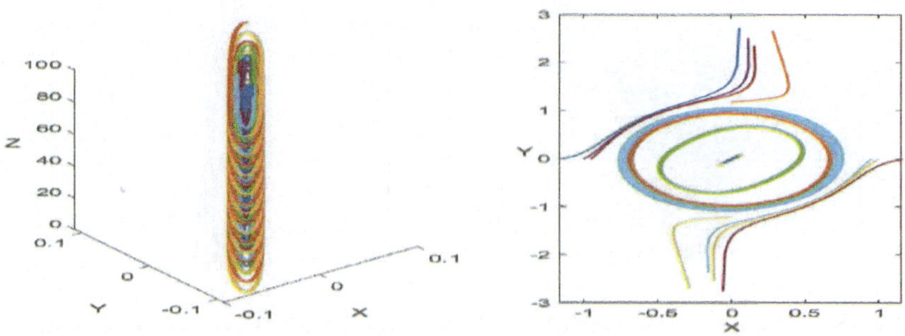

(b) $\alpha = -2,\ \beta = 2,\ a = 2,\ b = 1,\ R = 1,\ m = 2,\ \lambda = \sqrt{2},\ \mu = 2\sqrt{2}$

Fig. 4.100 $|\alpha| = \beta$

Table 4.5 The parameter range of the semi-stable limit cycle which is internally stable and externally unstable

α	β	a	b	m	R	ω	$\lambda\omega$	$\mu\omega$
$[0, 3]$	$[-3, 3]$ $\alpha + \beta \neq 0$	$(0, 400]$	$(0, 400]$	$[1.5, 2.11]$ $[3.5, 4.11]$ $[5.5, 6.13]$	$[0.8, 1.1]$	$\omega \geq$ 0.8	$\lambda\omega \geq$ $0.8\sqrt{\frac{b}{a}}$	$\mu\omega \geq$ $0.8\sqrt{\frac{a}{b}}$
$[-3, 0]$	$[0, 3]$ $\beta \neq \|\alpha\|$ $\|\alpha\| < 1.1\beta$	$(0, 400]$	$(0, 400]$	$[1.5, 2.11]$ $[3.5, 4.11]$ $[5.5, 6.13]$	$[0.8, 1.1]$	$\omega \geq$ 0.8	$\lambda\omega \geq$ $0.8\sqrt{\frac{b}{a}}$	$\mu\omega \geq$ $0.8\sqrt{\frac{a}{b}}$

In the three-dimensional space, the inside of the flow field tends to limit cycle C and increases with time t. The behavior of the limit cycle is shown in Table 4.7.

Similarly, the case of n water particles can be seen in Fig. 4.105.

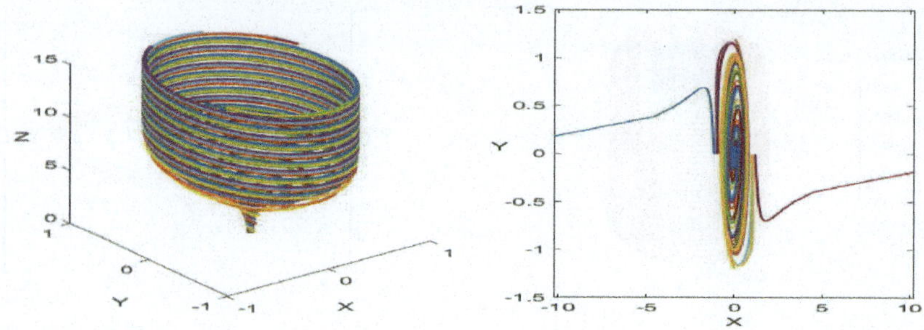

(a) $\alpha=1,\ \beta=-2.5,\ a=2,\ b=1,\ R=1,\ m=1,\ \lambda=2\sqrt{2},\ \mu=4\sqrt{2}$

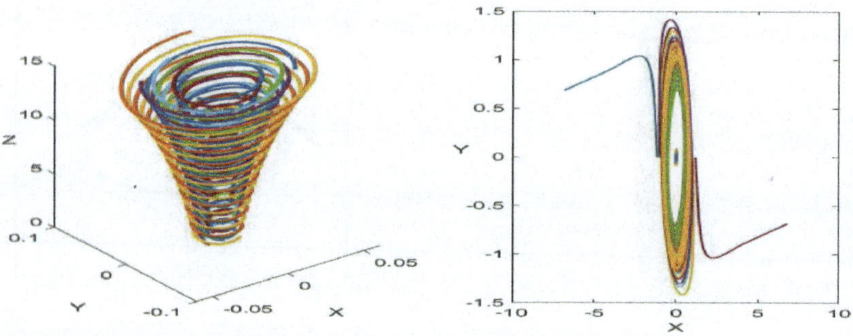

(b) $\alpha=1,\ \beta=-1.2,\ a=2,\ b=1,\ R=1,\ m=1,\ \lambda=2\sqrt{2},\ \mu=4\sqrt{2}$

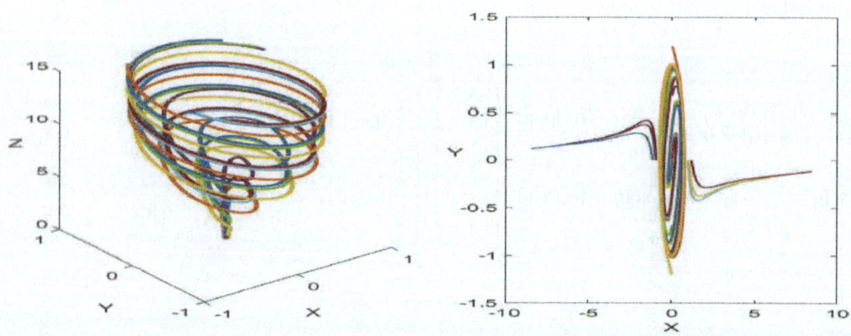

(c) $\alpha=1,\ \beta=-2,\ a=2,\ b=1,\ R=1,\ m=1,\ \lambda=0.75\sqrt{2},\ \mu=1.5\sqrt{2}$

Fig. 4.101 $|\beta| \neq \alpha$

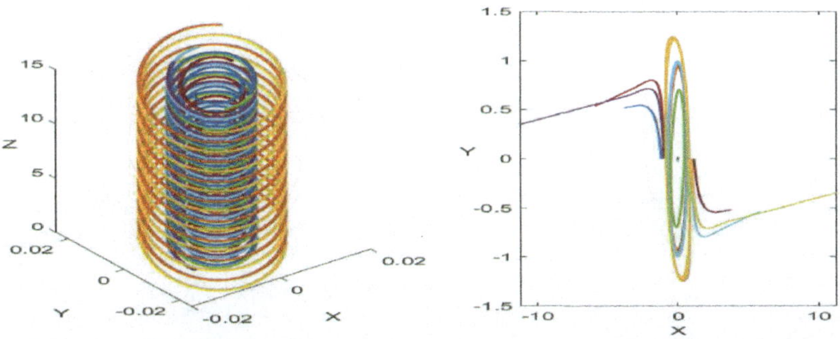

(a) $\alpha = 2,\ \beta = -2,\ a = 2,\ b = 1,\ R = 1,\ m = 1,\ \lambda = 2\sqrt{2},\ \mu = 4\sqrt{2}$

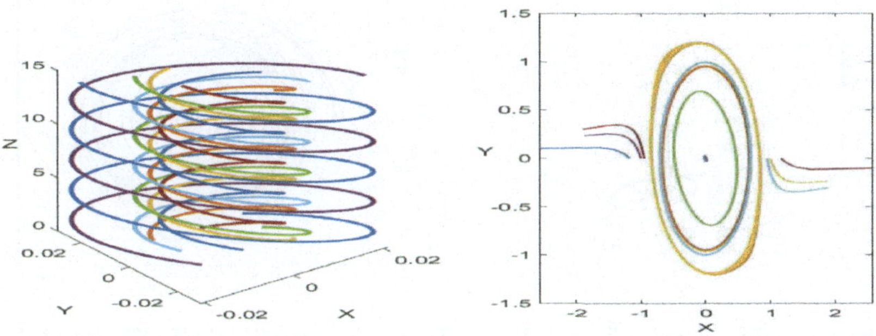

(b) $\alpha = 1,\ \beta = -1,\ a = 2,\ b = 1,\ R = 1,\ m = 2,\ \lambda = 0.75\sqrt{2},\ \mu = 1.5\sqrt{2}$

Fig. 4.102 $|\beta| \approx \alpha$

4.6 Externally Stable and Internally Unstable Semi-stable Limit Cycle

In Sect. 4.4, when parameters α, β, a, b, m, R, λ, μ, and ω meet Table 4.6, then the system has a semi-stable limit cycle that is internally stable but externally unstable.

Table 4.6 The parameter range of the semi-stable limit cycle which is internally unstable and externally stable

α	β	a	b	m	R	ω	$\lambda\omega$	$\mu\omega$
$[-3, 0]$	$[0, 3]$ $\lvert\alpha\rvert \geq 1.1\beta$	$(0, 400]$	$(0, 400]$	$[1.5, 2.11]$ $[3.5, 4.11]$ $[5.5, 6.13]$	$[0.8, 1.1]$	$\omega \geq$ 0.8	$\lambda\omega \geq$ $0.8\sqrt{\frac{b}{a}}$	$\mu\omega \geq$ $0.8\sqrt{\frac{a}{b}}$
$[-3, 0]$	$[-3, 0]$	$(0, 400]$	$(0, 400]$	$[1.5, 2.11]$ $[3.5, 4.11]$ $[5.5, 6.13]$	$[0.8, 1.1]$	$\omega \geq$ 0.8	$\lambda\omega \geq$ $0.8\sqrt{\frac{b}{a}}$	$\mu\omega \geq$ $0.8\sqrt{\frac{a}{b}}$

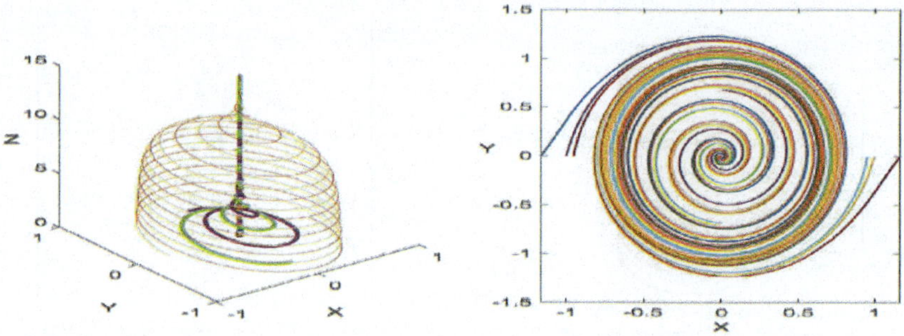

(a) $\alpha=-1,\ \beta=-1,\ a=2,\ b=1,\ R=1,\ m=2,\ \lambda=2\sqrt{2},\ \mu=4\sqrt{2}$

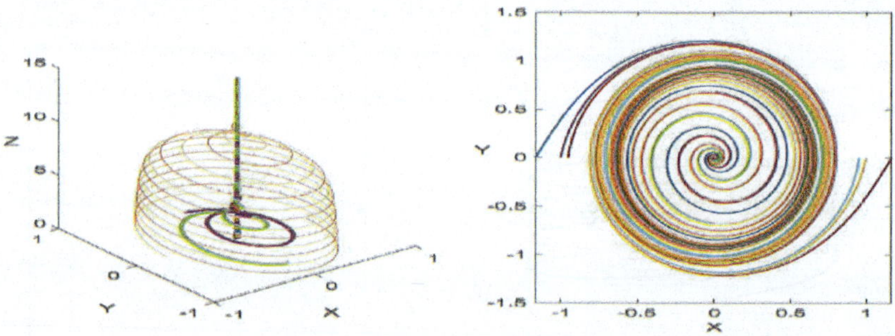

(b) $\alpha=-0.8,\ \beta=-1.5,\ a=2,\ b=1,\ R=1,\ m=2,\ \lambda=2\sqrt{2},\ \mu=4\sqrt{2}$

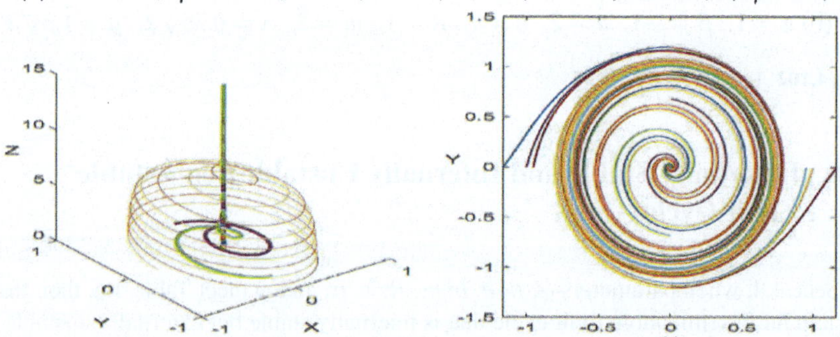

(c) $\alpha=-1.1,\ \beta=-2,\ a=2,\ b=1,\ R=1,\ m=2,\ \lambda=2\sqrt{2},\ \mu=4\sqrt{2}$

Fig. 4.103 α and β are negative

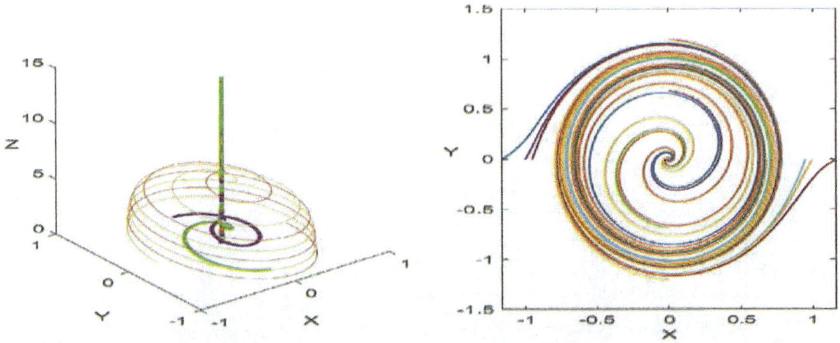

(d) $\alpha = -2.5,\ \beta = -1.3,\ a = 2,\ b = 1,\ R = 1,\ m = 2,\ \lambda = 2\sqrt{2},\ \mu = 4\sqrt{2}$

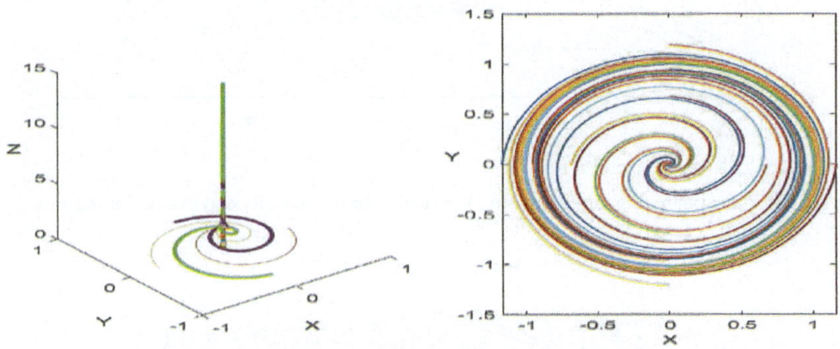

(e) $\alpha = -2.5,\ \beta = -1.3,\ a = 1,\ b = 1,\ R = 1,\ m = 2,\ \lambda = 4,\ \mu = 4$

Fig. 4.103 (continued)

For example, in system (3.3.16), take $a = b = R = 1, \alpha = \beta = -1, \lambda\omega = 1, \mu\omega = -1$, and $m = 2$. Then system (3.3.16) becomes

$$\begin{cases} \dot{x} = y - x(2x^2 + y^2 - 1)^2, \\ \dot{y} = -x - y(2x^2 + y^2 - 1)^2, \\ \dot{z} = \varepsilon(f). \end{cases} \tag{4.6.1}$$

According to the limit cycle theory [478, 479], system (4.6.1) has a semi-stable limit cycle. As shown in Table 4.6, it is internally unstable and externally stable, that is, in the flow field in two dimensions, the internal streamlines of the semi-stable limit cycle are far away from it, while the external streamlines tend to converge at the center, as shown in Fig. 4.106.

In the three-dimensional space, all the streamlines inside limit cycle C are far away from it and tend to converge at the center. So, as the time t increases, the mesoscale eddy rises, showing a "vortex." As shown in Table 4.8, the three-dimensional semi-stable limit cycle with internal instability and external stability space situation, based

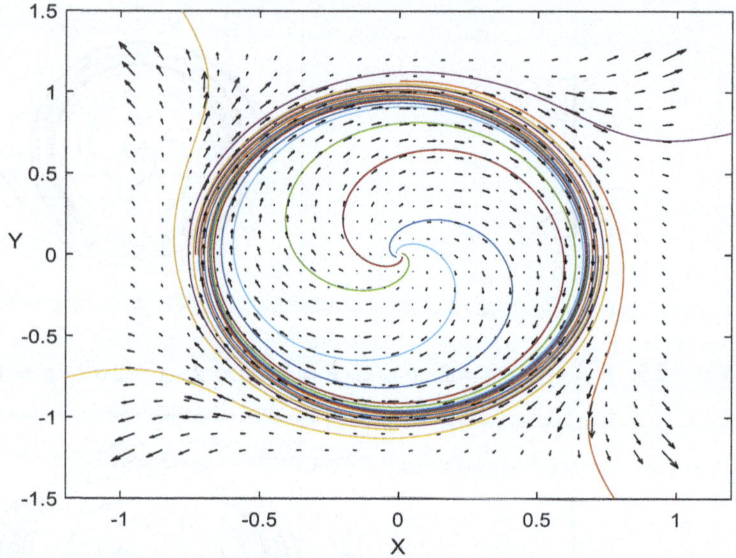

Fig. 4.104 Plane phase diagram of a limit cycle with internal stability and external instability

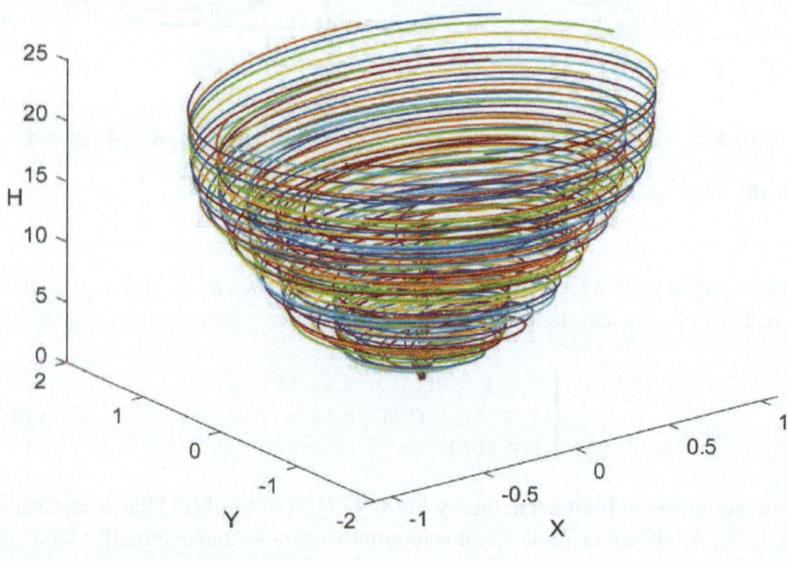

Fig. 4.105 Infinitely nested self-similar space fractal structure of n water particles for system (4.5.1)

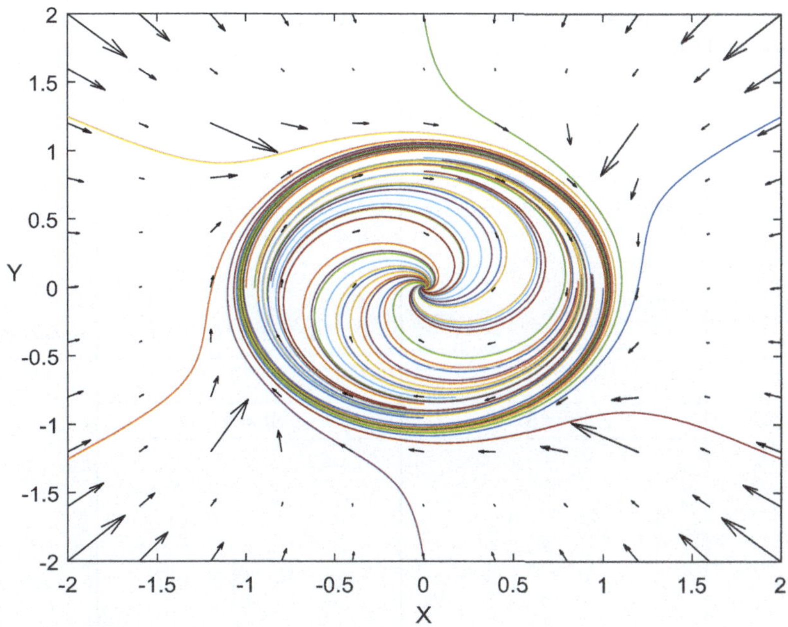

Fig. 4.106 The plane vortex formed by the trajectory of the limit cycle with internal instability and external stability

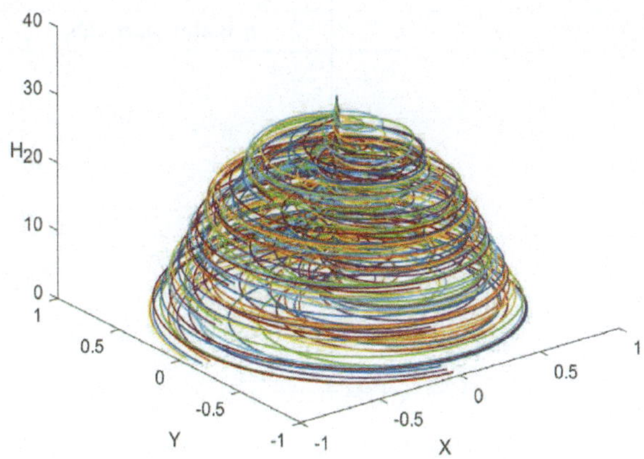

Fig. 4.107 Infinitely nested self-similar space fractal structure of n water particles for system (4.6.1)

Table 4.7 Three-dimensional space situation of semi-stable limit cycles with internal stability and external instability

$n=1$	$n=2$
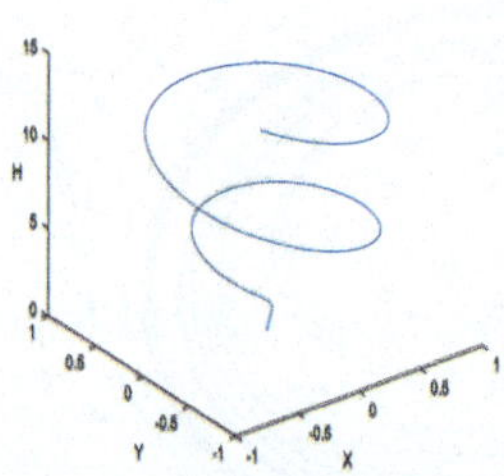	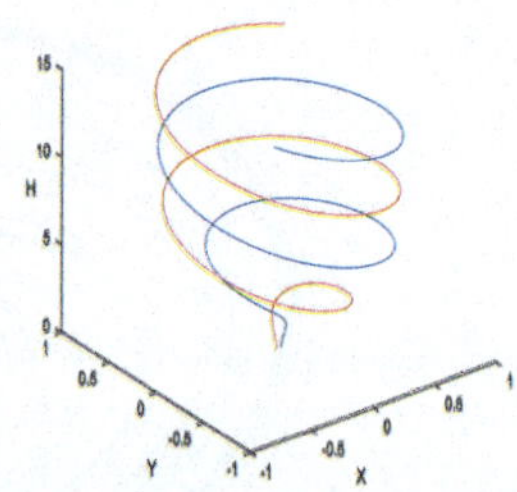
$n=3$	$n=4$
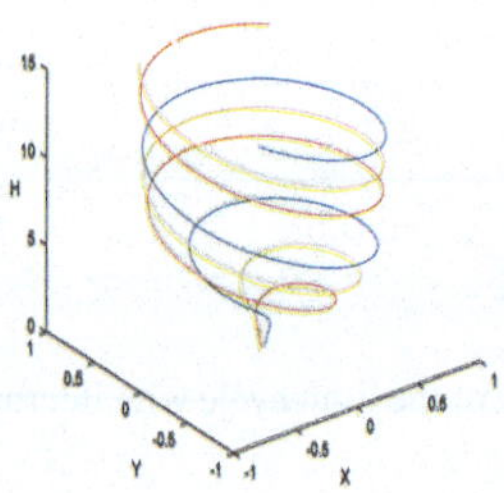	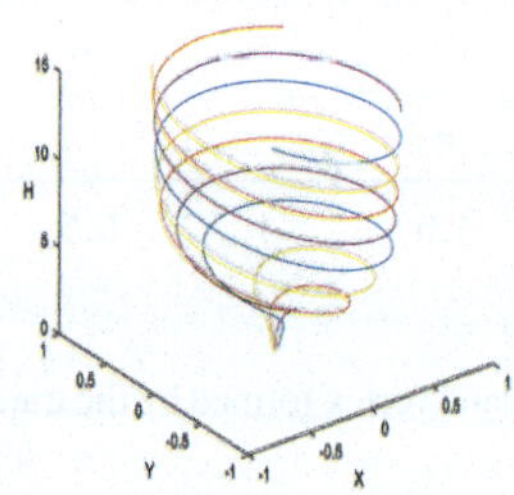
......	n water particles
......	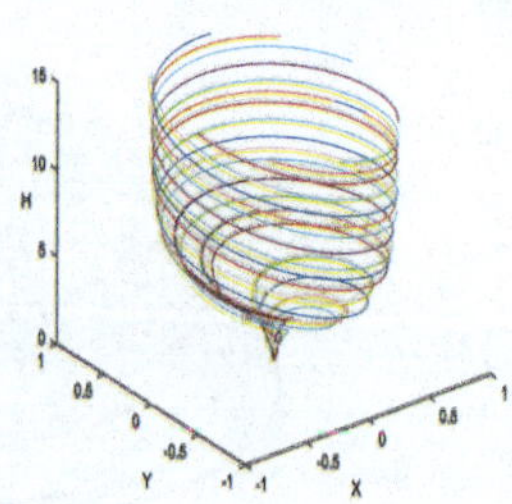

on the sequential increase of basic elements, the motion trajectories in the three-dimensional space are listed in Table 4.8.

Note that continuity of the system is infinite particles, but here is only a finite number of particles. The following infinitely nested self-similar space fractal structure is formed by n particles as shown in Fig. 4.107.

Table 4.8 Three-dimensional space situation of semi-stable limit cycles with internal instability and external stability

$n = 1$	$n = 2$
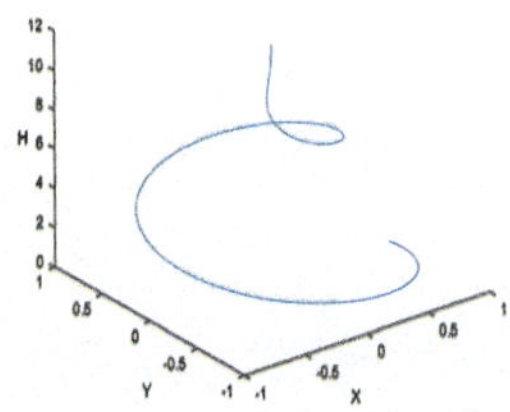	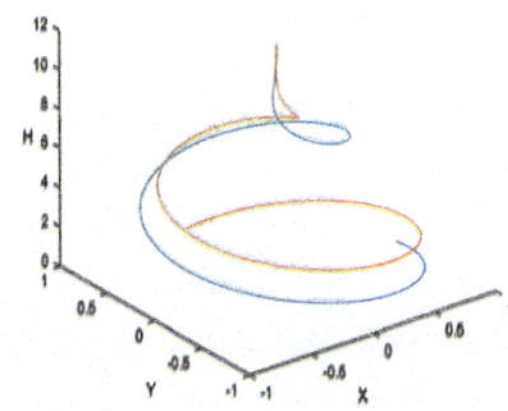
$n = 3$	$n = 4$
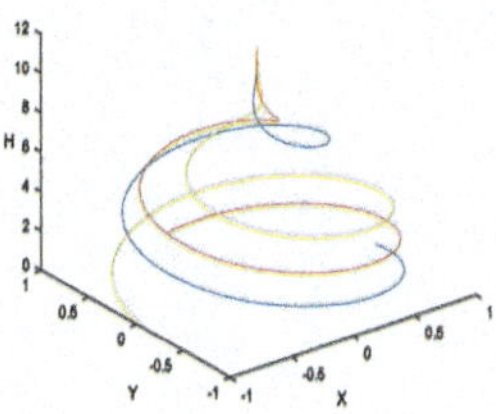	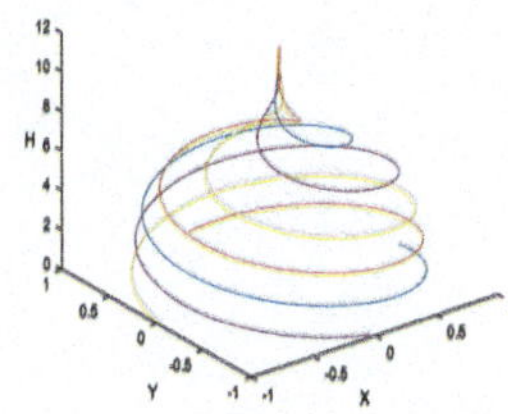
……	n water particles
……	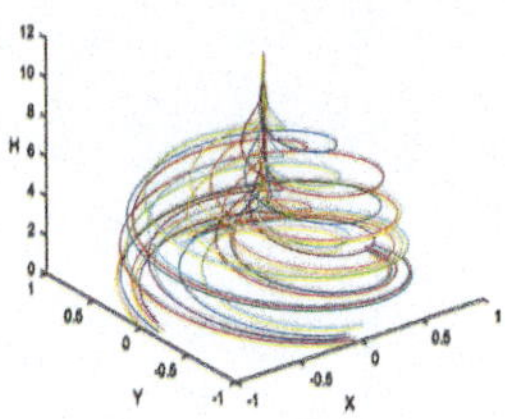

Chapter 5
Semi-stable Limit Cycles and Mesoscale Eddies

According to Sect. 4.3, there are two forms of semi-stability of nonlinear system (3.3.16), and the mesoscale eddy also has two forms: cold eddy and warm eddy. So they are exactly corresponding to each other, see Fig. 5.1.

5.1 Semi-stable Limit Cycles and Mesoscale Cold Eddies

System (3.3.16) is semi-stable: C is internally stable and externally unstable. According to the mesoscale eddy formation theory [498], when the sea surface in C has a negative anomaly, the pressure in C also has a negative anomaly, which causes the inward pressure gradient force to be balanced with the Coriolis force, thereby generating a cyclonic circulation. **According to Sverdrup transport theory [498], water within limit cycle C will gradually approach it, which generates a divergent mesoscale cold eddy.** See Fig. 5.2.

Nonlinear system (5.0a) is actually derived from system (3.3.16) when the perturbation function is taken as

$$\begin{cases} \omega p(x, y) = y - 0.1\left(\frac{1}{3}x^3 - x\right) - \alpha(ax^2 + by^2 - R)^m, \\ \omega q(x, y) = -x - \beta(ax^2 + by^2 - R)^m, \end{cases}$$

where the vertical direction is $\dot{z} = \varepsilon(f)t$.

S.-T. Liu et al., *Mathematical Principle and Fractal Analysis of Mesoscale Eddy*,
https://doi.org/10.1007/978-981-16-1839-0_5

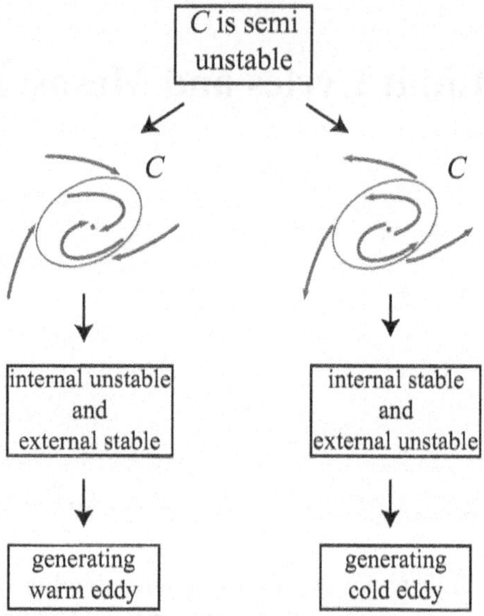

Fig. 5.1 Semi-stable limit cycles and mesoscale eddies

5.2 Semi-stable Limit Cycles and Mesoscale Warm Eddies

System (3.3.16) is semi-stable: C is internally unstable and externally stable. According to the mesoscale eddy formation theory [498], when the sea surface in C has a positive anomaly, the pressure in C also has a positive anomaly, which causes the outward pressure gradient force to be balanced with the Coriolis force, thereby generating an anticyclonic circulation. **According to Sverdrup transport theory [498], water within limit cycle C will gradually approach it, which generates a convergent mesoscale warm eddy**. See Fig. 5.3.

Nonlinear system (5.0b) is actually derived from system (3.3.16) when the perturbation function is taken as

$$\begin{cases} \omega p(x, y) = 2y - \alpha(ax^2 + by^2 - R)^m, \\ \omega q(x, y) = 0.8y(x^3 - x^2 - 1) - 2x - 3x^2 - \beta(ax^2 + by^2 - R)^m, \end{cases}$$

where the vertical direction is $\dot{z} = \varepsilon(f)t$.

1. The semi-stable limit cycle of a nonlinear dynamic system directly reveals the formation mechanism of the mesoscale eddy.
2. Since $\omega = \omega_0 f$ and f is the Coriolis force, system (3.3.16) describes the changes of the mesoscale eddy geostrophic fluid.
3. There are two forms of semi-stable limit cycles. A stochastic ellipse C divides the sea surface into two areas: internal stability and external instability as well as internal instability and external stability, which is corresponding to mesoscale cold eddy and mesoscale warm eddy water movement. **This shows that the**

C is a semi-stable limit circle with external instability and internal stability

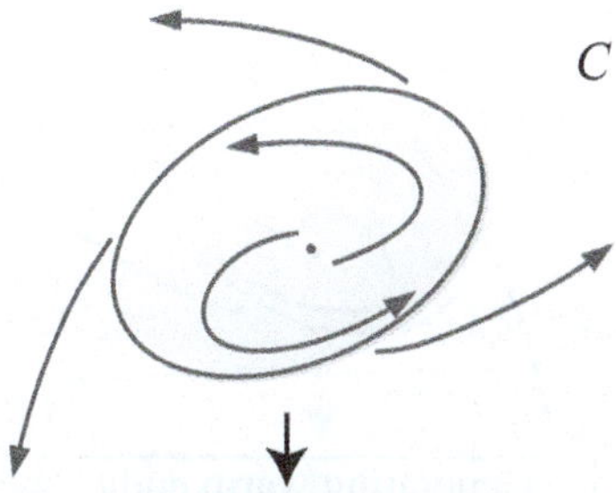

Generating cold eddy

$$\begin{cases} \dfrac{dx}{dt} = y - 0.1\left(\dfrac{x^3}{3} - x\right), \\ \dfrac{dy}{dt} = -x. \end{cases} \tag{5.0a}$$

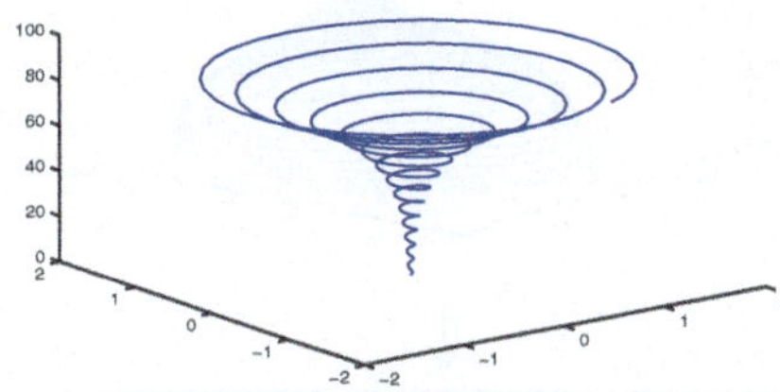

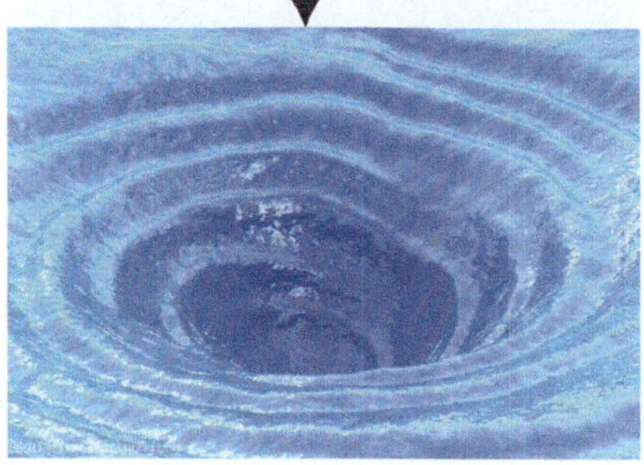

Fig. 5.2 A mesoscale cold eddy generated by semi-stable limit cycle which is internally stable and externally unstable

C is a semi-stable limit circle with internal instability and external stability

↓

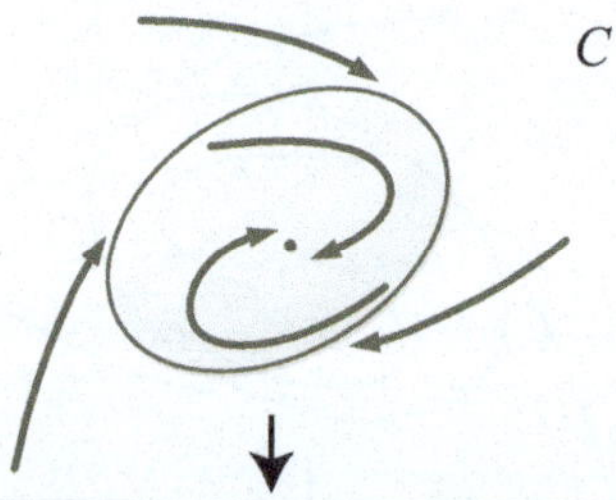

↓

Generating warm eddy

↓

$$\begin{cases} \dfrac{dx}{dt} = 2y, \\ \dfrac{dy}{dt} = 2x - 3x^2 + 0.8\left(x^3 - x^2 + y^2\right). \end{cases} \tag{5.0b}$$

↓

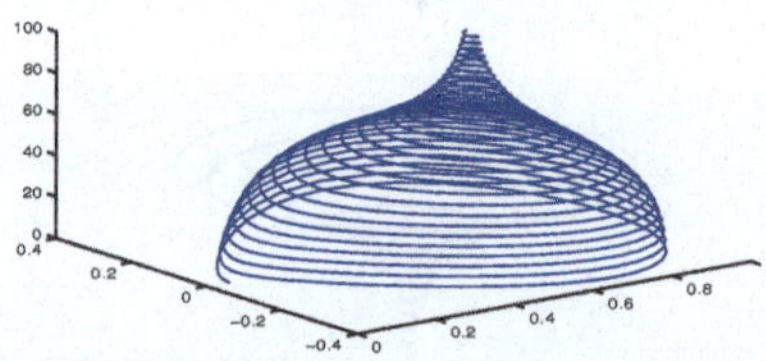

↓

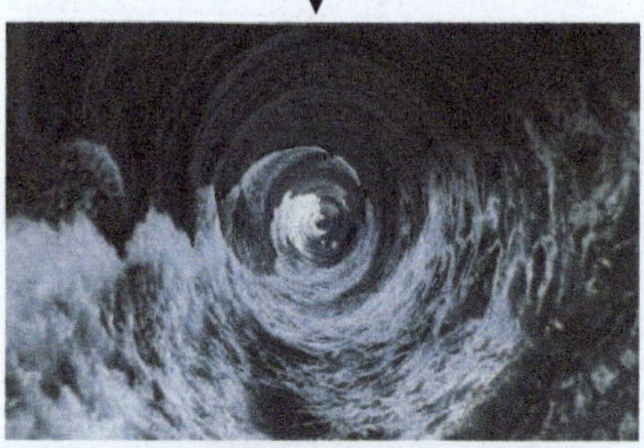

Fig. 5.3 A mesoscale warm eddy generated by semi-stable limit cycle which is internally unstable and externally stable

mathematical theory and the objective reality have reached perfect identity and symmetry.

4. **It shows that limit cycle theory in mathematics is a classic application in mesoscale eddies.**
5. From [193, 468], we know a and b greater than zero only affect the size of the limit cycle in the horizontal and vertical directions; λ and μ affect the particle track spacing and density (**It is related to the density, salinity, temperature, and velocity of seawater**); R affects the radius of the limit cycle; no matter how much m changes, there is no limit cycle with the opposite direction of convergence. Therefore, only α and β may affect the semi-stable limit cycle of the system, where other parameters are within the range previously obtained, if α and β are positive, a mesoscale cold eddy will definitely be generated; if α and β are negative, a mesoscale warm eddy will definitely be generated; other situations need to be discussed separately. These conclusions are also applicable to conventional systems and promotion systems.

Remark 5.1 We directly demonstrate theoretically that the mesoscale eddy is the nonlinear behavior of the nonlinear time-varying differential dynamic system. This problem is formed by the limit cycle of nonlinear dynamic systems [43], which confirms the use of a large number of statistical analysis — "*During the evolution of a eddy, the axisymmetric state is an attractive cycle of a system, and the system will spontaneously tend to this state. This also explains why the mesoscale eddy we see in the altimeter always appears as a circular finite-scale structure.*"

Remark 5.2 It is theoretically demonstrated the "circularity of mesoscale eddies on a limited scale" mentioned in [43].

Remark 5.3 It is noted that mesoscale eddies can be generated by nonlinear systems. The nonlinearity is usually strongly sensitive to the initial values. Therefore, it is a very important for the starting point of mesoscale eddies. Then the mesoscale cold eddy is generated from the sea surface from a sink to source while the mesoscale warm eddy is generated from the source to the sink at a certain depth of the ocean. The previous results are basically the analysis and statistical laws of tracking data records after the formation of mesoscale eddies. The starting point of mesoscale eddies has never been studied. However, it is precisely the nonlinear initial value that cannot be ignored for mesoscale eddies. This is an important issue because it involves external perturbation of mesoscale eddies and the transportation of water bodies.

Remark 5.4 Behavior in nature is in a relatively stable state of equilibrium, such as an acting force and reaction force. It is the same for mesoscale eddies, that is, the dipole problem. **From a theoretical point of view, a mesoscale eddy corresponds to a semi-stable limit cycle, which corresponds to internal stability and external instability (warm mesoscale eddy) as well as internal stability and external instability (cold mesoscale eddy). This reflects the symmetry of things in physics and mathematics, and reflects the symmetric beauty of the mathematical system!**

Remark 5.5 It can be seen from [382] that the behavior of mesoscale eddies is divided into disk-shaped eddies and cylindrical eddies. They are usually disk-shaped eddies (the diameter of the mesoscale eddy is always much greater than the distance from the eddy axis). That is why we choose the axial scale of a mesoscale eddy as $\dot{z} = \omega f$ when considering its vertical structure.

Chapter 6
Example Verification

6.1 Basic Method

In order to verify the correctness of the mathematical model of mesoscale eddies that we established, without loss of generality, we use nonlinear dynamic system (3.3.15), actual flow field data, and the genetic algorithm [479]. The process of the optimal solution of nonlinear programming with undetermined coefficients is shown in Fig. 6.1.

To simplify the problem, without losing its generality, the time-varying parameters are not considered in this section. As the dynamic system of mesoscale eddies (3.3.15), let

$$X = \begin{bmatrix} x \\ y \end{bmatrix}$$

and

$$f(X) = \begin{bmatrix} \lambda(y - y_0) + \alpha(x - x_0)[a(x - x_0)^2 + b(y - y_0)^2 - R]^m \\ -\mu(x - x_0) + \beta(y - y_0)[a(x - x_0)^2 + b(y - y_0)^2 - R]^m \end{bmatrix}.$$

We define $\theta = (x_0 \quad y_0 \quad R \quad a \quad b \quad \lambda \quad \mu \quad \alpha \quad \beta)$ as the vector of undetermined parameter, $\bar{X} = \{X_1, \ldots, X_N\}$ as the observation points, the velocity vector based on the observation points as $V(X_i) = (V_x(X_i) \quad V_y(X_i))^{\mathrm{T}}$, $i \in \{1, \ldots, N\}$ while N denote the number of observation points. Using the observation data, construct a nonlinear programming problem on parameter θ as

$$P_0 : \min \sum_{i=1}^{N} \|V(X_i) - f(X_i)\|_l, \quad \text{s.t.} \quad \theta_L \leq \theta \leq \theta_U, \tag{6.1.1}$$

where θ_L and θ_U are the lower and upper bounds of the parameter vector. While $i = 2$ denotes 2-norm. In the following, the objective function is constructed by 2-norm.

S.-T. Liu et al., *Mathematical Principle and Fractal Analysis of Mesoscale Eddy*,
https://doi.org/10.1007/978-981-16-1839-0_6

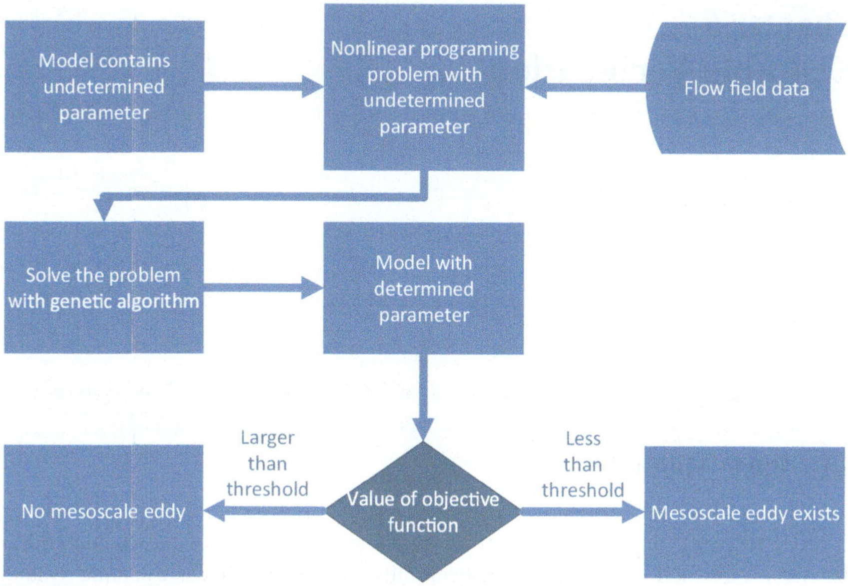

Fig. 6.1 Eddy identification process

We use model data[1] as experimental data, with the spatial resolution of 0.25°. Next, we will use the multi-objective NSGA-II algorithm [479] to solve problem (6.1.1) and define

$$Obj(X) = \begin{pmatrix} Obj_x & Obj_y \end{pmatrix}^{\mathrm{T}} = \sum_{i=1}^{N} \|V(X_i) - f(X_i)\|_l \tag{6.1.2}$$

as the objective function.

6.2 Numerical Experiment

To apply the model to the identification of mesoscale eddies, we mainly do the following work: Firstly, we simulate two mesoscale eddies (clockwise and anti-clockwise) according to the model with simplified parameters and complete parameters. By using of genetic algorithm, we can get the Pareto frontier of problem (6.1.1), therefore mesoscale eddies can be reconstructed. We can further obtain the criteria for identifying mesoscale eddies by analyzing the numerical results of the local area. Furthermore, we identify mesoscale eddies in the South China Sea based on the criteria, and then analyze the reliability of the model and the algorithm according to numerical results.

---

[1] http://apdrc.soest.hawaii.edu:80/dods/public_data/ECCO2/cube92.

6.2.1 Value in Special Circumstances

The genetic algorithm is used to estimate parameters in a local small area under the situation of simplified parameter that $a = b = 1$, $\lambda = \mu = \alpha = \beta = 1/\eta$ and $m = 2$, i.e., the parameter vector is $\theta = (x_0 \quad y_0 \quad R \quad \eta)^{\mathrm{T}}$. Based on the mathematical analysis, when $\eta > 0$, the solution of the model is a clockwise spiral, so it is called clockwise model; correspondingly, $\eta < 0$ denotes anti-clockwise model. Also we notice that the radius of the semi-stable limit cycle of the simplified model should be $\sqrt{R}$.

By using of the Geatpy library,[2] we implement the NSGA-II algorithm to solve the nonlinear programming problem (6.1.1). Set the population be 100 and the max generation be 1000.

Clockwise Model
Figure 6.2 illustrates the numerical result of the clockwise model with eddy center (284.90, 31.98), $R = 1.094$, and $\eta = 3.28$. We get the model as

$$\begin{cases} x' = \frac{1}{3.28}(y - 31.98) + \frac{1}{3.28}(x - 284.90) \\ \qquad \times[(x - 284.90)^2 + (y - 31.98)^2 - 1.094]^2, \\ y' = -\frac{1}{3.28}(x - 284.90) + \frac{1}{3.28}(y - 31.98) \\ \qquad \times[(x - 284.90)^2 + (y - 31.98)^2 - 1.094]^2. \end{cases}$$

As shown in Fig. 6.2b, the majority of the objective function value of feasible solutions in Pareto frontier is less than 4.5.

Table 6.1 shows the comparison between the numerical results and the flow field data.

Anti-clockwise Model
Figure 6.2 illustrates the numerical result of the clockwise model with eddy center (37.23, −21.51), $R = 1.37$, and $\eta = -8.714$. We get the model as

$$\begin{cases} x' = -\frac{1}{8.71}(y + 21.51) - \frac{1}{8.71}(x - 37.23) \\ \qquad \times[(x - 37.23)^2 + (y + 21.51)^2 - 1.37]^2, \\ y' = \frac{1}{8.71}(x - 37.23) - \frac{1}{8.71}(y + 21.51) \\ \qquad \times[(x - 37.23)^2 + (y + 21.51)^2 - 1.37]^2. \end{cases}$$

As shown in Fig. 6.3b, the majority of the objective function value of feasible solutions in Pareto frontier is less than 9.

Table 6.2 shows the comparison between the numerical results and the flow field data.

The main features of mesoscale eddies can be obtained with minor errors by using the simplified model to identify mesoscale eddies. The area of mesoscale eddies in each case only occupied 1/3 of the whole identification area, i.e., a large number of useless data were introduced and resulted in errors in the numerical value. Although

---

[2] http://geatpy.com/.

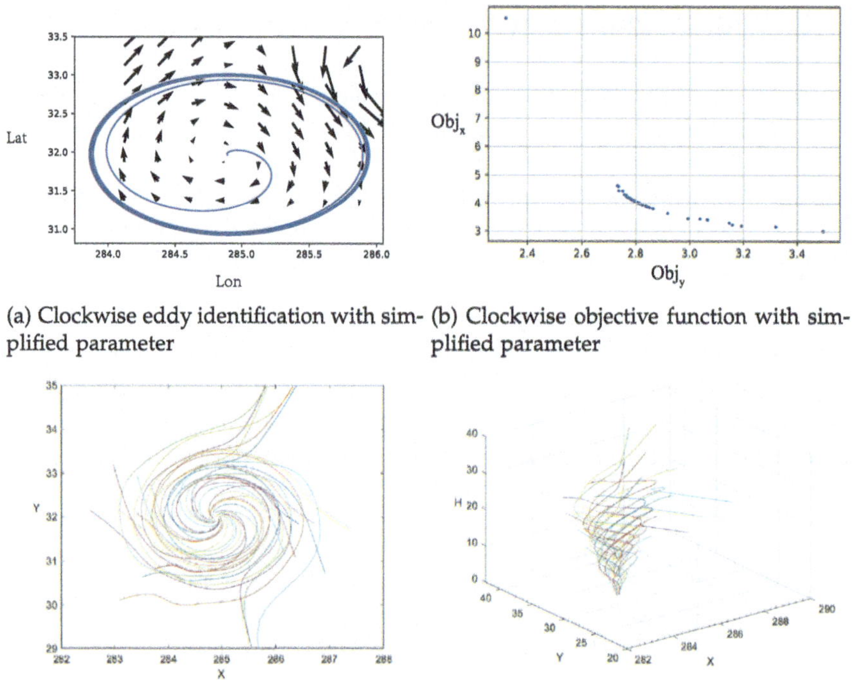

(a) Clockwise eddy identification with simplified parameter (b) Clockwise objective function with simplified parameter

(c) 2D phase portrait of simplified clockwise model (d) 3D phase portrait of simplified clockwise model

Fig. 6.2 Clockwise model

Table 6.1 Results comparison of clockwise model

	Radius	Eddy center	
		Longitude	Latitude
Numerical results	1.045	284.9	31.98
Flow field data	0.75	284.8	31.95

the objective function value of the feasible solution in Pareto frontier is not very small, the results still illustrated that this algorithm can be applied to the identification of mesoscale eddies. Moreover, the model can automatically distinguish the direction of the spiral with the sign of η, which is of profound significance to distinguish mesoscale cold eddies and mesoscale warm eddies.

In addition, due to the wide range of mesoscale eddies, this identification method is capable of controlling their behaviors.

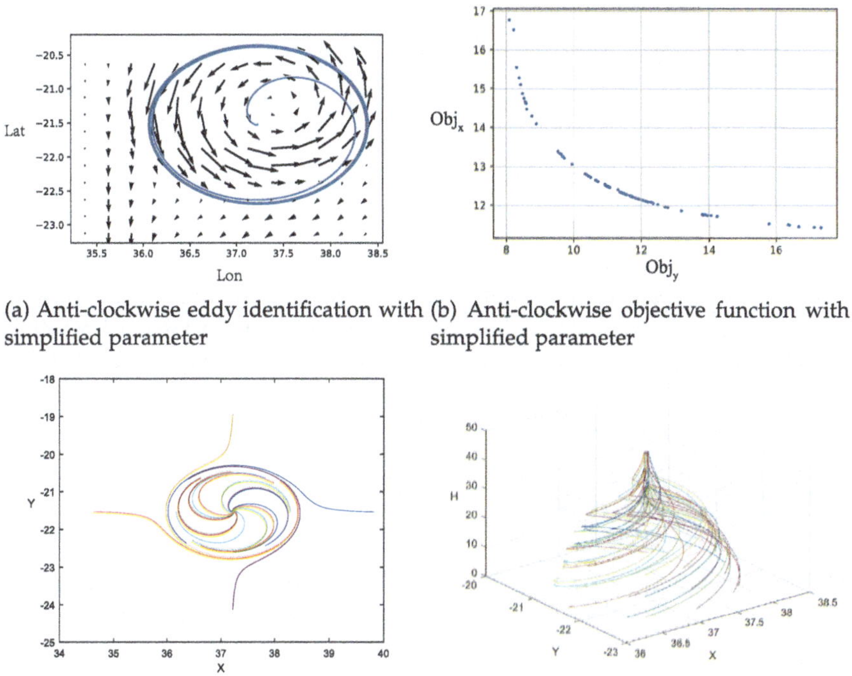

(a) Anti-clockwise eddy identification with simplified parameter

(b) Anti-clockwise objective function with simplified parameter

(c) 2D phase portrait of simplified anti-clockwise model

(d) 3D phase portrait of simplified anti-clockwise model

Fig. 6.3 Clockwise model

Table 6.2 Results comparison of anti-clockwise model

	Radius	Eddy center	
		Longitude	Latitude
Numerical results	1.17	37.23	−21.51
Flow field data	1.1	37.45	−21.4

6.2.2 *Full Parameter Case*

We take the complete parameter vector $\theta = (x_0 \quad y_0 \quad R \quad a \quad b \quad \lambda \quad \mu \quad \alpha \quad \beta)^{\mathrm{T}}$ in system (3.3.15) which has nine parameters to be estimated to solve problem (6.1.2). Compared with the special case in Sect. 6.2, we do not need to manually specify a clockwise model or a counterclockwise model. However, as the number of parameters increases, a larger initial population and more genetic algebra are required in the process of parameter estimation. Here we still choose the two examples in Sect. 6.2 for calculations. In order to facilitate the distinction, we still named them a clockwise model and counterclockwise model. But, they are unified in actual calculations.

Table 6.3 Range of undetermined parameter in genetic algorithm

x_0, y_0	Radius	a	b	λ	μ	α	β
Range of identification area	[0.1, 4]	[1, 4]	[1, 4]	[0.1, 10]	[0.1, 10]	[−3, 3]	[−3, 3]

To meet the concept of the mesoscale eddy and make the algorithm more efficient, we have adopted the parameter estimation range that is shown in Table 6.3, which is also the parameter range of the semi-stable limit cycle determined by the shooting method and the screening method.

6.2.3 Clockwise Model

In this case, the shape of the semi-stable limit cycle may be an ellipse, so according to the formula of the elliptical area, we define the radius of the mesoscale eddy as the radius of a circle, i.e., $\sqrt{R/\sqrt{ab}}$.

In the genetic algorithm, the number of population individuals is 400 and the maximum genetic generation number is 4000. The following results are obtained after calculation. See Table 6.4.

The corresponding model is determined as

$$\begin{cases} x' = 0.1(y - 31.78) + 0.2(x - 284.87) \\ \qquad \times[1.01(x - 284.87)^2 + (y - 31.78)^2 - 0.23]^2, \\ y' = -0.1(x - 284.87) + 0.46(y - 31.78) \\ \qquad \times[1.01(x - 284.87)^2 + (y - 31.78)^2 - 0.23]^2. \end{cases}$$

According to the results in Fig. 6.3, the algorithm is more accurate in identifying the position of the eddy center, but the estimated range of the mesoscale eddy is small. From the perspective of the objective function value of the Pareto front, it falls within the range of less than 3. Compare the objective function value in Sect. 6.2.1 is not large. At present, we believe that the main reason may be more parameters, resulting in many locally optimal solutions. The genetic algorithm is in the calculation process. It is easy to fall into a locally optimal solution.

Table 6.4 Numerical results of clockwise model

x_0	y_0	R	a	b	$\lambda\omega$	$\mu\omega$	α	β
284.84	31.78	0.23	1.01	1.00	0.10	0.10	0.20	0.46

Table 6.5 Results comparison of clockwise model

	Radius	Eddy center	
		Longitude	Latitude
Numerical results	0.47	284.85	31.75
Flow field data	0.75	284.8	31.95

Table 6.5 is a comparison of the main geometric parameters of the clockwise model calculation results and actual mesoscale eddies.

6.2.4 *Anti-clockwise Model*

The counterclockwise model follows the example in Sect. 6.2.1, and the results we calculated are as follows. See Fig. 6.5 and Table 6.6.

The corresponding model is determined as

$$\begin{cases} x' = 0.1(y - 21.35) - 0.06(x - 37.37) \\ \qquad \times[0.16(x - 37.37)^2 + 0.1(y - 21.35)^2 - 0.38]^2, \\ y' = -0.11(x - 37.37) - 0.18(y - 21.35) \\ \qquad \times[0.16(x - 37.37)^2 + 0.1(y - 21.35)^2 - 0.38]^2. \end{cases}$$

For the counterclockwise model, we can achieve better results. It can be seen from Fig. 6.4a that the model is more accurate in estimating the center and area of the eddy.

Table 6.7 shows the comparison between the calculation results of the counter-clockwise model and the main geometric parameters of the actual mesoscale eddy.

6.2.5 *Algorithm Parallelization and Model Checking in Global Oceans*

The content of the previous sections mainly analyzed the behavior of local sea areas. In this section, we will parallelize the above algorithms and apply them to identification and search of mesoscale eddies in wide range.

For the global flow field, if the search is carried out in the range $2° \times 2°$, it will compute the nonlinear programming problem about 20,000 times. For example, if we set the population as 50 and max generation as 200, with the computing platform i7-7700k 3.5GHz CPU the task will take about 13 hours. While the parallel algorithm in related literature only takes 0.13 hours (see Table 6.8). Therefore, it is necessary to design an appropriate parallel algorithm based on spatial division.

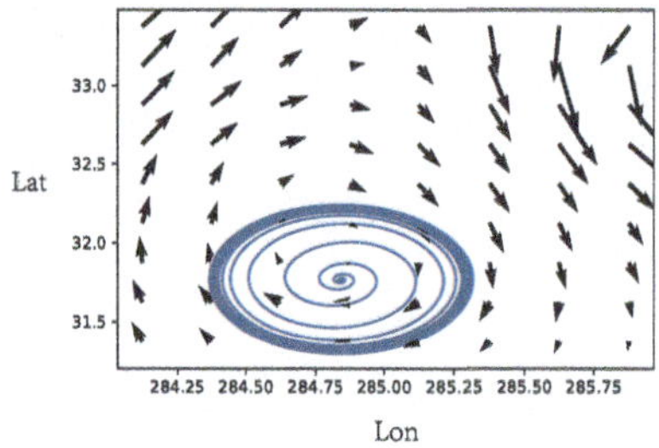

(a) Clockwise eddy identification with complete parameter

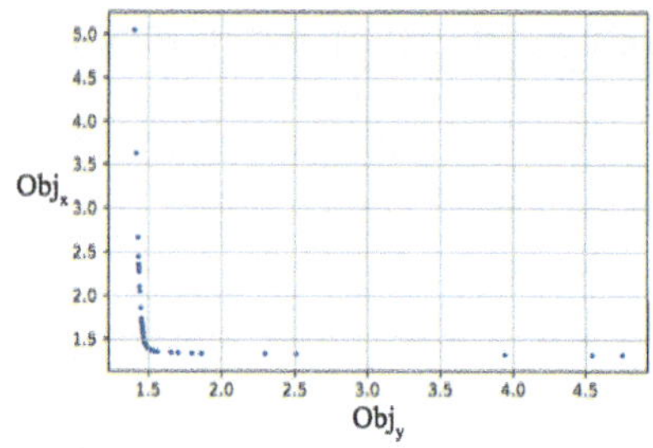

(b) Clockwise objective function with complete parameter

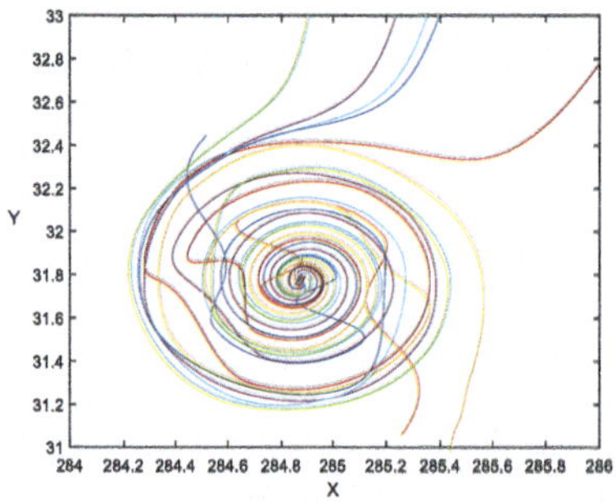

(c) 2D phase portrait of complete clockwise model

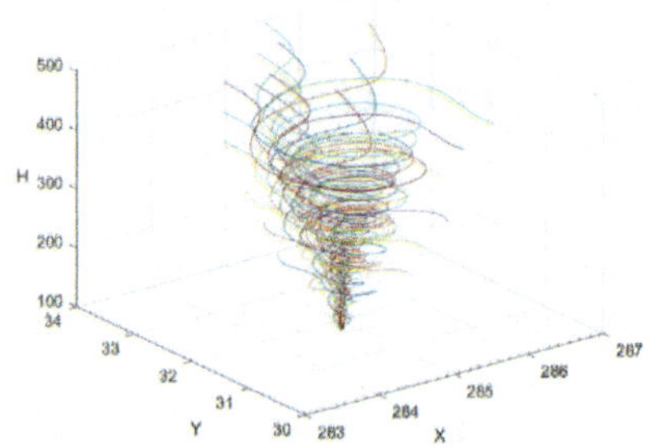

(d) 3D phase portrait of complete clockwise model

Fig. 6.4 Clockwise model

Figure 6.6 shows the idea of algorithm parallelization. A single rectangle (as shown in the green area) is the smallest computing unit, and all the grids constitute the global computing domain (as shown in the yellow and green areas). For instance, if we have five processes, we can assign the computing task as the division of the whole computing domain by the heavy lines.

On the other hand, it is reasonable to set a threshold of the objective function in the Pareto front as a criterion for the existence of mesoscale eddies in the search area. According to many experimental results we have carried out, it is reasonable to set the threshold value to 3 under the search accuracy of 1°. The results are shown in Fig. 6.7, in which the green marker is the clockwise model recognition result and the yellow marker is an anti-clockwise model recognition result.

Further, we use the parallel method to recognize mesoscale eddies in the South China Sea, see Fig. 6.8. The Lagrangian particle dynamic model and the genetic algorithm has the ability to obtain the coordinates and areas of mesoscale eddies and distinguish cold and warm ones according to the sign of parameters. The main errors come from the following two aspects. Firstly, for areas where streamlines are almost closed but unclosed, the algorithm will be cheated. Secondly, there is the phenomenon of identifying a single eddy repeatedly. For the first problem, related criteria will be

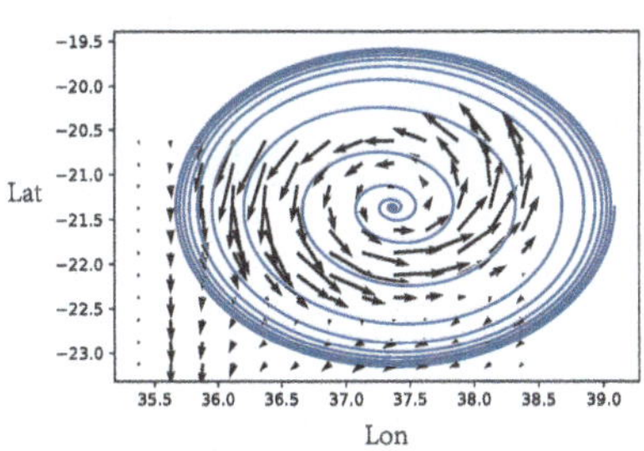

(a) Anti-clockwise eddy identification with complete parameter

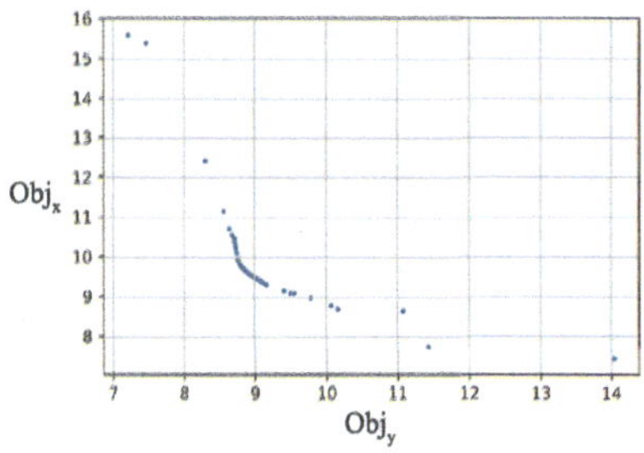

(b) Anti-clockwise objective function with complete parameter

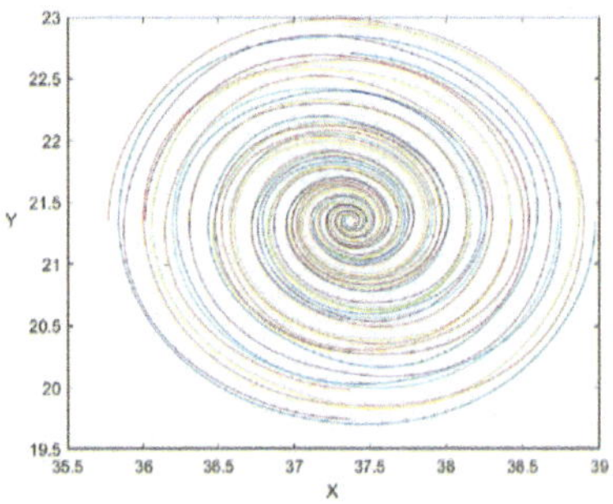

(c) 2D phase portrait of complete anti-clockwise model

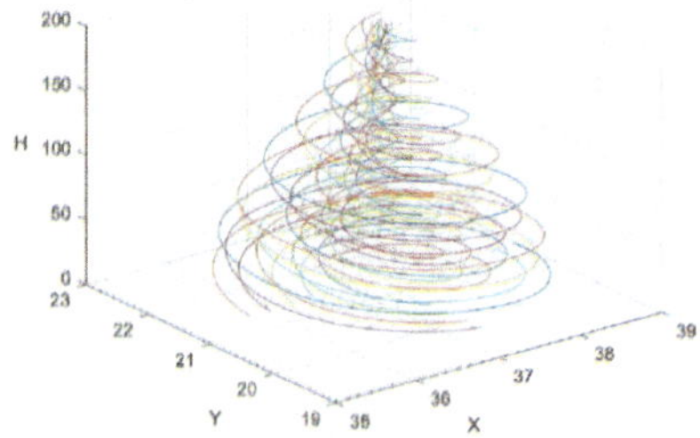

(d) 3D phase portrait of complete anti-clockwise model

Fig. 6.5 Anti-clockwise model

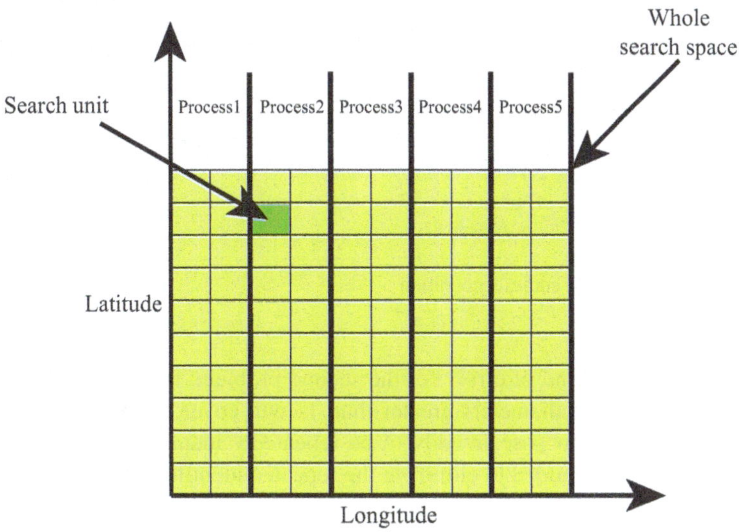

Fig. 6.6 Parallelization schema of algorithms

Table 6.6 Numerical results of anti-clockwise model

x_0	y_0	R	a	b	$\lambda\omega$	$\mu\omega$	α	β
37.37	21.35	0.38	0.16	0.10	0.10	0.11	−0.06	−0.18

Table 6.7 Results comparison of anti-clockwise model

	Radius	Eddy center	
		Longitude	Latitude
Numerical results	1.73	37.23	−21.51
Flow field data	1.1	37.45	−21.4

Table 6.8 Comparison of computing time

Computing method	Parallel SLA in [61]	This section
Parallel computing time (h)	0.13	13
Number of processes	8	1
Single process computing time	1.1	13

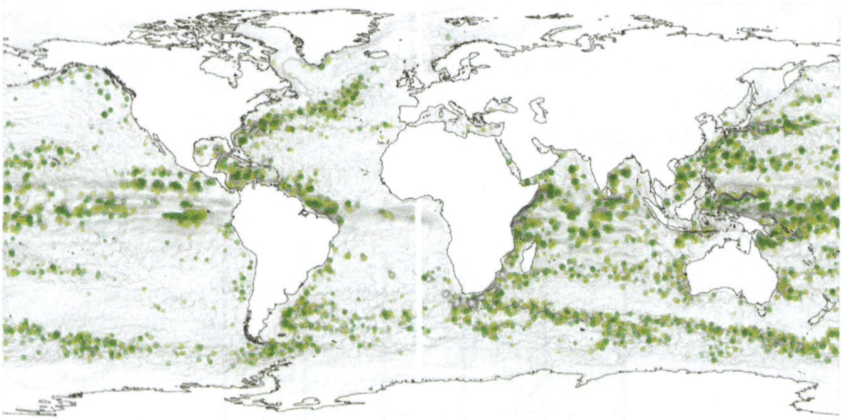

Fig. 6.7 Global mesoscale eddies recognition

added to reduce this kind of error. For the second problem, the main reason is that mesoscale eddies with diameters greater than 1° will cover multiple search areas, which will cause the mesoscale eddy to be repeatedly identified. The subsequent algorithm will also be added to eliminate the repeated identification results.

Remark 6.1 Universal mathematical models of plane structure and spatial structures are systems (3.3.15) and (3.3.16). They are all nonlinear systems with time-varying parameters. However, we select the constant-coefficient system with parameters vary-

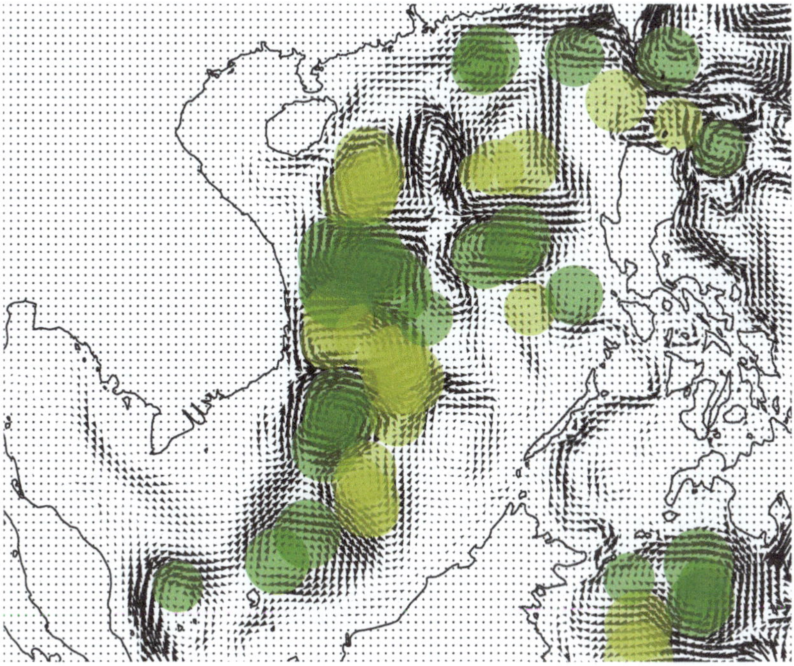

Fig. 6.8 Mesoscale eddies recognition in the South China Sea

ing in inappropriate range in actual operation, so there are some errors naturally, but these also basically describe the basic situation of mesoscale eddies.

Remark 6.2 After all, we have originally established a mathematical model with universal analytical form of mesoscale eddies, and used these models to identify mesoscale eddies. This is a subversive work, and the corresponding research on other issues will be a pioneering advance.

Chapter 7
Spatiotemporal Structure of Mesoscale Eddies: Self-similar Fractal Behavior

7.1 Spatiotemporal Fractal Structure of Mesoscale Warm Eddy

Attention to nonlinear system (3.3.16), we can take the initial value as $(x_{i0}, y_{i0}), i = 1, 2, \ldots, n$. For each initial value (x_{i0}, y_{i0}) starting from nonlinear system (3.3.16), n eddy lines will be generated, and each eddy line is set as $C(x_{i0}, y_{i0}), i = 1, 2, \ldots, n$. The mesoscale eddy formed by n points is denoted as E_n. Then, we have

$$E_n = \bigcup_{i=1}^{n} C(x_{i0}, y_{i0}), \tag{7.1.1}$$

which is a mesoscale eddy formed by system (3.3.16) with n water particles.

In practice, a mesoscale eddy should be generated by infinite water particles, which must satisfy that each water particle moves from east to west, and is subjected to approximately equal Coriolis force. Moreover, they are all continuum media with uniform consistency rules. Therefore, the actual mesoscale eddies should be

$$E_\infty = \bigcup_{i=1}^{\infty} C(x_{i0}, y_{i0}). \tag{7.1.2}$$

Noticing the actual flow field, model (7.1.2) should be the actual spatial infinitely nested self-similar fractal behavior. According to [47], it is consistent with the spatial structure of mesoscale warm eddies, as shown in Fig. 7.1a–c.

According to limit cycle theory [192], system (3.3.16) is internally unstable and externally stable, that is, system (3.3.16) can generate a mesoscale warm eddy.

Remark 7.1 Considering that the $f_i(\cdot)$ on the right side of dynamic system (3.3.16) contains salinity, density, temperature, and other connotative issues. These will seriously affect the behavior of the particle movement.

S.-T. Liu et al., *Mathematical Principle and Fractal Analysis of Mesoscale Eddy*,
https://doi.org/10.1007/978-981-16-1839-0_7

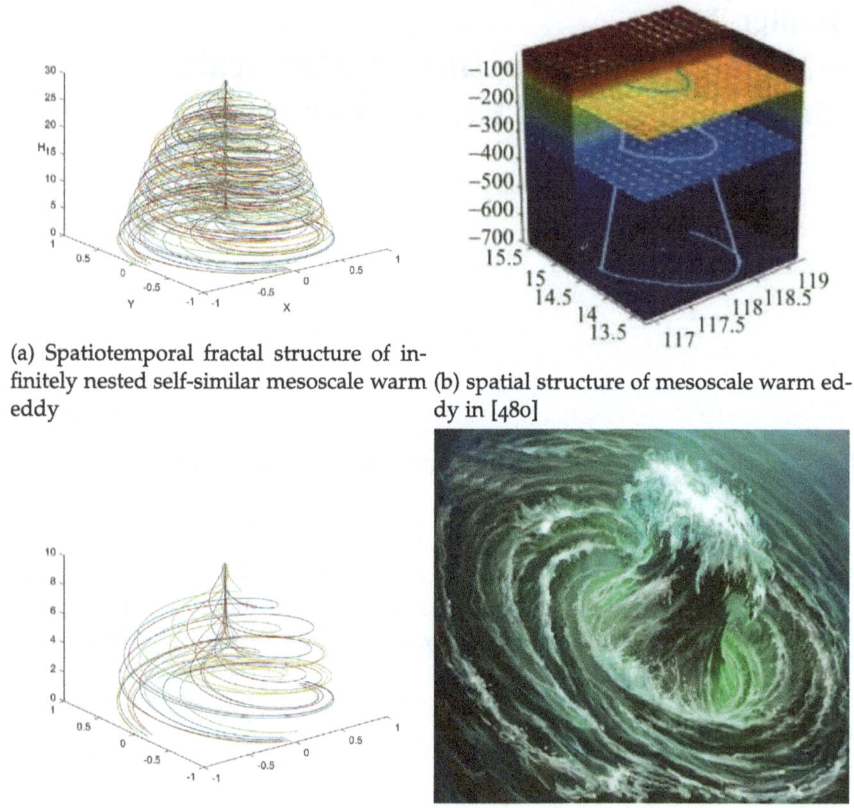

(a) Spatiotemporal fractal structure of infinitely nested self-similar mesoscale warm eddy

(b) spatial structure of mesoscale warm eddy in [480]

(c) Mesoscale warm eddy with n particles

(d) Actual mesoscale warm eddy

Fig. 7.1 The consistency between the universal model of mesoscale warm eddy and the historical data

7.2 Spatiotemporal Fractal Structure of Mesoscale Cold Eddy

In the same way, we take the initial value

$$(\overline{x}_{i0}, \overline{y}_{i0}), \quad i = 1, 2, \ldots, n.$$

System (3.3.16) is also a mesoscale eddy formed by n particles as

$$\overline{E}_n = \bigcup_{i=1}^{n} \overline{C}(\overline{x}_{i0}, \overline{y}_{i0}), \tag{7.2.1}$$

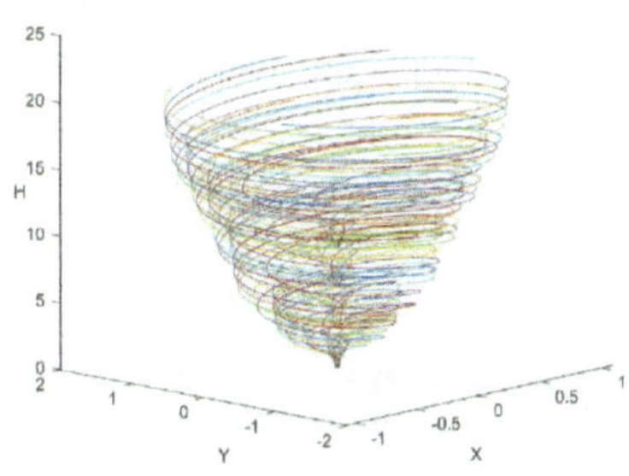

(a) Spatiotemporal fractal structure of infinitely nested self-similar mesoscale cold eddy

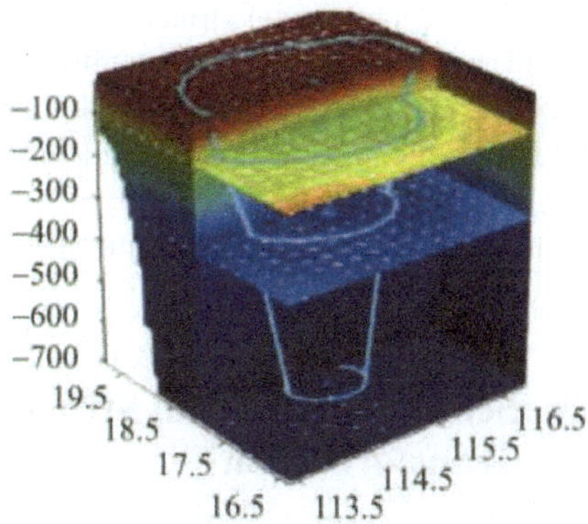

(b) spatial structure of mesoscale cold eddy in [480]

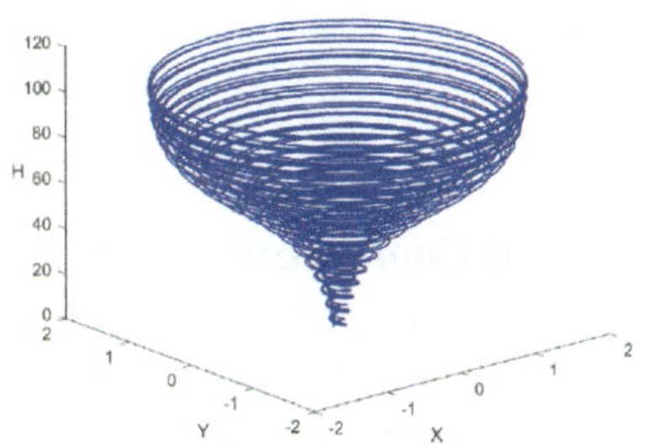

(c) Mesoscale cold eddy with n particles

(d) Actual mesoscale cold eddy

Fig. 7.2 The consistency between the universal model of mesoscale cold eddy and the historical data

where $\overline{C}(\overline{x}_{i0}, \overline{y}_{i0})$ is the i-th eddy curve generated by $(\overline{x}_{i0}, \overline{y}_{i0})$. Figure 7.2a is the complete eddy E_n. Like model (7.1.2), mesoscale cold eddies also are

$$\overline{E}_{\infty} = \bigcup_{i=1}^{\infty} \overline{C}(\overline{x}_{i0}, \overline{y}_{i0}). \tag{7.2.2}$$

According to the actual flow field, model (7.2.1) should be the actual spatial infinitely nested self-similar fractal behavior. According to the spatial structure demonstrated by the statistical theory in [480], model (7.2.1) is consistent with the spatial structure of mesoscale cold eddies, as shown in Fig. 7.2.

Remark 7.2 The fractal structure of a mesoscale eddy corresponds exactly to the "bowl" and "inverted cone" in its three-dimensional structure obtained from observation data in [480], which fully demonstrates that the three-dimensional fractal structure of the universal mathematical model established in this work is consistent with the reality. Meanwhile, it proves the correctness of our persistent use of the Lagrangian particle dynamics limit cycle theory to deal with mesoscale eddies.

7.3 Self-similar Fractal Structure under Affine Transformation

The spatial structure of mesoscale eddies corresponding to systems (7.1.2) and (7.2.2) is composed of eight water particles, see Figs. 4.104 and 4.106, respectively. Next, we will use two parts to prove that mesoscale eddies should have a self-similar fractal structure in mechanism. It is noted that from the eight water particles to the continuous medium infinite water particles E_∞ and $\overline{E}_\infty$ in (7.1.2) and (7.2.2), respectively. For any $p(x, y, z) \in E_\infty$ (or $\overline{E}_\infty$), it moves to $\overline{p}(\overline{x}, \overline{y}, \overline{z})$, that is,

$$P(x, y, z) \rightarrow \overline{P}(\overline{x}, \overline{y}, \overline{z}).$$

7.3.1 Transformation Relations of Spatial Coordinates

For any $p(x, y, z)$, it usually undergoes a finite number of basic transformations such as translation, rotation, scaling, staggering, and flipping. For the behavior of a mesoscale eddy, a clockwise mesoscale eddy will not flip to a counterclockwise one. So there is no need to flip the transformation but only the other four transformations, and their transformation matrices are shown in Table 7.1.

From the original point $P(x, y, z)$ to the target point $\overline{P}(\overline{x}, \overline{y}, \overline{z})$ is obtained by a finite number of transformations as

Table 7.1 Matrix transformation

Translation	Rotation
$T = \begin{bmatrix} 1 & 0 & 0 \\ 0 & 1 & 0 \\ d & n & 1 \end{bmatrix}$	$R = \begin{bmatrix} \cos\theta & \sin\theta & 0 \\ -\sin\theta & \cos\theta & 0 \\ 0 & 0 & 1 \end{bmatrix}$
Scale	**Shear**
$S_1 = \begin{bmatrix} k & 0 & 0 \\ 0 & k & 0 \\ 0 & 0 & k \end{bmatrix}$	$S_2 = \begin{bmatrix} 1 & \tan\phi & 0 \\ \tan\phi & 1 & 0 \\ 0 & 0 & 1 \end{bmatrix}$

$$(\overline{x}, \overline{y}, \overline{z})^{\mathrm{T}} = \sigma_n \sigma_{n+1} \ldots \sigma_2 \sigma_1 (x, y, z)^{\mathrm{T}},$$

where $\sigma_n \sigma_{n+1} \ldots \sigma_2 \sigma_1$ denotes the continuous transformation of finite translation, rotation, scaling, and shearing.

7.3.2 Spatial Structure

The spatial structures shown in Figs. 4.103 and 4.106 are based on E_∞ and $\overline{E}_\infty$. Any point $Q(x, y, z)$ on E_∞ or $\overline{E}_\infty$ can reach point $\overline{Q}(\overline{x}, \overline{y}, \overline{z})$ through affine transformation

$$Q(x, y, z) \longrightarrow \overline{Q}(\overline{x}, \overline{y}, \overline{z}) = (\alpha x, \beta y, \gamma z + t), \tag{7.3.1}$$

where the general transformation formula is

$$\begin{bmatrix} x_{n+1,i} \\ y_{n+1,i} \\ z_{n+1,i} \end{bmatrix} = \begin{bmatrix} \alpha_n & 0 & 0 \\ 0 & \beta_n & 0 \\ 0 & 0 & \gamma_n \end{bmatrix} \begin{bmatrix} x_{n,i} \\ y_{n,i} \\ z_{n,i} \end{bmatrix} + \begin{bmatrix} 0 \\ 0 \\ t_{n,i} \end{bmatrix}. \tag{7.3.2}$$

Note that transformations (7.3.1) and (7.3.2) are an iterative self-similar one. The self-similar transformation of the iterative function determines that Figs. 7.1(a) and 7.2(a) are two infinitely nested self-similar fractal sets. Furthermore, the basic properties of the iterative function determine its own unique attractor, which naturally contains strange attractors. This will lead to an infinitely nested self-similar structure, that is, a fractal structure [472], which confirms that the mesoscale eddy is a fractal structure of self-similar space, see Fig. 7.2(a).

Chapter 8
Mesoscale Eddies: Disk and Columnar Shapes

It can be seen from Sect. 7.3 that mesoscale eddies are usually obtained through a limited number of basic transformation processes such as translation, rotation, scaling, cross-cutting, and flipping (sometimes twisting and twisting nesting at different depth levels of seawater are used). It is also worth noting that there are two forms of mesoscale eddies, namely, disk-shaped mesoscale eddies whose diameter is larger than that of the mesoscale eddy axis and cylindrical vortices whose diameter is smaller than that of the mesoscale eddy axis. Usually, the mesoscale eddy is a disk-shaped one, which is also our focus.

8.1 The Specific Implementation Process of Water Particle Motion Transformation of Disk-Shaped Mesoscale Cold Eddy

8.1.1 Specific Transformation

According to the parameter distribution interval of the mesoscale cold eddy that in Tables 4.5 and 4.6, after taking appropriate parameters, model (3.3.16) is transformed into

$$\begin{cases} \frac{\mathrm{d}x}{\mathrm{d}t} = x(2x^2 + y^2 - 1)^2 + \frac{8\sqrt{2}}{2}y, \\ \frac{\mathrm{d}y}{\mathrm{d}t} = y(2x^2 + y^2 - 1)^2 - 8\sqrt{2}x. \end{cases} \tag{8.1.1}$$

In this regard, the following specific transformations and simulations are implemented.

(a1) Translation. Set $\overline{\mathbf{X}} = (\overline{x}, \overline{y}, \overline{z})$. Then

S.-T. Liu et al., *Mathematical Principle and Fractal Analysis of Mesoscale Eddy*,
https://doi.org/10.1007/978-981-16-1839-0_8

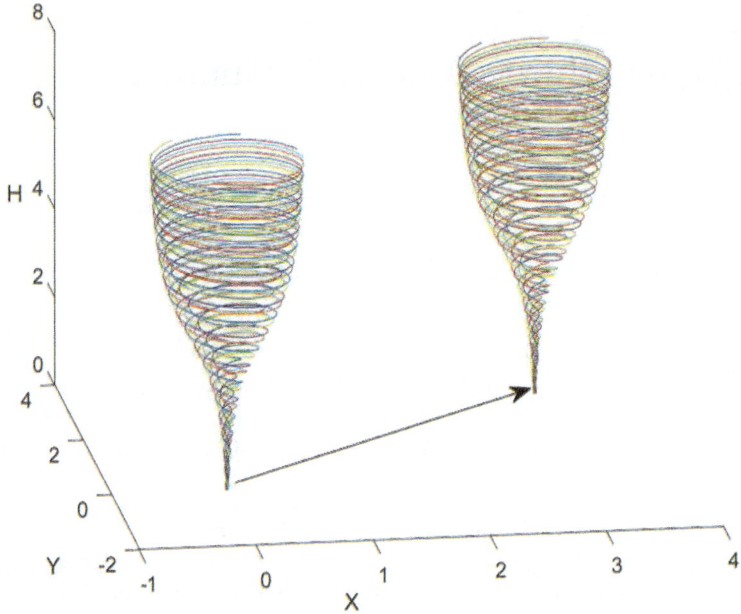

Fig. 8.1 Translation

$$\begin{bmatrix} \overline{x} \\ \overline{y} \\ \overline{z} \end{bmatrix} = \begin{bmatrix} x \\ y \\ z \end{bmatrix} + \begin{bmatrix} x_0 \\ y_0 \\ z_0 \end{bmatrix}.$$

Shift coordinate axes x and y by three units, and substitute $\overline{\mathbf{X}}$ into system (8.1.1). The simulation is shown in Fig. 8.1.

(a2) Rotation. Consider

$$\begin{bmatrix} \overline{x} \\ \overline{y} \\ \overline{z} \end{bmatrix} = \begin{bmatrix} \cos\theta & \sin\theta & 1 \\ -\sin\theta & \cos\theta & 1 \\ 0 & 0 & 1 \end{bmatrix} \begin{bmatrix} x_0 \\ y_0 \\ z_0 \end{bmatrix}.$$

Take $\theta = 45°$, and substitute $\overline{\mathbf{X}}$ into system (8.1.1). The simulation is shown in Fig. 8.2.

(a3) Scale. Consider

$$\begin{bmatrix} \overline{x} \\ \overline{y} \\ \overline{z} \end{bmatrix} = \begin{bmatrix} k & 0 & 0 \\ 0 & k & 0 \\ 0 & 0 & k \end{bmatrix} \begin{bmatrix} x_0 \\ y_0 \\ z_0 \end{bmatrix}.$$

Take $k = 1.5$, and substitute $\overline{\mathbf{X}}$ into system (8.1.1). The simulation is shown in Fig. 8.3.

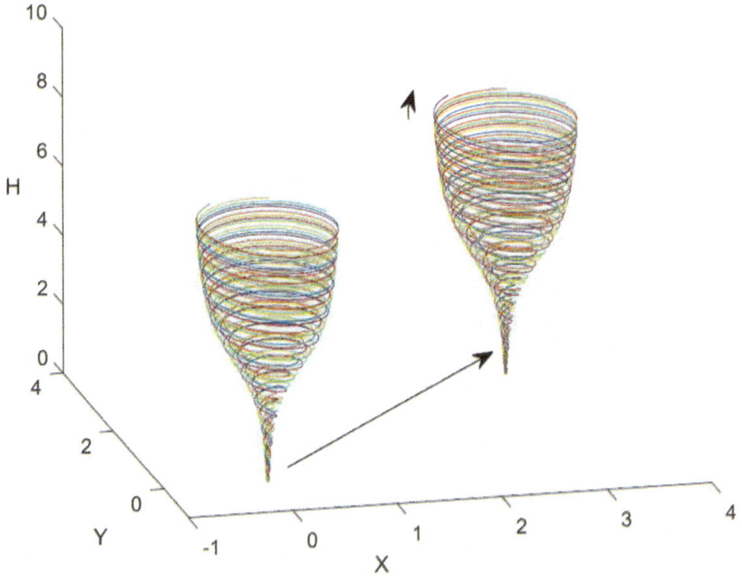

Fig. 8.2 Rotation

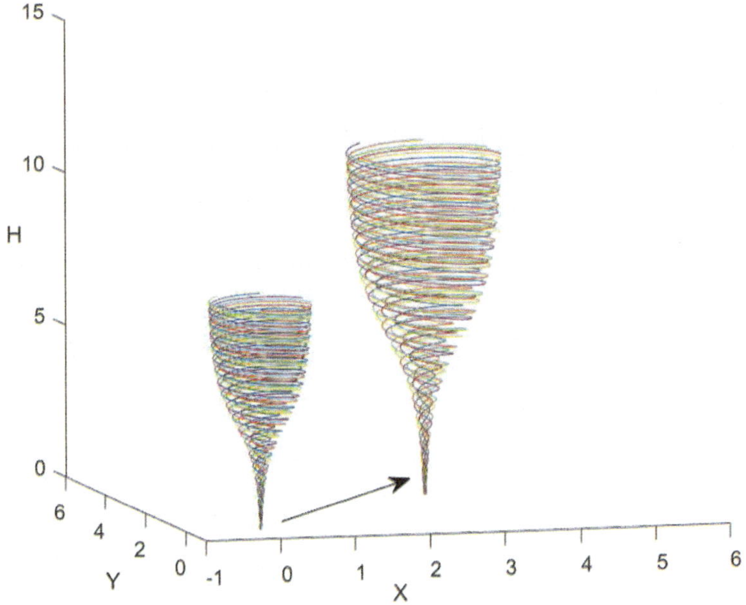

Fig. 8.3 Scale

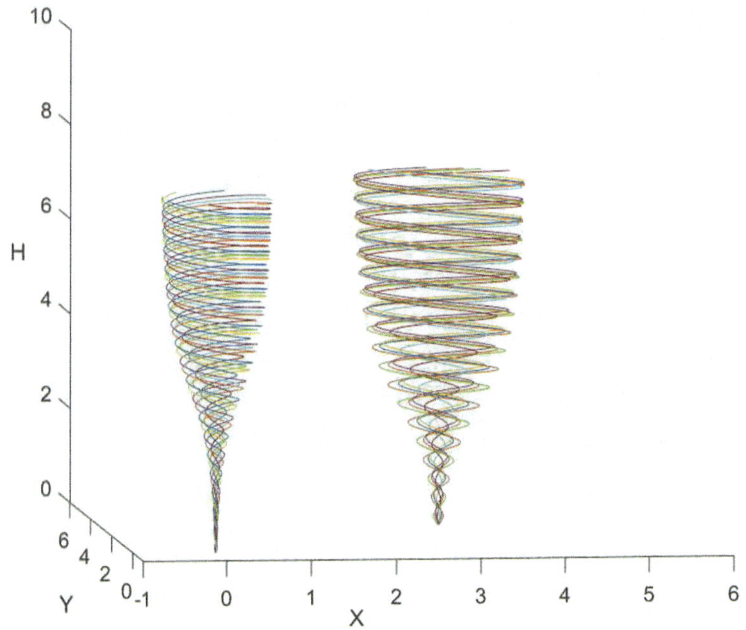

Fig. 8.4 Shear

(a4) Shear. Consider

$$\begin{bmatrix} \overline{x} \\ \overline{y} \\ \overline{z} \end{bmatrix} = \begin{bmatrix} 1 & \tan\phi & 0 \\ \tan\phi & 1 & 1 \\ 0 & 0 & 1 \end{bmatrix} \begin{bmatrix} x_0 \\ y_0 \\ z_0 \end{bmatrix}.$$

Take $\phi = 60°$, and substitute $\overline{\mathbf{X}}$ into system (8.1.1). The simulation is shown in Fig. 8.4.

In addition to the above four transformations, mesoscale eddies will also have tilt and distortion that are more consistent with the reality of mesoscale eddies.

(a5) Tilt. Set $\widehat{\mathbf{X}} = (\widehat{x}, \widehat{y}, \widehat{z})$. Then

$$\begin{cases} \widehat{x} = x + k_1 z, \\ \widehat{y} = y + z, \\ \widehat{z} = z. \end{cases}$$

Take $k_1 = 0.2$ and $k_2 = 0.22$. The simulation is shown in Fig. 8.5.

(a6) Distortion. Distortion occurs on the basis of the tilt of the mesoscale eddy, and its essence is to produce disturbance. Changes (a1)–(a5) are all transformations under ideal conditions, but in the actual ocean, there is no ideal mesoscale eddy, and they are all inclined and distorted. Therefore, to perturb the inclined mesoscale

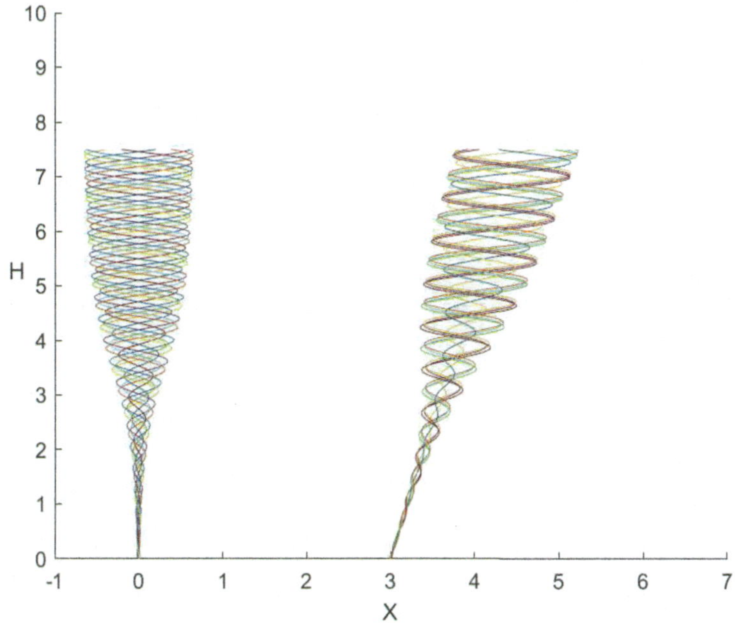

Fig. 8.5 Tilt

eddy $\widehat{E}$, in fact, the axis of the mesoscale eddy has a depth. Naturally, we perturb mesoscale eddy $\widehat{E}$ piecewise, which also takes into account the actual stratification of the ocean. Suppose mesoscale eddy $\widehat{E}$ is divided into eight layers, which are $\widehat{E}_1,\ \widehat{E}_2, \ldots, \widehat{E}_8$. Then, add random disturbance to each.

$$\begin{cases} \overline{E}_1 = \widehat{E}_1 + \gamma_1, \\ \overline{E}_2 = \widehat{E}_2 + \gamma_2, \\ \cdots \\ \overline{E}_8 = \widehat{E}_8 + \gamma_8, \end{cases}$$

where $\gamma_1,\ \gamma_2, \ldots, \gamma_8$ are very small randomly distributed value. The simulation is shown in Fig. 8.6.

(a7) Twisted nest. Actual mesoscale eddies are an infinitely nested self-similar fractal behavior of space. The mesoscale cold eddy formed by n space limit cycles is recorded as $\widetilde{E}_n$ defined as model (7.2.1). The simulation is shown in Fig. 8.7.

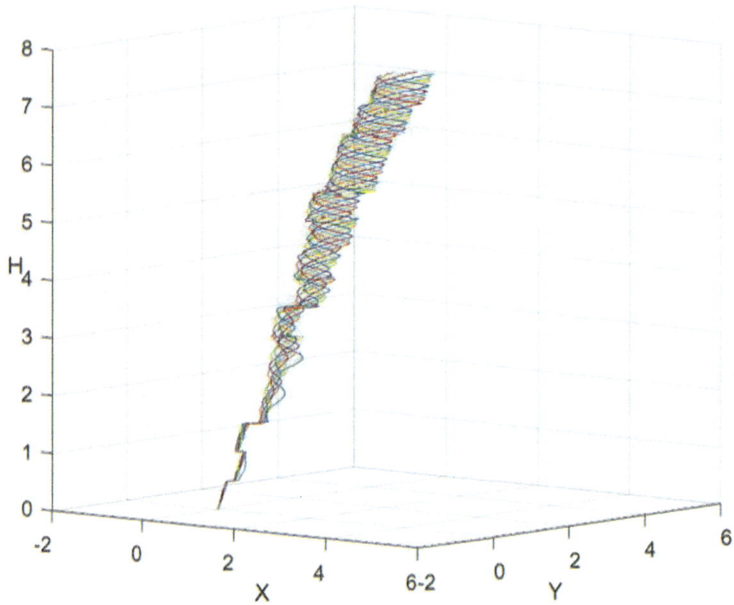

Fig. 8.6 Distortion

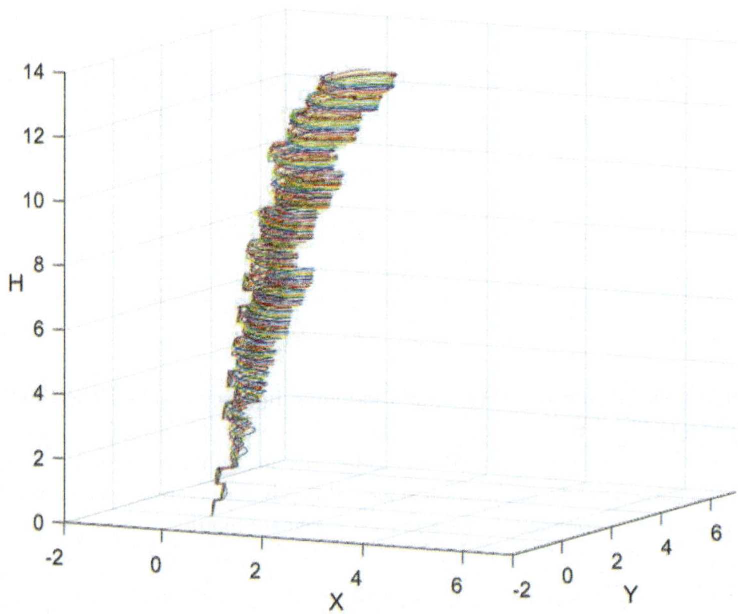

Fig. 8.7 Twisted and nested mesoscale cold eddy

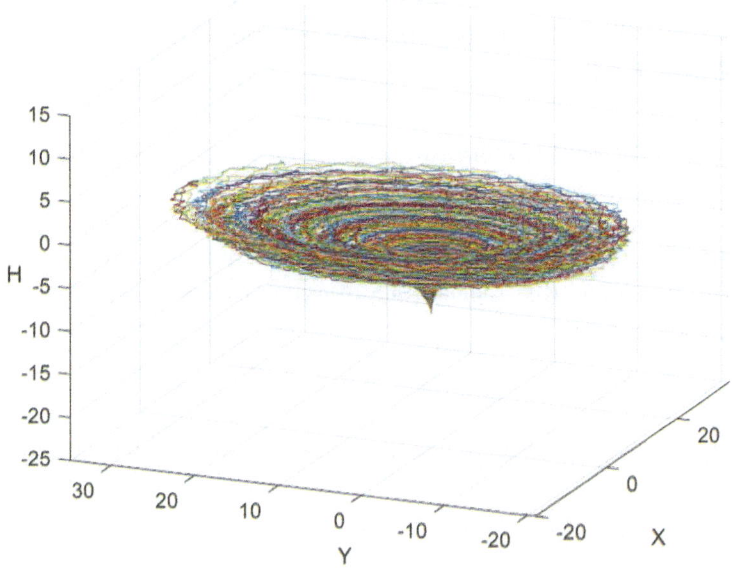

Fig. 8.8 Disk-shaped mesoscale cold eddy

8.1.2 Disk-Shaped Mesoscale Cold Eddy

The actual mesoscale eddies are mostly "shallow disk-shaped," because the horizontal scale of the ocean is much larger than the vertical scale. To get the actual result, we need to enlarge the horizontal scale of mesoscale eddy $\overline{E}$ on the basis of model (7.2.1) tilted and twisted nested mesoscale eddy $\widetilde{E}$, or reduce mesoscale eddy $\widetilde{E}$. For vertical scale, both can also be done simultaneously. Let disk-shaped mesoscale eddy $E_d(x_d, y_d, z_d)$ be

$$\begin{cases} x_d = k_3\widetilde{x}, \\ y_d = k_4\widetilde{y}, \\ z_d = k_5\widetilde{z}, \end{cases}$$

where $k_3 > 1$, $k_4 > 1$, and $k_5 < 1$. The simulation is shown in Fig. 8.8.

8.2 Specific Implementation Process of Water Particle Motion Transformation of Disk-Shaped Mesoscale Warm Eddy

8.2.1 Specific Transformation

According to the parameter distribution interval of mesoscale warm eddy that in Tables 4.5 and 4.6, after taking $\alpha = -1$, $\beta = -1$, $a = 1.01$, $b = 1.01$, $m = 2$, $\omega = 8.5$, $\lambda = 1$, $\mu = 1$, and $R = 1$, model (3.3.16) is transformed into

$$\begin{cases} \frac{dx}{dt} = -x(1.01x^2 + 1.01y^2 - 1)^2 + 8.5y, \\ \frac{dy}{dt} = -y(1.01x^2 + 1.01y^2 - 1)^2 - 8.5x. \end{cases} \tag{8.2.1}$$

In this regard, the following specific transformations and simulations are implemented.

(b1) Translation. Set $\overline{\mathbf{X}} = (\overline{x}, \overline{y}, \overline{z})$. Then

$$\begin{bmatrix} \overline{x} \\ \overline{y} \\ \overline{z} \end{bmatrix} = \begin{bmatrix} x \\ y \\ z \end{bmatrix} + \begin{bmatrix} x_0 \\ y_0 \\ z_0 \end{bmatrix}.$$

Shift coordinate axes x and y by three units, and substitute $\overline{\mathbf{X}}$ into system (8.2.1). The simulation is shown in Fig. 8.9.

(b2) Rotation. Consider

$$\begin{bmatrix} \overline{x} \\ \overline{y} \\ \overline{z} \end{bmatrix} = \begin{bmatrix} \cos\theta & \sin\theta & 1 \\ -\sin\theta & \cos\theta & 1 \\ 0 & 0 & 1 \end{bmatrix} \begin{bmatrix} x_0 \\ y_0 \\ z_0 \end{bmatrix}.$$

Take $\theta = 45°$, and substitute $\overline{\mathbf{X}}$ into system (8.2.1). The simulation is shown in Fig. 8.10.

(b3) Scale. Consider

$$\begin{bmatrix} \overline{x} \\ \overline{y} \\ \overline{z} \end{bmatrix} = \begin{bmatrix} k & 0 & 0 \\ 0 & k & 0 \\ 0 & 0 & k \end{bmatrix} \begin{bmatrix} x_0 \\ y_0 \\ z_0 \end{bmatrix}.$$

Take $k = 1.6$, and substitute $\overline{\mathbf{X}}$ into system (8.2.1). The simulation is shown in Fig. 8.11.

(b4) Shear. Consider

$$\begin{bmatrix} \overline{x} \\ \overline{y} \\ \overline{z} \end{bmatrix} = \begin{bmatrix} 1 & \tan\phi & 0 \\ \tan\phi & 1 & 1 \\ 0 & 0 & 1 \end{bmatrix} \begin{bmatrix} x_0 \\ y_0 \\ z_0 \end{bmatrix}.$$

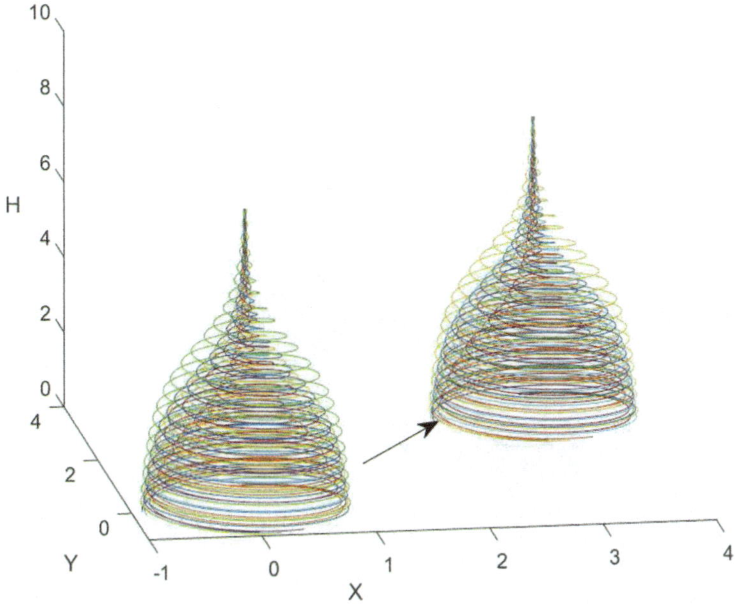

Fig. 8.9 Translation

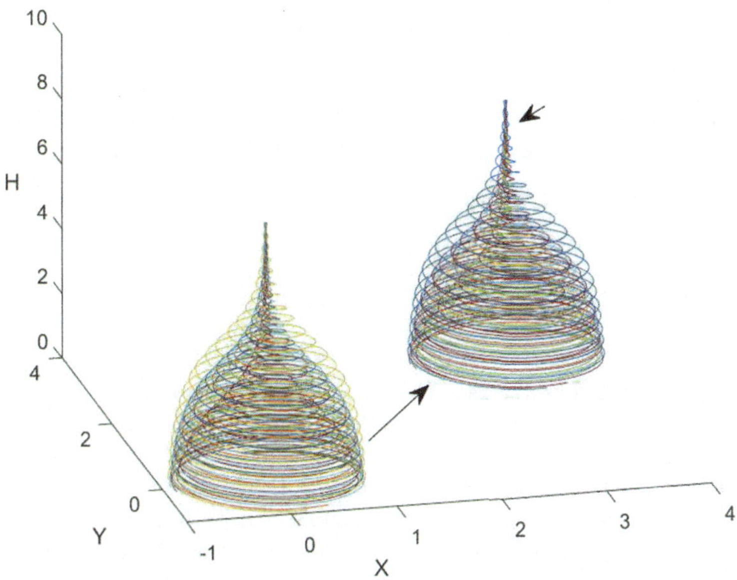

Fig. 8.10 Rotation

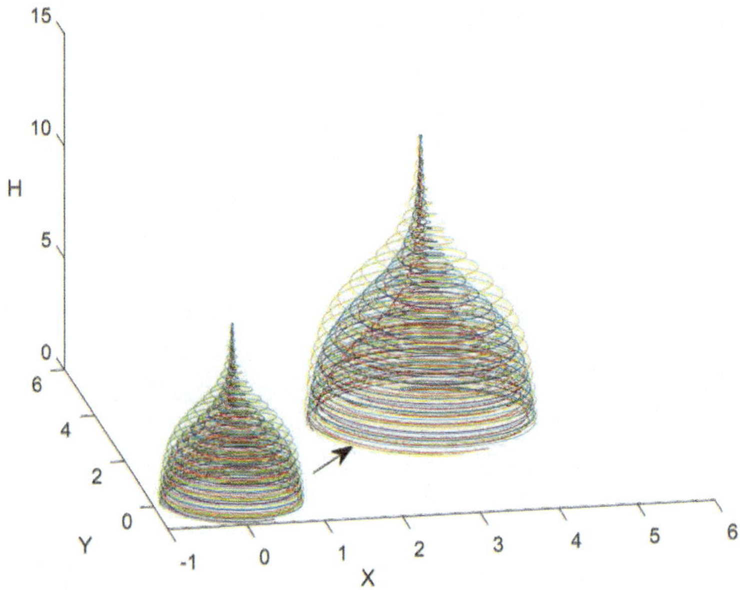

Fig. 8.11 Scale

Take $\phi = 50°$, and substitute $\overline{\mathbf{X}}$ into system (8.2.1). The simulation is shown in Fig. 8.12.

In addition to the above four transformations, mesoscale eddies will also have tilt and distortion that are more consistent with the reality of mesoscale eddies.

(b5) Tilt. Set $\widehat{\mathbf{X}} = (\widehat{x}, \widehat{y}, \widehat{z})$. Then

$$\begin{cases} \widehat{x} = x + k_1 z, \\ \widehat{y} = y + z, \\ \widehat{z} = z. \end{cases}$$

Take $k_1 = 0.21$ and $k_2 = 0.219$. The simulation is shown in Fig. 8.13.

(b6) Distortion. Distortion occurs on the basis of the tilt of the mesoscale eddy, and its essence is to produce disturbance. Changes (b1)–(b5) are all transformations under ideal conditions, but in the actual ocean, there is no ideal mesoscale eddy, and they are all inclined and distorted. Therefore, to perturb the inclined mesoscale eddy $\widehat{E}$, in fact, the axis of the mesoscale eddy has a depth. Naturally, we perturb mesoscale eddy $\widehat{E}$ piecewise, which also takes into account the actual stratification of the ocean. Suppose mesoscale eddy $\widehat{E}$ is divided into seven layers, which are $\widehat{E}_1,\ \widehat{E}_2, \ldots, \widehat{E}_7$. Then, add random disturbance to each.

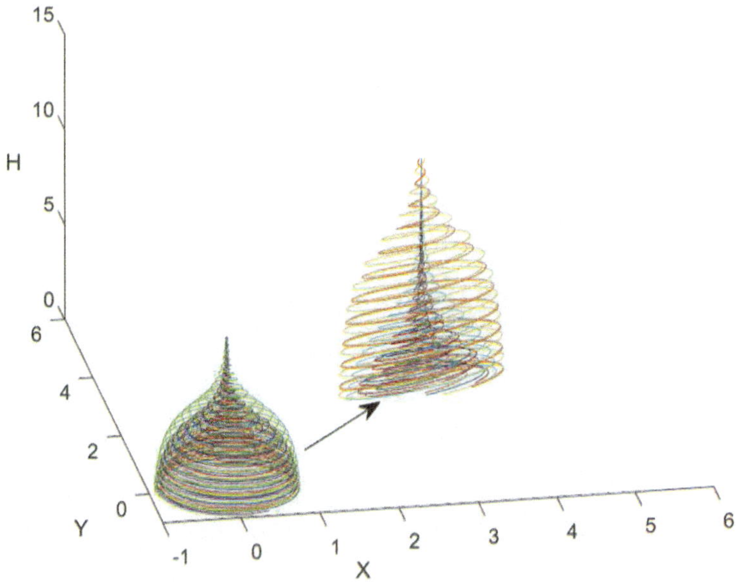

Fig. 8.12 Shear

$$\begin{cases} \overline{E}_1 = \widehat{E}_1 + \gamma_1, \\ \overline{E}_2 = \widehat{E}_2 + \gamma_2, \\ \cdots \\ \overline{E}_7 = \widehat{E}_7 + \gamma_7, \end{cases}$$

where $\gamma_1, \ \gamma_2, \ldots, \gamma_7$ are very small randomly distributed value. The simulation is shown in Fig. 8.14.

(b7) Twisted nest. Actual mesoscale eddies are an infinitely nested self-similar fractal behavior of space. The mesoscale warm eddy formed by n space limit cycles is recorded as $\widetilde{E}_n$ defined as model (7.1.1). The simulation is shown in Fig. 8.15.

8.2.2 Disk-Shaped Mesoscale Warm Eddy

Similar to Sect. 8.1.2, by synthesizing (b1)–(b7), the simulation of a disk-shaped mesoscale warm eddy is shown in Fig. 8.16.

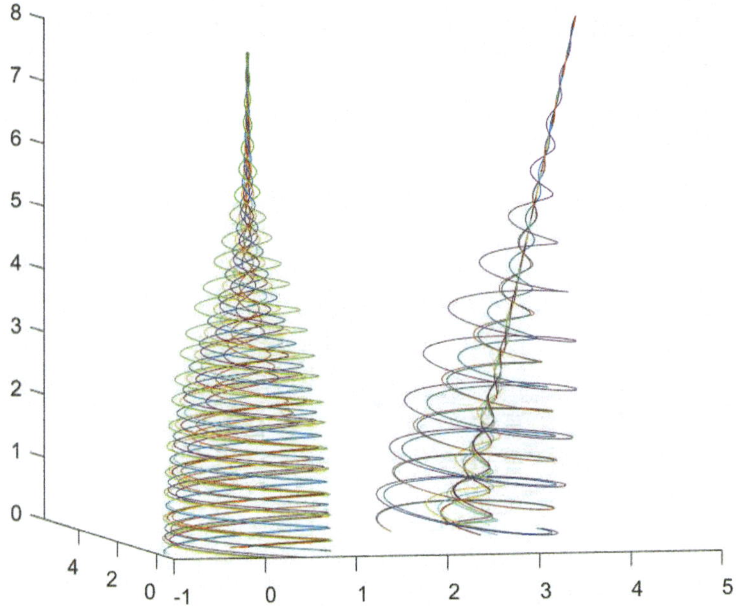

Fig. 8.13 Tilt

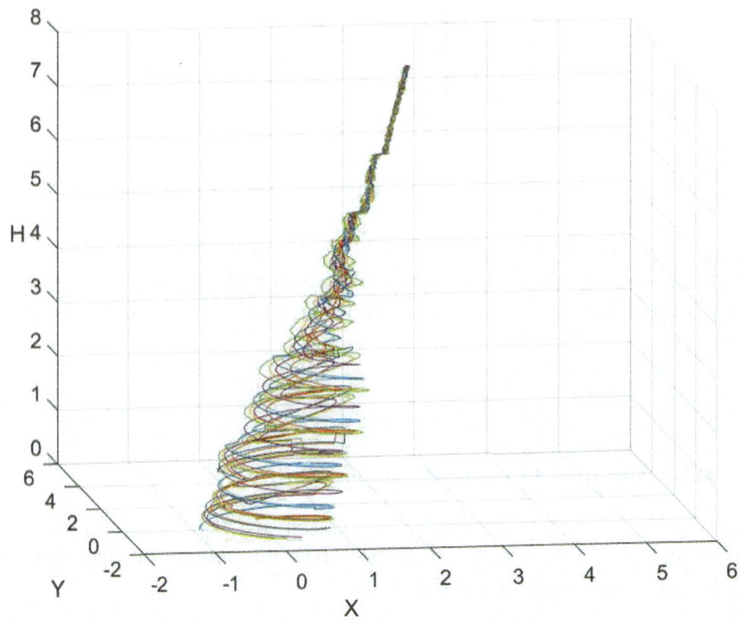

Fig. 8.14 Distortion

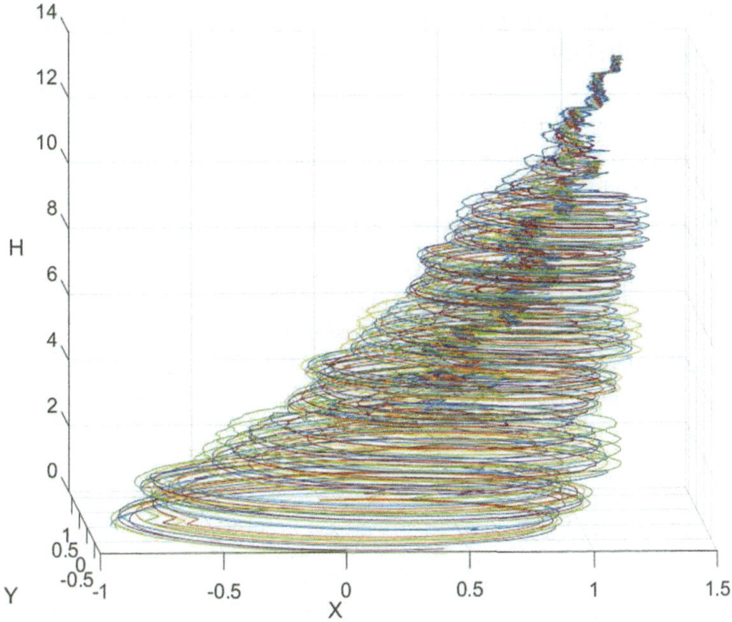

Fig. 8.15 Disk-shaped mesoscale warm eddy

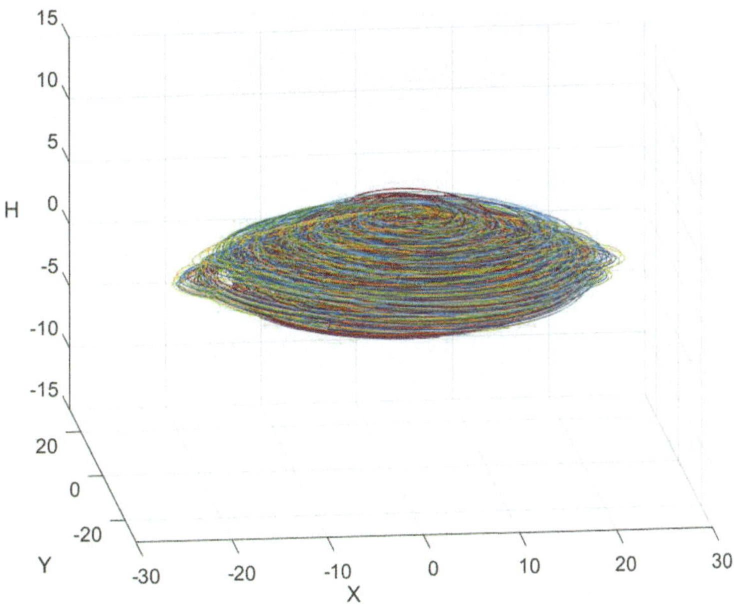

Fig. 8.16 Disk-shaped mesoscale cold eddy

8.3 Approximate Approximation of Mesoscale Disk-Shaped Mesoscale Eddy

In order to be more in line with the actual situation and make the disk-shaped mesoscale eddy gradually approach the real one, we retake the (1)–(7) transformation process and re-verify the control parameters that are more in line with the actual situation of the mesoscale eddy, such as within a reasonable range.

1. Increase or decrease the angle of rotation;
2. Increase or decrease the angle of the tangent;
3. Increase or decrease the coefficient of expansion;
4. Increase or decrease the degree of distortion;
5. Increase or decrease the tilt and distortion values of different levels.

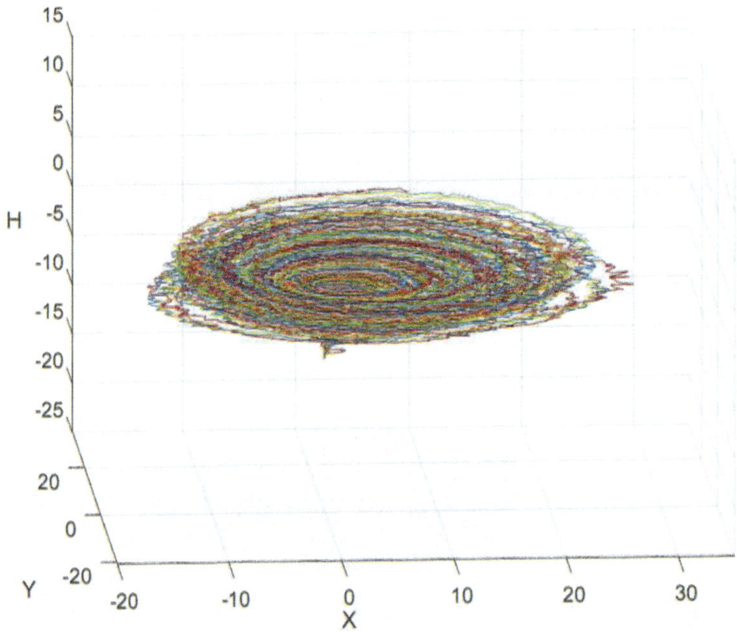

Fig. 8.17 Numerically approximated a mesoscale cold eddy

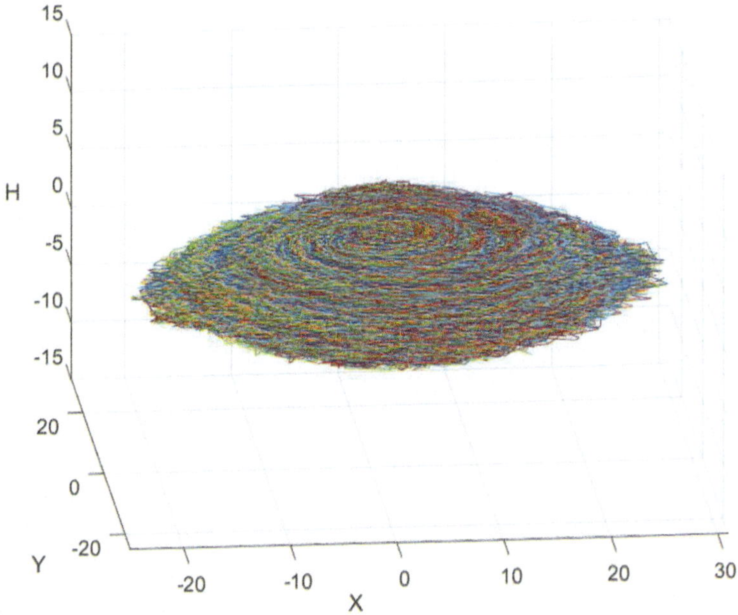

Fig. 8.18 Numerically approximated a mesoscale warm eddy

After a series of screening procedures with different parameters, we will get closer and closer to the reality of the mesoscale eddy. See Figs. 8.17 and 8.18. This actually involves the correlation between the mathematical model of the mesoscale eddy and the actual prediction, and it is also a practical application problem.

Chapter 9
Fractal Analysis and Prediction for Spatiotemporal Complexity of Mesoscale Eddy

9.1 Spatiotemporal Structure of Mesoscale Eddies Based on Universal Model

9.1.1 Mesoscale Cold Eddy

The mesoscale cold eddy in Fig. 4.105 is obtained from nonlinear system (3.3.16) through affine transformations (7.3.1) and (7.3.2). For the complexity of the spatial structure in Fig. 4.105, since the mesoscale cold eddy is a self-similar fractal structure, the fractal dimension can be used to describe its complexity. To this end, we take the vertical cross-section of the mesoscale cold eddy and a finite number of vertical intercepts to obtain a series of cross-sections. When the distance between each other is infinitely small, it can be regarded as a composition of multiple infinitely small intervals. So, the complexity of the three-dimensional structure of the mesoscale cold eddy can be revealed according to the fractal dimension of the cross-section. Notice the axial symmetry of mesoscale cold eddies, that is, the geometric structures of the parts $x < 0$ and $x > 0$ are the same in area $|x| < 2$. So only the part of $0 \leq x < 2$ is intercepted. Take the vertical cross-section to get Fig. 9.1.

From the perspective of fractal geometry, fractal dimension can describe the variation degree of cross-section, that is, the change of cross-section will cause the change of fractal dimension, which is the corresponding relation with cross-section. By adjusting parameters, fractal graphs of different eddy cross-sections can be obtained. According to the box dimension method, 42 cross-sections (Fig. 9.1 are four examples 42 groups) are taken, and the values of their fractal dimension and the number of boxes are calculated, as shown in Table 9.1, where $\log_2 N$ represents the number of small squares covering the edge of the mesoscale cold eddy cross-section with a side length of 0.001, and fractal dimension D describes the complexity of the mesoscale cold eddy.

S.-T. Liu et al., *Mathematical Principle and Fractal Analysis of Mesoscale Eddy*,
https://doi.org/10.1007/978-981-16-1839-0_9

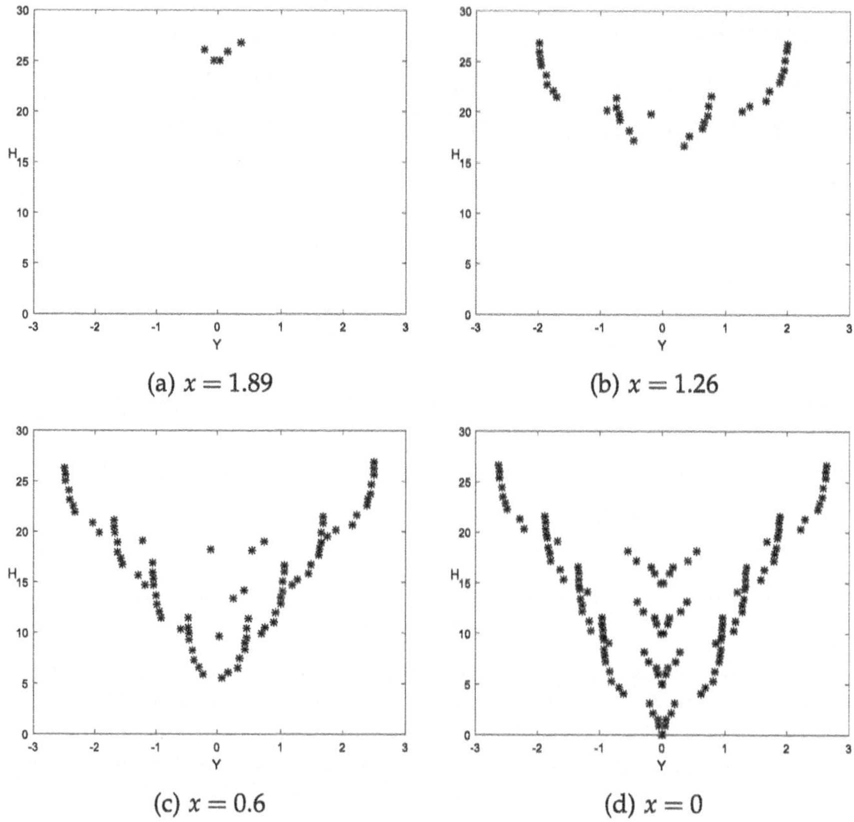

Fig. 9.1 Vertical section of mesoscale cold eddy

Table 9.1 Fractal dimension D and boxes ($\log_2 N$) of mesoscale cold eddy

	1	2	3	4	5	6	7	...	42
D	1.0335	1.0402	1.0432	1.0456	1.0473	1.0469	1.0435	...	1.3064
$\log_2 N$	11.372	11.511	11.549	11.569	11.581	11.578	11.551	...	13.612

Figure 9.1 corresponds to the 1, 15, 29, and 42 sets of data in Table 9.1. These four legends describe the fractal dimension and the number of boxes increases one by one, reflecting the fractal of the 42 sets of the mesoscale cold eddy from its edge cross-section to the center cross-section. The fractal dimension and the number of boxes gradually increase, that is, the complexity of the mesoscale cold eddy cross-section changes from weak to strong.

We use MATLAB to fit the data in Table 9.1 and get the numerical curve shown in Fig. 9.2, from which we can see that 42 cross-section data points from the eddy edge to the center in Table 9.1 are relatively evenly distributed on the fitting curve. Fractal

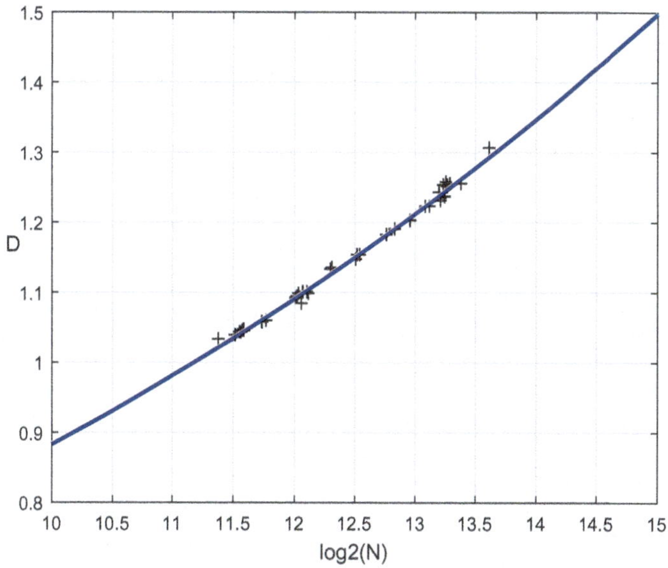

Fig. 9.2 The relation curve between the fractal dimension (D) and the number of boxes ($\log_2 N$) of mesoscale cold eddy ("*": data points; "-": fitting curve)

dimension D of the mesoscale cold eddy is positively correlated with the number of boxes $\log_2 N$. Fractal dimension D increases with the increase of the number of boxes $\log_2 N$ and the complexity of the spatial structure of eddies also increases gradually. The fractal dimension of the eddy cross-section is mostly in (1, 1.35) and the complexity of the mesoscale cold eddy is also in this range. Mathematical relation between fractal dimension D and number of boxes $\log_2 N$ of the mesoscale cold eddy is

$$D = 0.3079e^{0.1055 \log_2 N}. \tag{9.1.1}$$

It depicts the complexity of the spatial structure of mesoscale cold eddies.

9.1.2 Mesoscale Warm Eddy

Similar to the above, we focus on nonlinear system (3.3.16) and obtain mesoscale warm eddies in Fig. 4.107 in space by affine transformation (7.3.2). We also note the symmetry of eddies. For example, the vertical cross-sections for $|x < 3|$ are shown in Fig. 9.3.

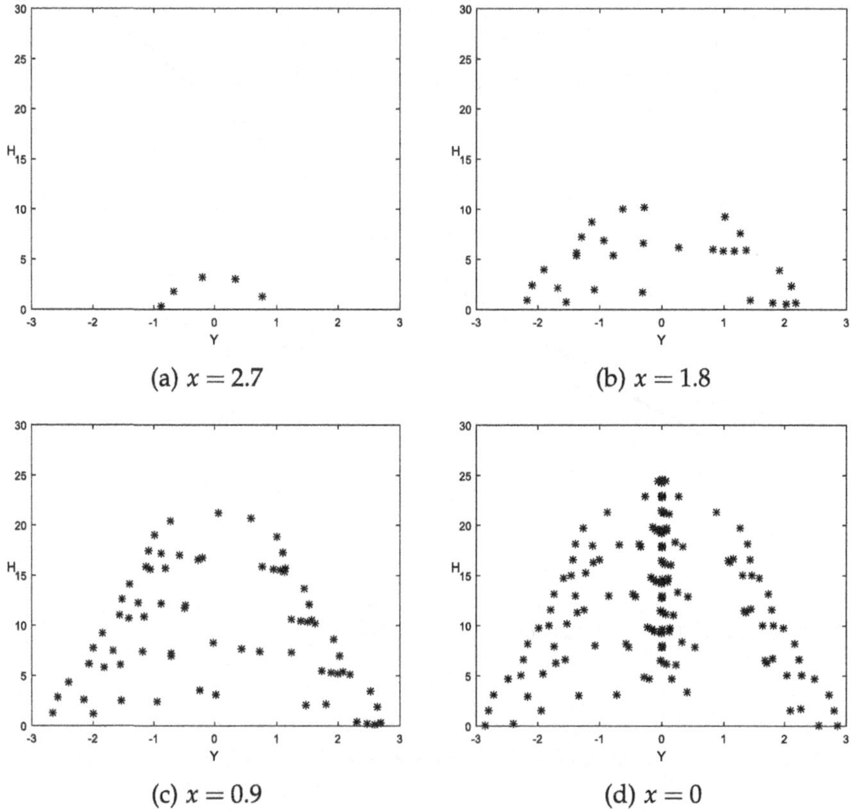

Fig. 9.3 Vertical section of mesoscale warm eddy

Table 9.2 Fractal dimension D and boxes ($\log_2 N$) of mesoscale warm eddy

	1	2	3	4	5	6	7	...	61
D	1.0363	1.0390	1.0753	1.0729	1.0859	1.1110	1.1111	...	1.4233
$\log_2 N$	11.413	11.435	11.929	11.953	12.051	12.305	12.304	...	14.785

We take 61 sections of the mesoscale warm eddy (Fig. 9.3 are four of the 61 groups) and then calculated their fractal dimensions and the number of boxes, as shown in Table 9.2, where $\log_2 N$ represents the number of small squares covering the eddy edge cross-section with a side length of 0.001. Fractal dimension D describes the complexity of the mesoscale warm eddy.

Figure 9.3 corresponds to the 1, 21, 41, and 61 sets of data in Table 9.2. These four legends describe the fractal dimension and the number of boxes increase one by one, reflecting the fractal of the 61 sets of the mesoscale warm eddy from its

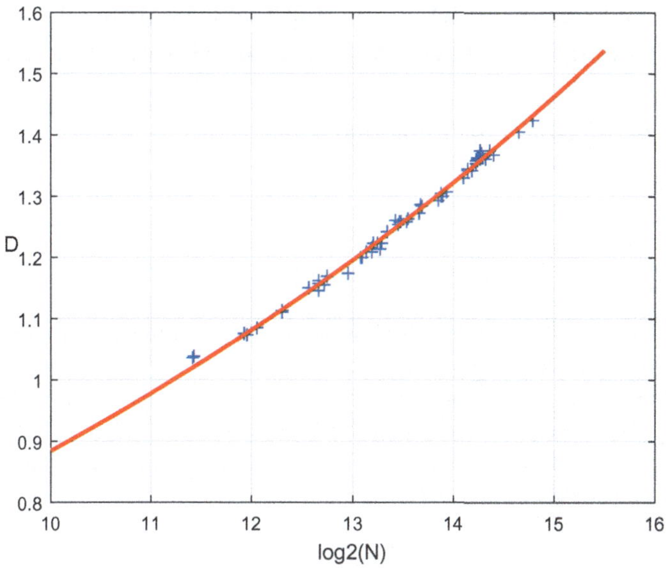

Fig. 9.4 The relation curve between the fractal dimension (D) and the number of boxes ($\log_2 N$) of mesoscale warm eddy ("*": data points; "-": fitting curve)

edge cross-section to the center cross-section. The fractal dimension and the number of boxes gradually increase, that is, the complexity of the mesoscale warm eddy cross-section changes from weak to strong.

We use MATLAB to fit the data in Table 9.2 seeing [481–483] to obtain the numerical curve shown in Fig. 9.4. We can see from Fig. 9.4 that 61 cross-section data points from the eddy edge to the center in Table 9.2 are evenly distributed on the fitting curve. The fractal dimension d of mesoscale warm eddy is positively correlated with the number of boxes $\log_2 N$. Fractal dimension D increases with the increase of the number of boxes $\log_2 N$ and the complexity of spatial structure of the mesoscale eddy also gradually increases. The fractal dimension of eddy cross-section is mostly in $(1, 1.45)$ and the complexity of the mesoscale warm eddy is also in this range. Expression of relation between fractal dimension D and number of boxes $\log_2 N$ of the mesoscale warm eddy is

$$D = 0.3244e^{0.1004 \log_2 N}, \tag{9.1.2}$$

which describes the complexity of mesoscale warm eddies.

9.2 Mathematical Model and Complexity Analysis of Spatiotemporal Fractal Structure of Mesoscale Eddies

9.2.1 *Fractal Model of Snowflake*

Snowflakes are regular fractals in nature. Consider

$$\mathscr{F}(\mathscr{L}_0) = (g \circ f^n)(\mathscr{L}_0) = g\{f^n(\mathscr{L}_0)\}, \tag{9.2.1}$$

where $\mathscr{F} = g \circ f^n$, $\mathscr{L}_0$ is a straight line segment, f is a functional relation of an equilateral triangle with a gap formed by dividing $\mathscr{L}_0$ into three equal parts and removing the middle part and adding two edges of an equilateral triangle, and g is a transformation of an equilateral triangle with variable $f^n(\mathscr{L}_0)$ as its edge.

The recursive relationship (9.2.1) is a composite function relationship with independent variables. Particularly, select

$$\mathscr{L}_{n+1} = f^n(\mathscr{L}_0), \quad n = 0, 1, 2, 3, 4, 5. \tag{9.2.2}$$

Then the recursive relation (9.2.2) generates Koch curves shown in Fig. 9.5.

Let $\mathscr{L}_5 = f^4(\mathscr{L}_0)$. Function $g(\mathscr{L}_5)$ produces the Koch snowflake shown in Fig. 9.6.

It can be seen from the above that model (9.2.1) is the mathematical model of snowflakes in nature, that is, the snowflakes obtained by Swedish mathematician Koch in 1904 by using regular fractals. Especially in model (9.2.1), the straight line segment $\mathscr{L}_0$ is the independent variable of compound function $g\{f^n(\cdot)\}$, also known as generator.

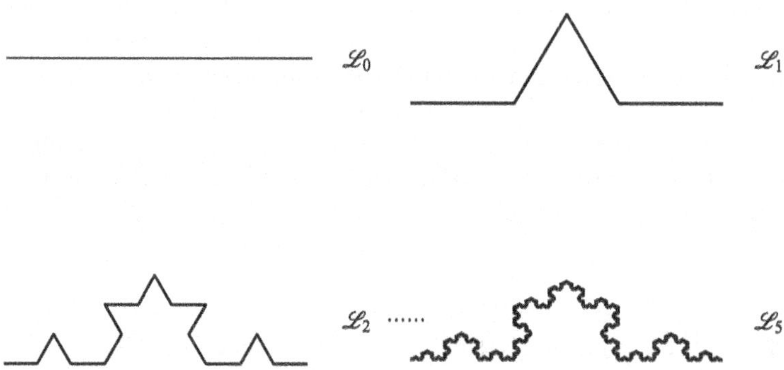

Fig. 9.5 Koch curve

Fig. 9.6 Koch snowflake

9.2.2 Fractal Model of Random Snowflake

Let

$$\mathscr{F}(\mathscr{L}_0) = g\left(f_\omega^n(\tilde{\mathscr{L}_0})\right), \tag{9.2.3}$$

where $\tilde{\mathscr{L}_0}$ is a straight line segment; f_ω with $\omega \in \{0, 1\}$ is a functional relation that divides $\tilde{\mathscr{L}_0}$ into three equal parts, removes the middle part and adds two sides of an equilateral triangle to form a polyline composed of four equal-length line segments; and g is the transformation function of an equilateral triangle with variable $f_\omega^n(\tilde{\mathscr{L}_0})$. Thus, model (9.2.3) is a compound function relationship with the independent variable $\tilde{\mathscr{L}_0}$. In particular, take

$$\mathscr{L}_{n+1} = f(\mathscr{L}_n), \quad \tilde{\mathscr{L}}_{n+1} = f_\omega(\tilde{\mathscr{L}}_n), \quad n = 0, 1, 2, 3, 4, 5. \tag{9.2.4}$$

Then the recursive relation (9.2.4) with $\tilde{\mathscr{L}}$ generates random Koch curves shown in Fig. 9.7.

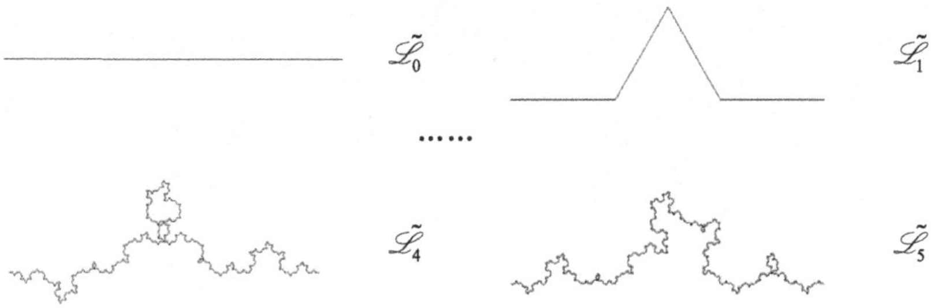

Fig. 9.7 Random Koch curve

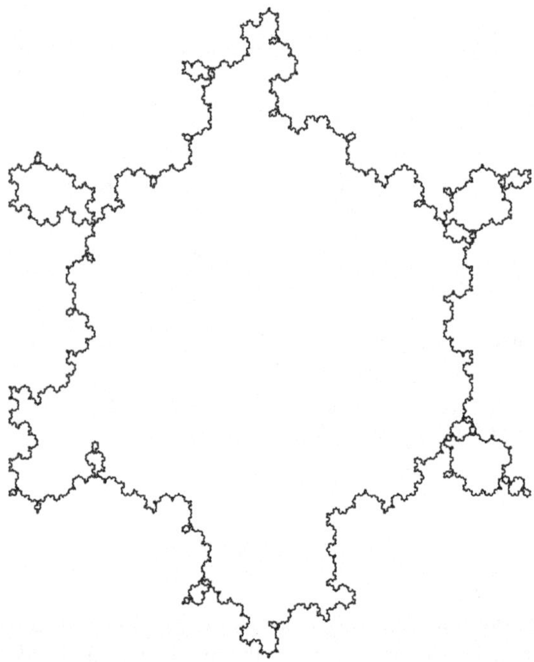

Fig. 9.8 Random Koch snowflake

Let $\mathscr{L}_5 = f^4(\mathscr{L}_0)$. Function $g(\mathscr{L}_5)$ produces a random Koch snowflake shown in Fig. 9.8.

Similarly, the generators of different snowflakes can be obtained by different methods, as long as the independent variable $\mathscr{L}_0$ is a curve and the iterated function is F_ω which is random.

9.2.3 Mesoscale Eddies and Spatiotemporal Fractal Structures of Cantor Self-Similar Fractal Sets

For convenience, we first consider the case of m-particles. Consider a mesoscale cold eddy defined as model (7.1.1) and a mesoscale warm eddy defined as model (7.2.1). Notice that E_m and $\bar{E}_m$ are both closed subsets of $D \subseteq \mathbb{R}^3$. Thus, the mathematical structure of mesoscale eddies is a stochastic fractal behavior of compression map in [192].

In fact, in $\mathbb{R}^3$, let D be an area of the ocean and $E_m, \bar{E}_m \subset D$. For arbitrary $p(x, y)$ in E_m or $\bar{E}_m$, take the continuous map $S : D \to D$, such that

$$|S(x) - S(y)| \begin{cases} < c\,|x - y|\,, & \text{compression of water particle,} \\ = c\,|x - y|\,, & \text{similarity of water particle,} \end{cases}$$

where $0 < c < 1$ and S is a random similarity map. So S turns E_m or $\bar{E}_m$ into a geometric similarity set, that is, S is similar.

Pay attention to the formation process of Figs. 4.105 and 4.107. For arbitrary $p(x, y) \in E_m$ or $\bar{E}_m$, we can transform $p(x, y)$ into a mesoscale eddy ε by means of a series maps

$$\varepsilon = E_n = f_n f_{n-1} \cdots f_2 f_1(E_m). \tag{9.2.5}$$

Therefore, ε is a fractal structure, and map (9.2.5) is mathematical expression of the spatial structure of mesoscale eddies and it is similar to the model of snowflake. In a particular case, when $f_j = f_0$, $j = 1, 2, \ldots, n$, we can transform map (9.2.5) into

$$\varepsilon = {f_0}^n(E_m). \tag{9.2.6}$$

Thus map (9.2.6) is an iterative behavior, which is also an infinitely nested self-similar behavior. Similarly, there are also

$$\overline{\varepsilon} = {f_0}^n(\overline{E}_m). \tag{9.2.7}$$

Using the idea of the mathematical model of snowflake, it is the basic element (independent variable) for the generation of mesoscale eddies. The geometric behavior of mesoscale eddies is the continuous evolution of geometric bodies, and the process is the transformation of map of stochastic fractals. Thus we have the fractal model of spatial structure as (9.2.6) and (9.2.7). From this, we have the following conclusions.

Theorem 1 *Mesoscale eddies ε and $\bar{\varepsilon}$ have a Cantor self-similar fractal structure.*

Proof By use of the particle trajectories of planar phase shown in Figs. 4.105 and 4.107, we can construct fractal structure as shown in Fig. 9.9.

First, we set $E_0 = E_m$ as the basic element of a mesoscale eddy with fractal structure. In order to study its fractal structure, we used to transform the coordinates

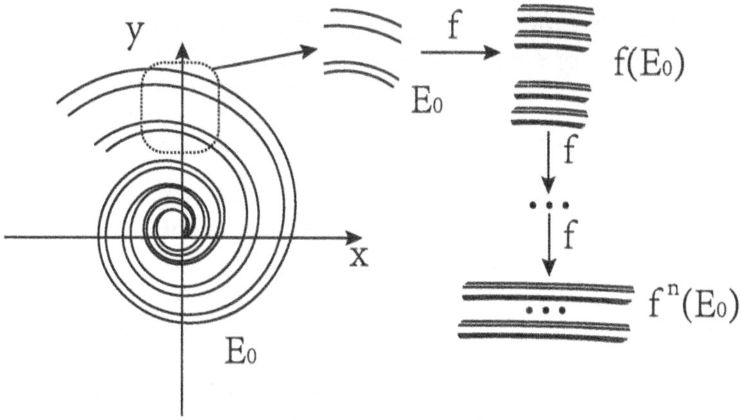

Fig. 9.9 Planar phase diagram of m-particles mesoscale eddy

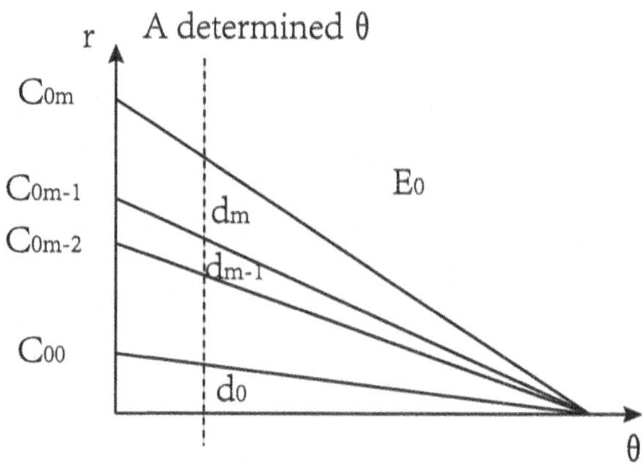

Fig. 9.10 Streamline in $\theta - r$-plane

from $x - y$-plane to $\theta - r$-plane. For the convenience of research, we might as well convert the obtained E_0 to the polar coordinate plane of $\theta - r$ and the following form can be obtained as shown in Fig. 9.10.

Next, we can convert the spiral trajectory in $x - y$-plane into an approximate linear trajectory in $\theta - r$-plane. For each fixed θ, we define the distance between points on the trajectory in E_0 as $d_i(\theta), i = 1, \ldots, m$. Here we use the idea of constructing Cantor three-division set (map a line segment into two through compression map line segment). We can set a list of points determined by each fixed $\{p_{0i}(\theta)\}$, where 0 means the point is the point in E_0. Each of the points P is mapped into n points proportional to the distance of each point in the point column $\{p_{0i}(\theta)\}$.

For example, if $m = 3$, Fig. 9.11 is the result of the transformation of the constructed map. Thus, the mesoscale eddy with complex structure is transformed into an intuitive point sequence problem with Cantor self-similar fractal structure. Then the mesoscale eddy $\overline{\varepsilon}$ is expressed as a form with self-similar fractal structure, and

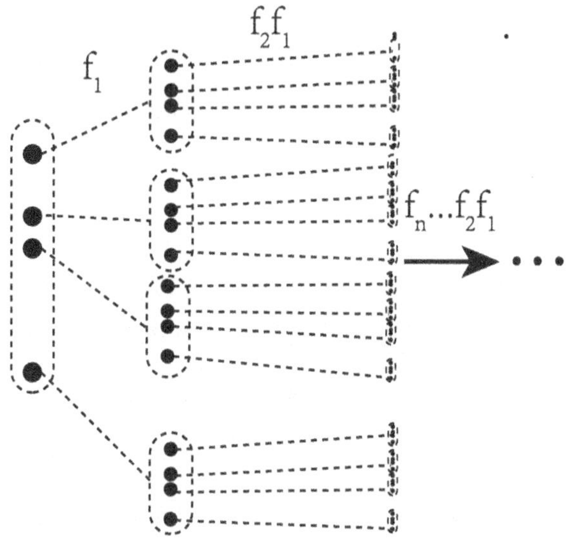

Fig. 9.11 $m = 3$

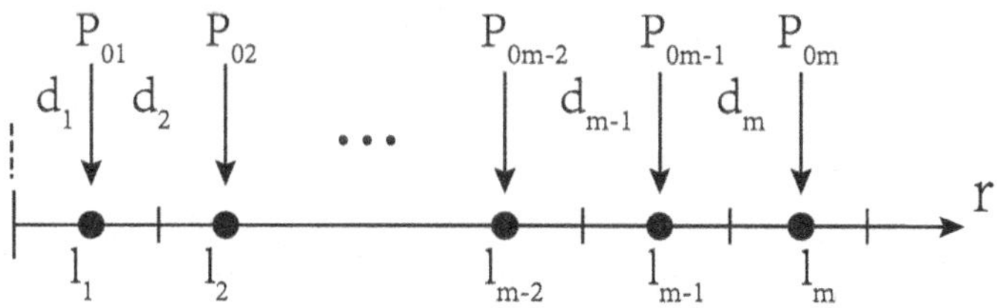

Fig. 9.12 Distribution of points and distances on r-axis

Table 9.3 Fractal generation coefficient l_i of mesoscale eddy

i	1	$[2, m-1]$	m
l_i	$d_1 + \frac{d_2}{2}$	$\frac{d_i+d_{i+1}}{2}$	d_m

the map is derived. The specific expression form of S iteration process is shown in Fig. 9.12.

In Fig. 9.13, for m points determined by the fixed θ, they occupy the range of motion of $R(\theta) = \frac{d_1+d_m}{2} + \sum_{i=2}^{m} d_i$ in the mesoscale eddy. The coordinates of this point are determined by $r_i(\theta)$, $i = 1, \ldots, m$. The distance between m points is expressed by $d_i(\theta)$, $i = 1, \ldots, m$. The connection between m points is composed of $1, 2, 3, 4, \ldots, m-1, m$ line segments, and the length of each line segment as l_i, $i = 1, \ldots, m$ satisfies the relationship in Table 9.3.

For each point, the range obtained by mapping it will be limited to the range of its line segment. For m points, we need m maps to form a map family within its defined

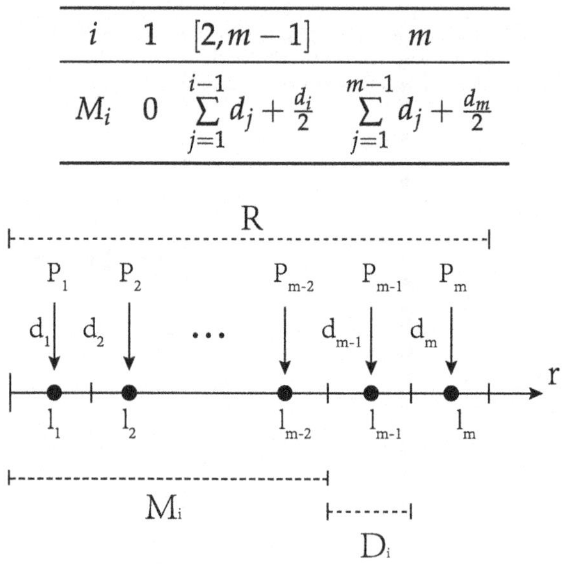

Fig. 9.13 Determination rules of related parameters

Table 9.4 Fractal generation coefficient of mesoscale eddy M_i

i	1	$[2, m-1]$	m
M_i	0	$\sum_{j=1}^{i-1} d_j + \frac{d_i}{2}$	$\sum_{j=1}^{m-1} d_j + \frac{d_m}{2}$

range map. For the points on the line segment 0-1, we can get the corresponding map as $f_1(r) = a\frac{l_1}{R(\theta)}r$, where r is the coordinates of the corresponding point, $a \in (0, 1)$ is called random fractal coefficient, and can be used to calculate the fractal dimension. For the point at the line segment 1-2, the corresponding map is $f_2(r) = a\frac{l_2}{R(\theta)}r + d_1 + \frac{d_2}{2}$. By analogy, we have

$$f_i(r) = a\frac{D_i}{R}r + M_i,$$

where M_i satisfies the relationship in Table 9.4.

Further, we found that in the process of using compressed map $S = \{f_i\}$ iterative process, although the overall map ideas and methods are always the same for each iteration $S = \{f_i\}$, the various parameters are changed. This theoretical conclusion can be obtained through a set of recursive formulas. For i-th map $\{f_i\}$ in n iteration, the relevant parameters can be calculated by

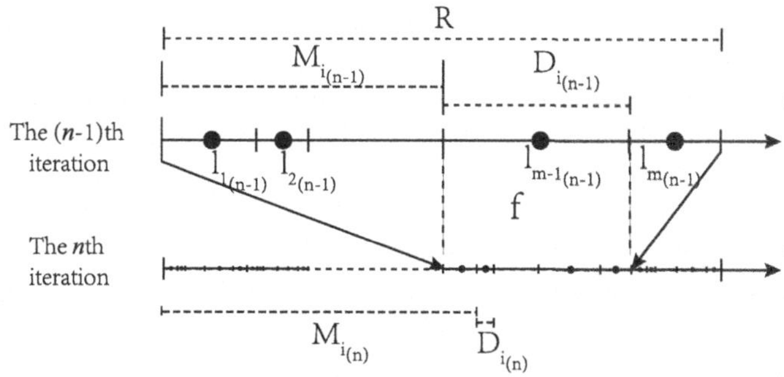

Fig. 9.14 Fractal generation rules of mesoscale eddy

$$\begin{cases} M_{i_{(n)}}(n) = M_{i_{(n-1)}}(n-1) + D_{i_{(n-1)}}(n-1)\frac{l_{i_{(n)}}}{R}, \\ D_{i_{(n)}}(n) = D_{i_{(n-1)}}(n-1)\frac{l_{i_{(n)}}}{R}. \end{cases}$$

So far, we have obtained a method of generating a point sequence with a self-similar structure from a point sequence. When we apply map $S = \{f_i\}$ to the set E_0 on $\theta - r$ plane, we can get ε and $\overline{\varepsilon}$. After several iterations, a family of curves with fractal structure $\varepsilon = f^n(E_m)$ can be approximated. Convert it from $\theta - r$-plane to $x - y$-plane. Then get the mesoscale eddy with E_m as the generator and a self-similar fractal structure. See Fig. 9.14.

It can be noted that the operation here is only on $x - y$-plane, and the trajectory is not compressed and translated on the vertical axis. Because the trajectory contained in E_m can fully represent the movement trend of water particles in the mesoscale eddy, the fractal structure that we mapped and presented is only for the complex structure of it to be more in line with the realistic particle motion. **So far we have proved that the mesoscale eddy is an ocean surging behavior with Cantor self-similar fractal structure.**

In summary, nonlinear dynamic system (3.3.16) is a semi-stable attractor. It is again a chaotic system, and thus a semi-strange attractor. In fact, system (3.3.16) is a semi-stable attraction set [484], and all initial points that make up the attraction set are the basin of attraction (also called attraction domain) of the semi-stable strange attractor. It is also concerned that it is a self-similar fractal structure, so that the mesoscale eddy is actually a fractal basin of semi-strange attractors.

9.3 Spatiotemporal Fractal Analysis and Prediction of the Complexity of Mesoscale Eddies

From the current point of view, whether a mesoscale eddy is obtained from the Lagrange particle dynamics universal mathematical model or from a satellite altimeter, its essence is a geometric gradual behavior. And this geometric gradient is fractal behavior! An important indicator to describe the scale of geometric gradients is the fractal dimension.

Starting from fractal theory, this section uses the SSH-based eddy tracking method (py-eddy-tracker) provided by Evan Mason to analyze the statistics, distribution characteristics, motion characteristics, shape, and nonlinear characteristics of mesoscale eddies in the global ocean. See Figs. 9.15 and 9.16. We found that the shape of the mesoscale eddy is asymmetrical, not smooth, and has fractal geometric characteristics. Therefore, we use the calculation of the fractal dimension to study the change and severity of the shape of the mesoscale eddy.

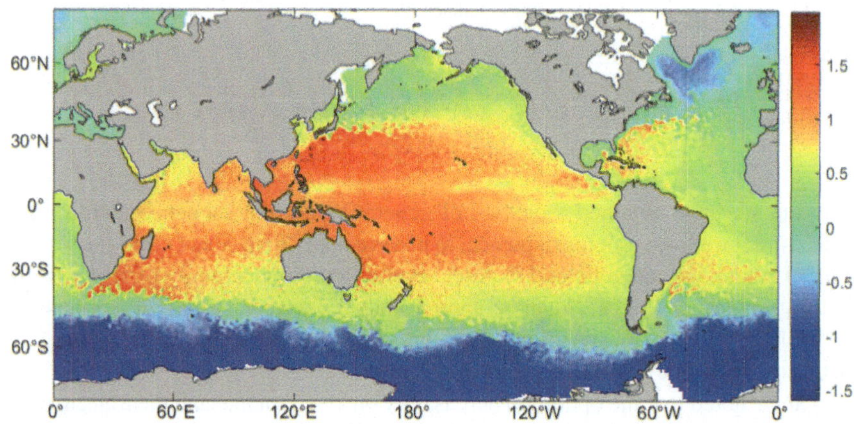

Fig. 9.15 Raw SSH data of sea height

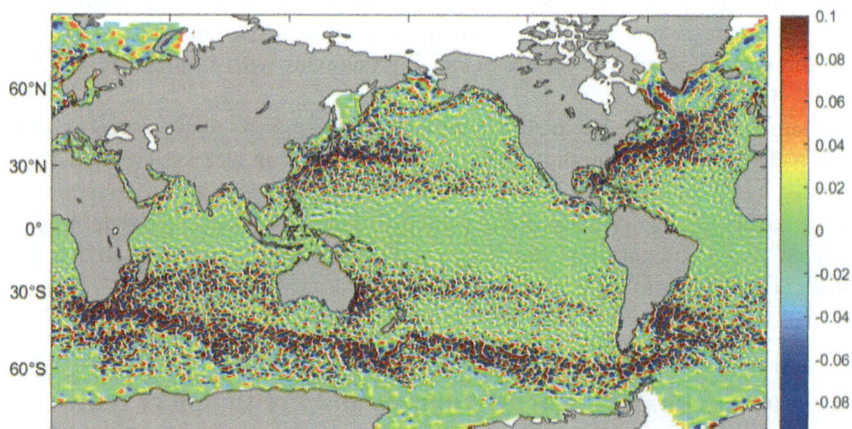

Fig. 9.16 Filtered SSH data of sea height

9.3.1 Data

Based on the analysis of SSH data, the py-eddy-tracker eddy tracking detection method[1] is used to identify and automatically track mesoscale eddies observed by satellite altimetry in the user-specified area of the global ocean. As input, the code requires sea level abnormal data, such as the data provided by AVISO data archiving, verification, and interpretation. The output format is

1. a data file containing eddy current characteristics, including position, radius, amplitude, and geostrophic velocity;
2. displays a sea height map covered with active eddy centers and tracks.

SSH data is used as eddy current identification data, we consider the mesoscale eddy current detection data released by py-eddy-tracker, its horizontal resolution is 0.25.

For the sea surface data SSH, we use the py-eddy-tracker eddy tracking detection method to identify all mesoscale eddies in the global ocean. In Fig. 9.17, blue is the mesoscale cold eddy, and red is the mesoscale warm eddy. There is a total of 7086 mesoscale eddies, which are divided into two types according to their polarity: mesoscale cold eddies and mesoscale warm eddies. Among them, there are 3628 mesoscale cold eddies and 3458 mesoscale warm eddies.

By analyzing the 7086 mesoscale eddies identified above, the radius distribution of mesoscale cold eddies and mesoscale warm eddies is obtained. Mesoscale warm eddies are consistent with the numerical distribution of the radius, amplitude, and rotation speed. The number of mesoscale cold eddies is slightly more than the number of mesoscale warm eddies. The most radius value is in (30, 50), followed by (50, 70). This means that the radius of mesoscale eddies is the most distributed in (30, 70); the most number of amplitude values is about 2cm; the most number of rotation speeds is 0.1m/s. That is to say, the measured global mesoscale eddy features are 30 km–50 km in radius, 2 cm amplitude, and 0.1 m/s rotation speed.

For the study of mesoscale eddies, people usually regard their shapes on the sea surface as a circle or ellipse for identification and tracking. In fact, the mesoscale eddy on the surface is not a smooth circle or ellipse. It is an irregular, asymmetric, non-smooth, random geometric shape generated by the deformation of a random circle or ellipse under the action of the flow field. The approximate one is called a stochastic ellipse. See Figs. 9.18 and 9.19.

9.3.2 Fractal Dimension of Mesoscale Eddy

In order to facilitate the study and avoid the error of results caused by different scales, we put all the identified mesoscale eddies in a square area with latitude and longitude of $4.2^\circ \times 4.2^\circ$ and then calculate the fractal dimension of mesoscale eddies by the box dimension method. The result is obtained in Table 9.5.

[1] https://bitbucket.org/emason/py-eddy-tracker.

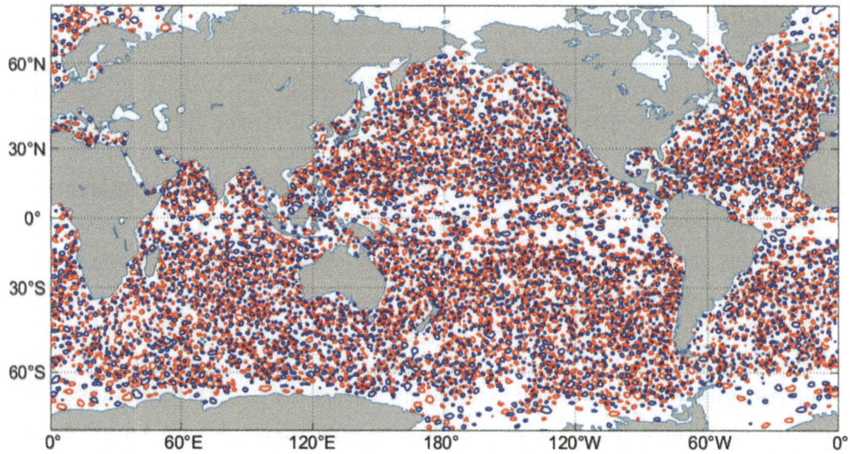

Fig. 9.17 Global mesoscale eddy distribution

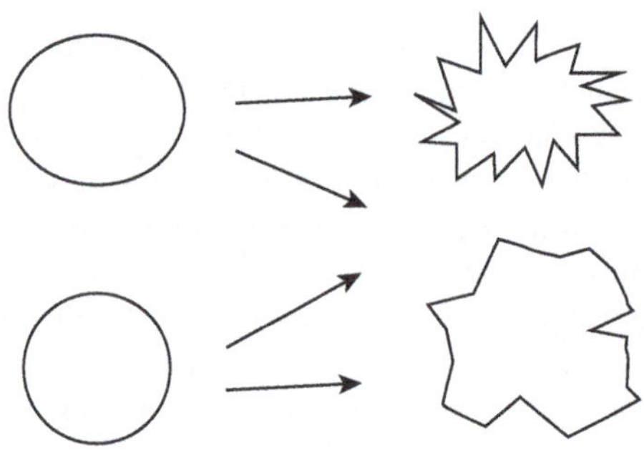

Fig. 9.18 Circle and ellipse deformation

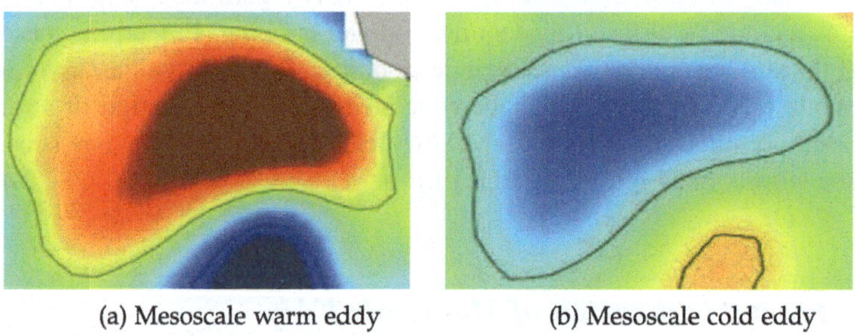

Fig. 9.19 Eddy shape

Table 9.5 Correspondence between shape, radius *R*, and fractal dimension *D* of mesoscale eddy

R and fractal dimension D	Cold eddy	R and fractal dimension D	Warm eddy
R = 126.55 D = 1.1722	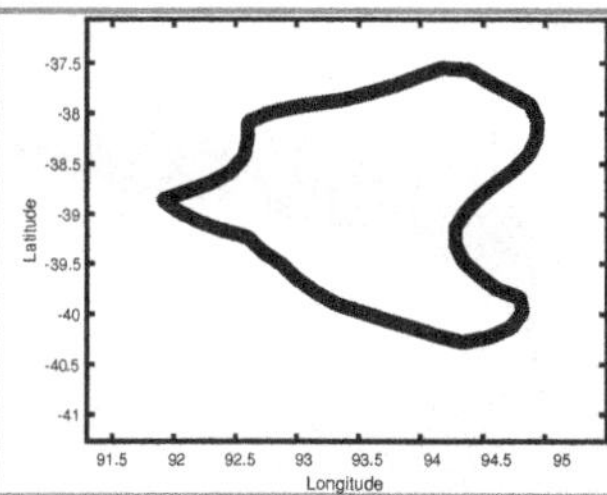	R = 153.55 D = 1.1911	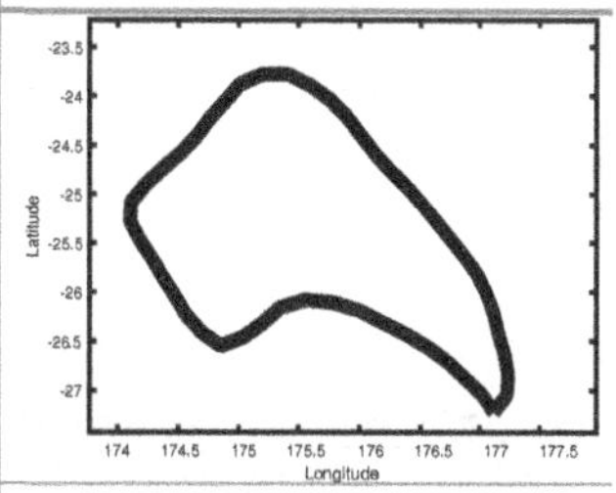
R = 105.45 D = 1.1441	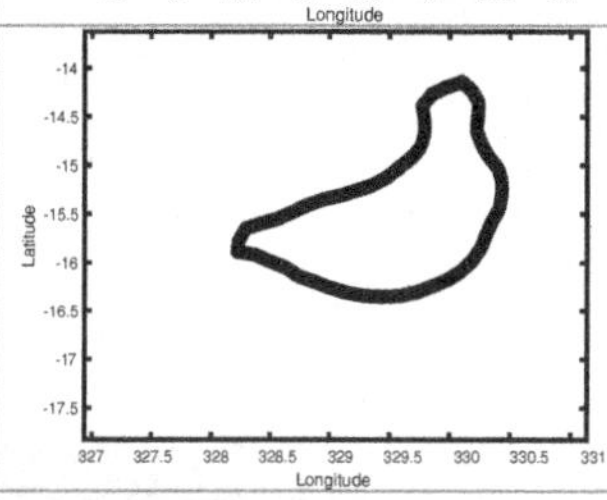	R = 49.70 D = 1.0742	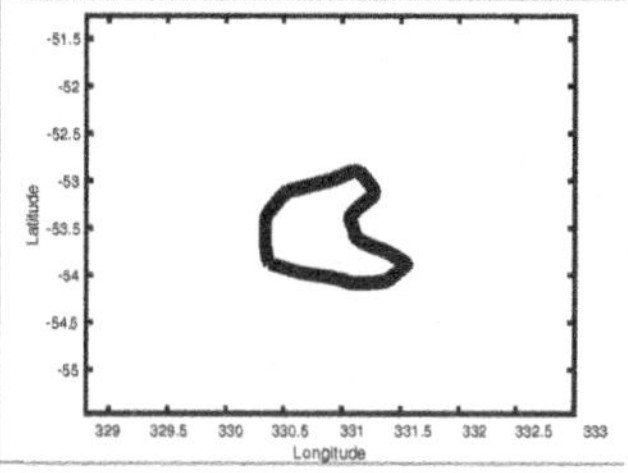
…	…	…	
R = 82.80 D = 1.1076	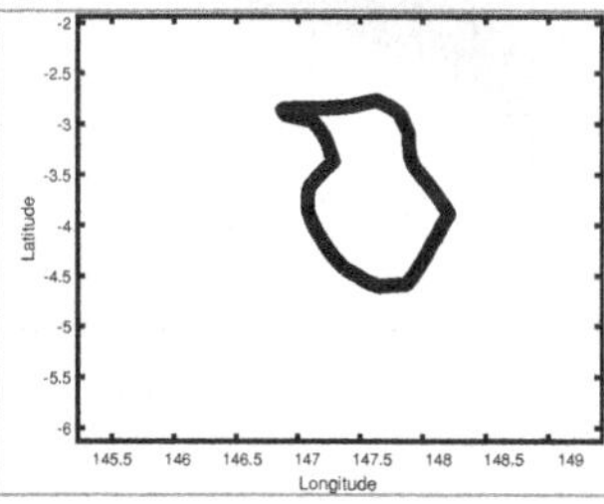	R = 112.5 D = 1.1648	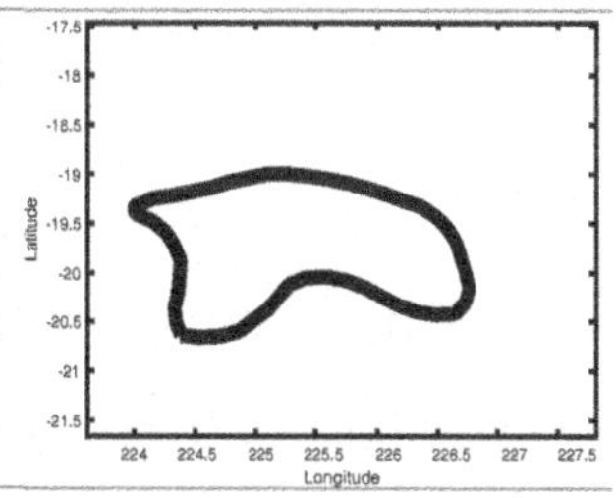

To study the complexity of the shape of the mesoscale eddy when its radius *R* of the global ocean is different, we rearrange radius *R* of all mesoscale eddies according to the size. The study found that radius *R* changes from small when it is large. And fractal dimension *D* of the mesoscale eddy will gradually increase, the same changing trend for both mesoscale cold eddies and mesoscale warm eddies.

Figure 9.20 shows that the fractal dimension *D* increases when radius *R* increases. The uniform and consistent relationship between all radii and dimensions of the geometrical shape of the mesoscale eddy is used for data fitting, with radius *R* as the independent variable and fractal dimension *D* as the dependent variable. The result is obtained in Fig. 9.21.

In order to more clearly and quantitatively analyze the relationship between the radius *R* and fractal dimension *D*, we use MATLAB software to simulate the identified radius *R* with fractal dimension *D* data. Together, the numerical curve shown in Fig. 9.21 is obtained. From Fig. 9.21, it can be seen that the measured data points are more evenly distributed on the fitting curve, the fractal dimension *D* and radius

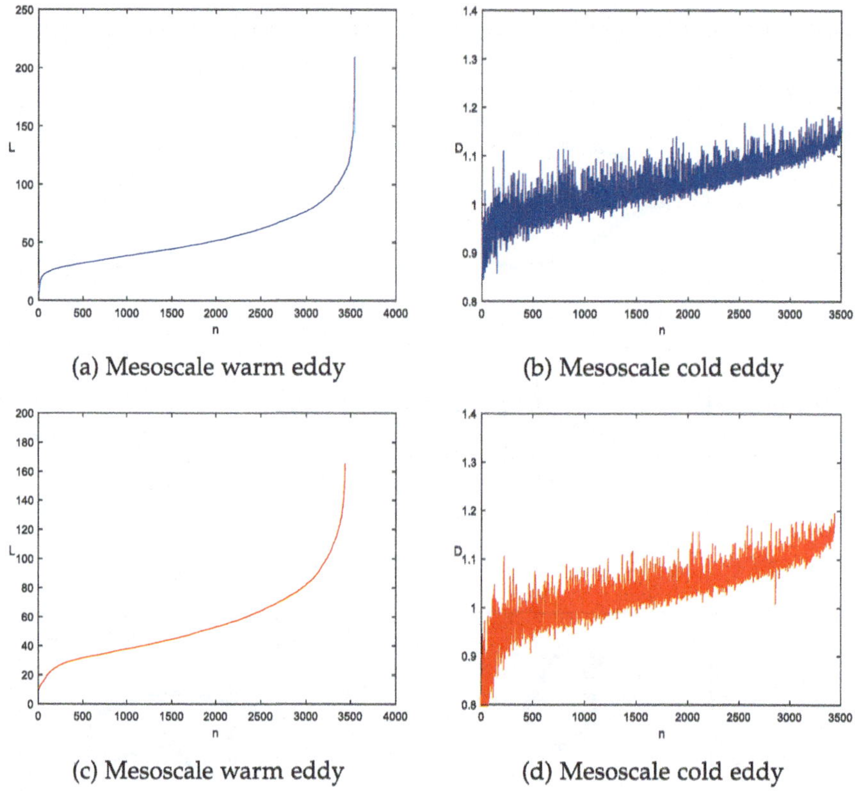

Fig. 9.20 Variation trend of the radius and fractal dimension of the mesoscale eddy

R are positively correlated, fractal dimension D increases with the increase of radius R, and the geometric complexity of the mesoscale eddy also gradually increases. It could be said that larger the fractal dimension, the larger the radius of the mesoscale eddy. Since we are studying the mesoscale eddy on a two-dimensional plane, the geometric fractal dimension of the mesoscale eddy is $D < 2$, specifically in the interval of $(1, 1.21)$, and the complexity of the mesoscale eddy is also in this range. The relationship between the fractal dimension D and radius R is

$$D = \alpha R^{\beta} + \gamma,$$

where $\alpha = -1.376$, $\beta = -0.2689$, and $\gamma = 1.582$. Through the above analysis, we have obtained the unified relationship between radius R and fractal dimension D of a mesoscale eddy. Next, we separately analyze the respective conditions of the mesoscale eddy, and study the relationship between the respective radius R and fractal dimension D.

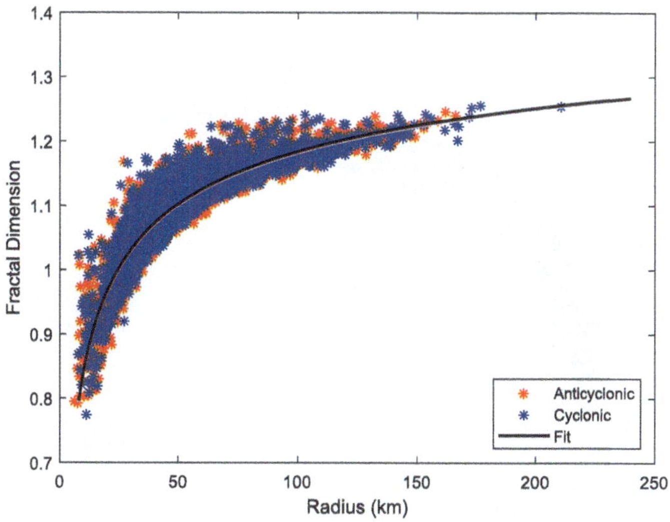

Fig. 9.21 Curves of radius R and fractal dimension D (red "*" is mesoscale warm eddy, blue "*" is mesoscale cold eddy, "-" is the fitted curve)

Based on the unification of mesoscale eddies above, the relationship between radius R and fractal dimension D of the geometric shape of mesoscale eddies is analyzed by data analysis and curve fitting, with radius R as the independent variables and fractal dimension D as the dependent variable, respectively. Similarly, we imitated the uniformity of the mesoscale eddy using MATLAB to fit the identified radius R with fractal dimension D data, and get Figs. 9.22 and 9.23.

Separate the study and the unified situation to obtain the same result. Therefore, the relationship between fractal dimension D and radius R of mesoscale eddies is unified. The situation is the same as

$$D = \alpha R^{\beta} + \gamma,$$

where $\alpha = -1.376$, $\beta = -0.2689$, and $\gamma = 1.582$.

The fractal dimension distribution is shown in Fig. 9.24, from which it can be concluded that fractal dimension D is in (1, 1.22) interval. So the complexity of the mesoscale eddy is mostly in this range.

Fit the distribution of the fractal dimension of all mesoscale eddies. The normal distribution is the most common distribution in nature. This distribution consists of two parameters—the mean μ and the variance σ. Determine the probability density function $f(x)$ of the normal probability distribution model. The normal distribution has many good properties. We will use the normal distribution to study the distribution of the fractal dimension D, see Fig. 9.25.

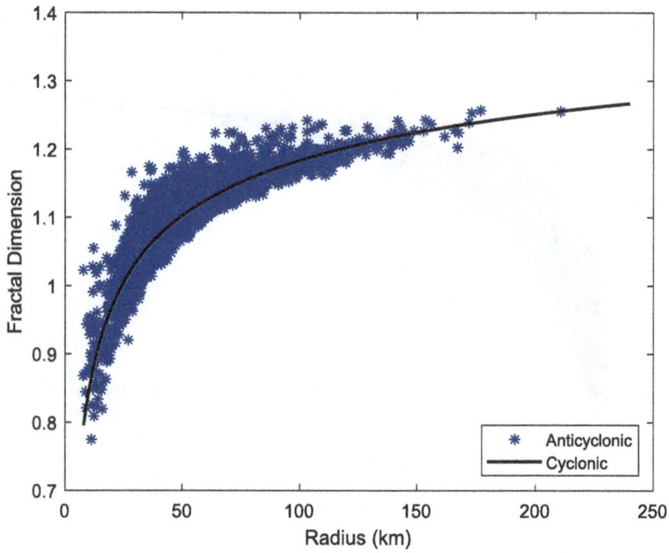

Fig. 9.22 Radius R and fractal dimension D of the mesoscale cold eddy ("*" is the measured data, "-" is the fitted curve)

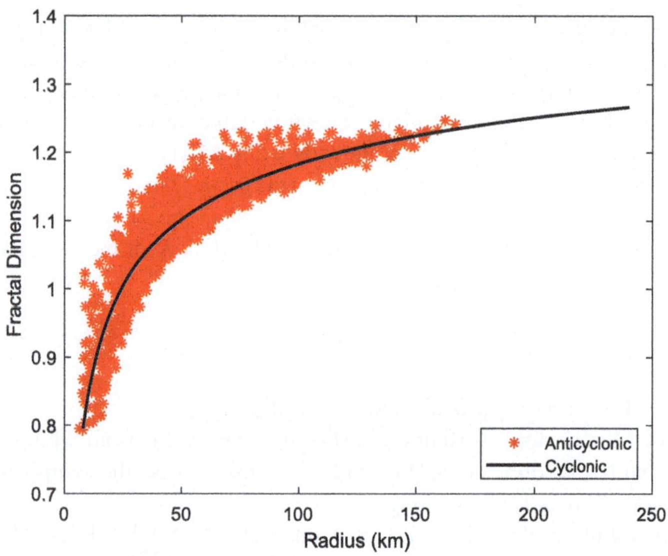

Fig. 9.23 Radius R and fractal dimension D of the mesoscale warm eddy ("*" is the measured data, "-" is the fitted curve)

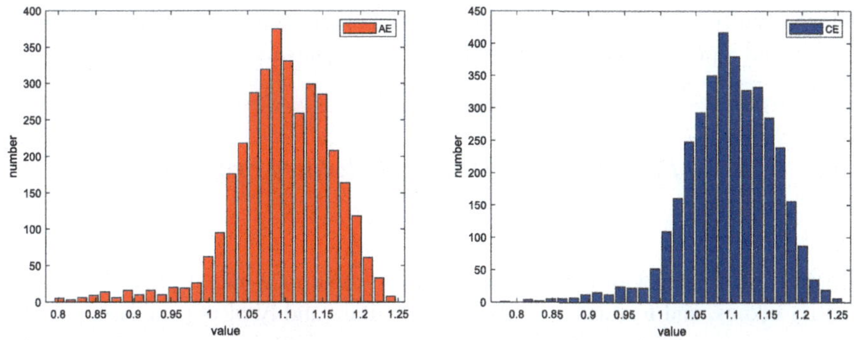

(a) Mesoscale warm eddy (b) Mesoscale cold eddy

Fig. 9.24 Fractal dimension distribution of mesoscale eddies

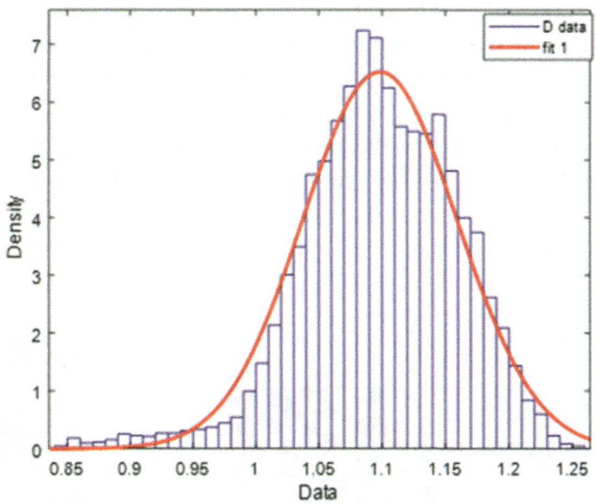

Fig. 9.25 Fitting of fractal dimension size distribution of mesoscale eddies

The following normal distribution function is fitted to the fractal dimensions of all mesoscale eddies

$$f(x) = \frac{1}{\sigma\sqrt{2\pi}}e^{-\frac{(x-u)^2}{2\sigma^2}},$$

where $\mu = 1.0983$ and $\sigma = 0.0611$. The normal fit passes the test of significance level 0.05. So the fractal dimension of the mesoscale eddy conforms to the normal distribution.

From the above chart analysis, it can be concluded that the fractal dimension of the mesoscale eddy is mostly in (1, 1.22) and the complexity of the mesoscale eddy is also in this range. The larger the fractal dimension, the more complex the shape of the mesoscale eddy, and the the larger its contact area with the outside. That is, greater the influence of mesoscale eddy on nutrients, chlorophyll, density, pressure, velocity, and other substances in its surrounding environment, the greater its influence on the survival and reproduction of surrounding marine organisms.

9.3.3 Fractal Processing of Mesoscale Eddies Profile of the Ocean

In order to compare with the method of using sea surface height anomaly data to identify eddies, and to further verify the scientific validity of our method, we use image processing of sea surface height anomalies to obtain the edge profile of the mesoscale eddy and its fractal dimension.

The sea surface height data image is processed to extract the edge lines of the selected mesoscale eddy. Take the one in Fig. 9.26 as an example. The results are obtained in Fig. 9.27. Calculate the fractal dimension of the edge of the mesoscale eddies in the image processing example above. Results are obtained in Table 9.6.

Applying the method of using sea surface height data to identify mesoscale eddies in the South China Sea, we get as shown in Fig. 9.28. Similarly, the fractal dimension of the edge of mesoscale eddies is calculated and shown in Table 9.7.

By comparing the two methods to get the fractal dimension of mesoscale eddies, we can get that the rough shape of the eddy edge obtained by image processing is similar to that obtained by SSH data recognition. See Table 9.8. However, the former

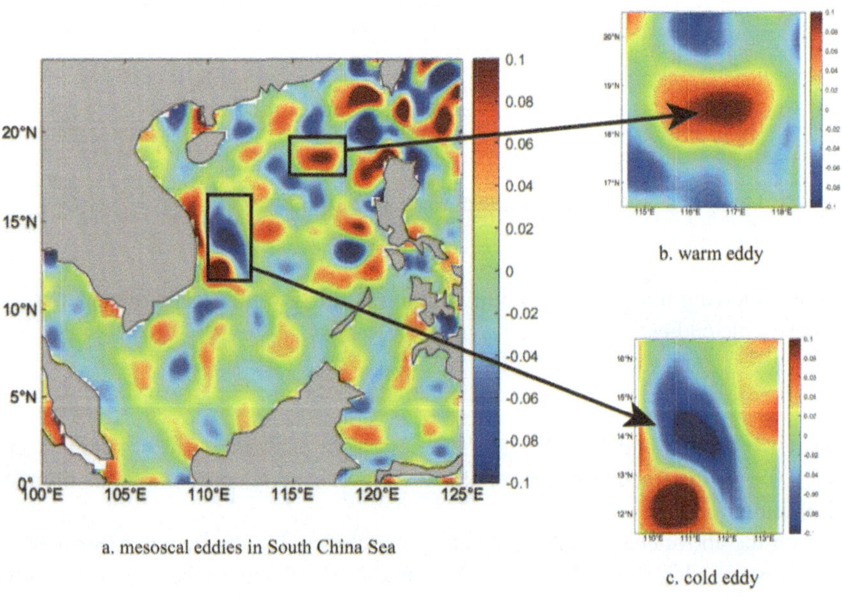

Fig. 9.26 Legends of mesoscale eddies in the South China Sea

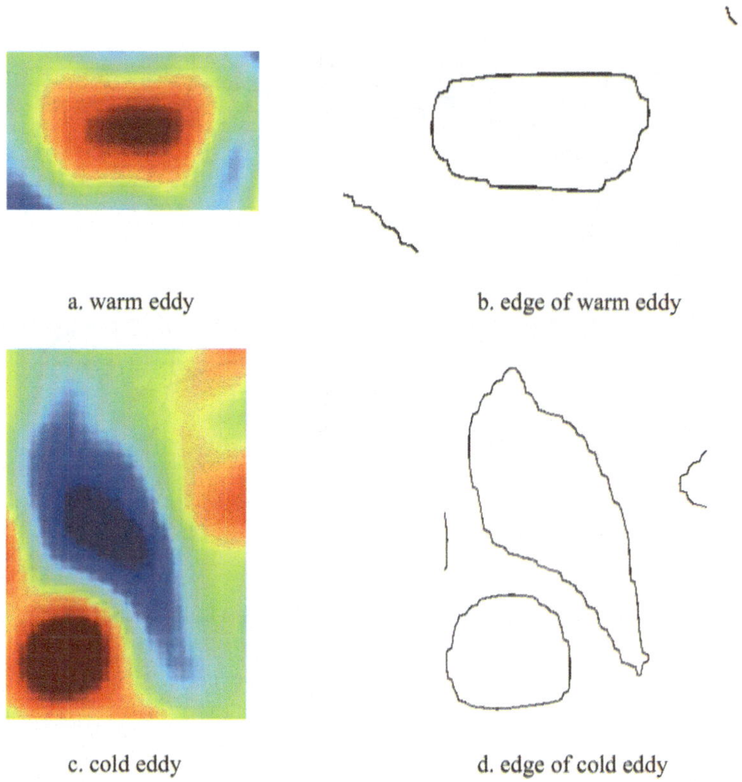

Fig. 9.27 Identifying the edges of mesoscale eddies

Table 9.6 The fractal dimension of the edge of mesoscale eddies in Fig. 9.27

Mesoscale warm eddy	Mesoscale cold eddy
1.1032	1.2010

is rougher and the latter is relatively smooth, but the fractal dimension of the edge line obtained by the two methods has little difference. The dimension of the edge line of the mesoscale eddy obtained by image processing is larger than that obtained by SSH data recognition. When the mesoscale eddy changes, its fractal dimension will also change, and the changing trend of the two methods is the same, but the edge of the mesoscale eddy obtained by image processing is rough. Therefore, the py-eddy-tracker method based on SSH is selected to obtain the mesoscale eddy, and the fractal dimension of its edge line is used to characterize its complexity.

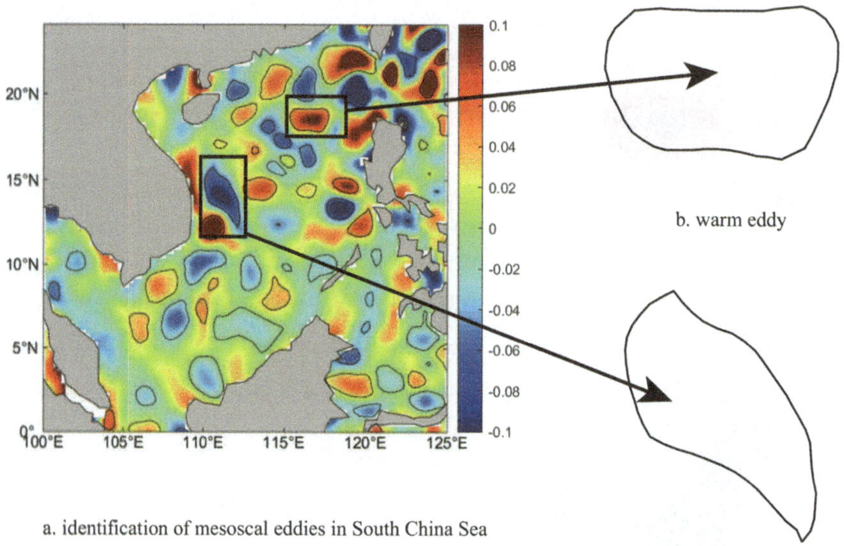

a. identification of mesoscal eddies in South China Sea

b. warm eddy

c. cold eddy

Fig. 9.28 Distribution of mesoscale eddies in the South China Sea based on SSH data

Table 9.7 The fractal dimension of the edges of mesoscale eddies in Fig. 9.28

Mesoscale warm eddy	Mesoscale cold eddy
1.0925 ~ 1.1032	1.1688 ~ 1.2010

9.3.4 Three-Dimensional Fractal Structure of Abnormal Salinity

Since satellites cannot provide any information on the subsurface depth of a mesoscale eddy, we can quantitatively study the three-dimensional structure characteristics of the temperature and salt inside the mesoscale eddy by synthetic analysis of the Argo profile data. Different from the traditional vertical profile of salinity anomaly, we analyze the complexity of salinity of a mesoscale eddy from fractal perspective. Take a mesoscale cold eddy whose center is $(118.45° E, 13.37° N)$ in the South China Sea for an example. Then the salinity anomaly boundaries of the mesoscale eddy at different depths are obtained, see Figs. 9.29 and 9.30. Besides, dimension of the salinity anomaly boundary is calculated to obtain the relationship as shown in Fig. 9.31.

From Figs. 9.29, 9.30, and 9.31, it can be seen that the deeper the depth of the mesoscale cold eddy, the smaller the fractal dimension. And, the more shallow it is, the more complex it is. The fractal dimension of salinity anomaly on the surface of a mesoscale cold eddy is not the largest, but the dimension of salinity anomaly between

Table 9.8 Two methods for the eddy edge and comparison about fractal dimension

Mesoscale eddy	Identification method	Edge of eddy	Fractal dimension
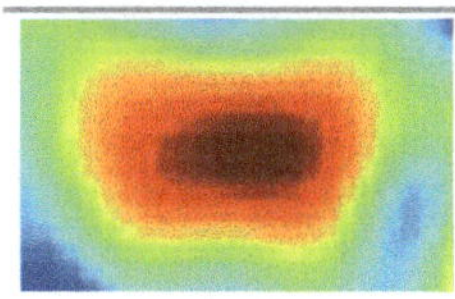	Image processing		1.1032
	py-eddy-tracker(SSH based)	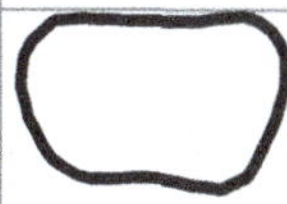	1.0925
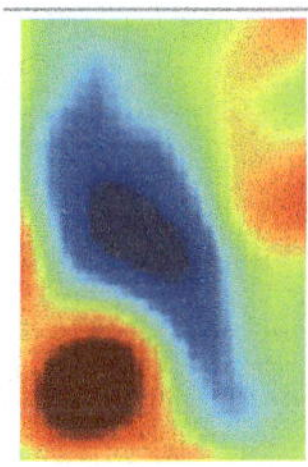	Image processing		1.2010
	py-eddy-tracker(SSH based)	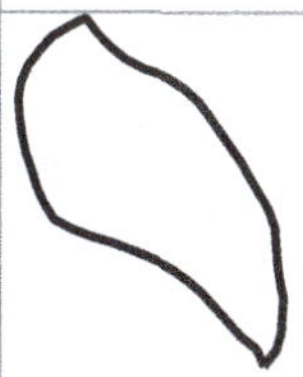	1.1688

200 km and 400 km is larger, which is consistent with the observed data. In short, the fractal dimension of three-dimensional structure can reflect the salinity anomaly of the mesoscale eddy: the larger the dimension, the more severe it becomes.

9.3.5 *Comprehensive Analysis*

By calculating the fractal dimension of mesoscale eddies, the relationship between fractal dimension of the mesoscale eddy and its radius is analyzed, which is compared with the mesoscale eddy profile obtained by image processing. The results show that the surface fractal dimension of the mesoscale eddy is positively correlated with the radius. The larger the radius is, the more intense the mesoscale eddy is. Moreover, the

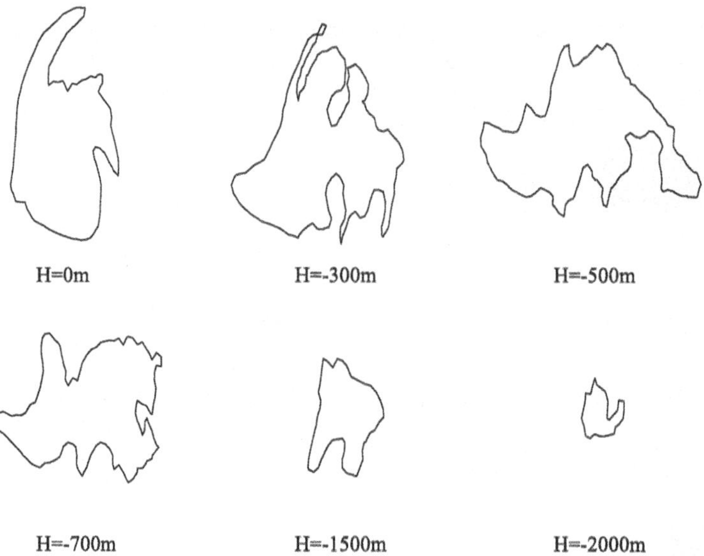

Fig. 9.29 The depths of the mesoscale cold eddy salinity anomaly contour line

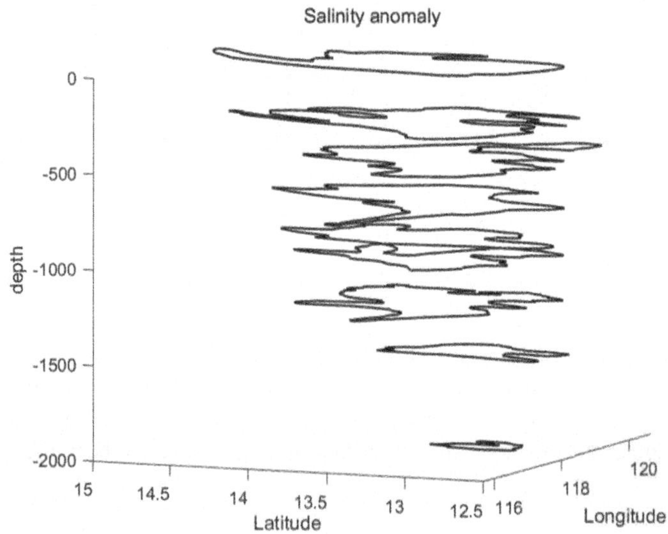

Fig. 9.30 Three-dimensional mesoscale cold eddy salinity anomalies

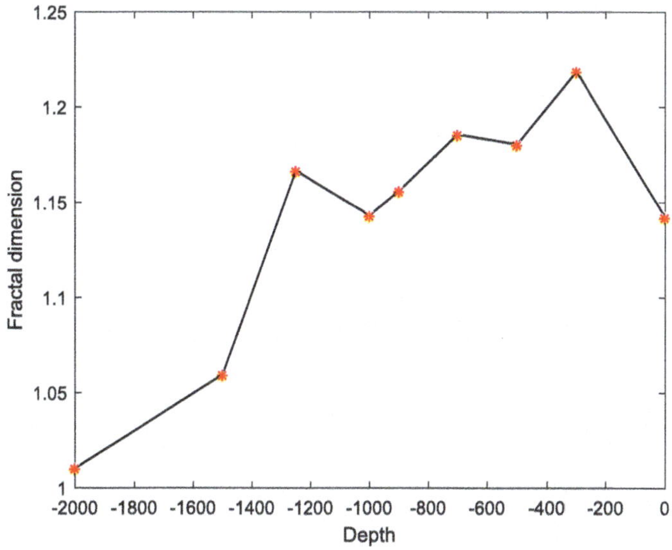

Fig. 9.31 The relationship between fractal dimension and depth of abnormal salinity profile

fractal dimension of the mesoscale eddy obeys normal distribution, and the fractal dimension of its edge profile obtained by image processing is smaller. From the three-dimensional analysis of the salinity anomaly of the mesoscale eddy, it is found that the deeper the mesoscale cold eddy depth, the smaller its fractal dimension. The shallower it is, the more complicated it is.

Chapter 10
Nonlinear Characteristics of Universal Mathematical Model of Mesoscale Eddy

10.1 Dissipation of Nonlinear Systems

Under the constraints of parameters, the flow field curl $div(\mathbf{V})$ in nonlinear dynamic system (3.3.16) becomes

$$\begin{aligned}&div(\mathbf{V}) = \tfrac{\partial \dot{x}}{\partial x} + \tfrac{\partial \dot{y}}{\partial y}\\&= \left[\sqrt{((2m+1)\alpha+\beta)a}x + \sqrt{(\alpha+(2m+1)\beta)b}y\right]^2 (ax^2+by^2-R)^{m-1}\\&-[2\sqrt{((2m+1)\alpha+\beta)a(\alpha+(2m+1)\beta)b}xy + (\alpha+\beta)R](ax^2+by^2-R)^{m-1}\\&< 0.\end{aligned}$$

It is known from [507] that the system is a dissipative system. A D-dimensional dissipative system after a long period of gradual motion must be on a system structure lower than D-dimensional system. This structure is called attractor [493, 494]. From the shooting method and the screening method [505, 515], if parameters satisfy Table 4.4, nonlinear dynamic system (3.3.16) has a semi-stable limit cycle.

10.2 Chaotic Behavior of Universal Nonlinear System of Mesoscale Eddy

For system (3.3.16), set

$$F(x, y) = [f(x, y), g(x, y)]^{\mathrm{T}},$$

where

$$\begin{cases} f(x, y) = \alpha x(ax^2+by^2-1)^m + \lambda\omega y, \\ g(x, y) = \beta y(ax^2+by^2-1)^m - \mu\omega x. \end{cases}$$

S.-T. Liu et al., *Mathematical Principle and Fractal Analysis of Mesoscale Eddy*,
https://doi.org/10.1007/978-981-16-1839-0_10

Then the Jacobian matrix of the model at the initial value (x_0, y_0) is

$$J_F(x_0, y_0) = \left.\frac{\partial(f, g)}{\partial(x, y)}\right|_{(x_0, y_0)} = \begin{bmatrix} J_{11} & J_{12} \\ J_{21} & J_{22} \end{bmatrix},$$

where

$$\begin{cases} J_{11} = \alpha(ax_0^2 + by_0^2 - 1)^{m-1}(a(2m+1)x_0^2 + by_0^2 - 1), \\ J_{12} = 2\alpha mbx_0y_0(ax_0^2 + by_0^2 + 1)^{m-1} + \lambda, \\ J_{21} = 2\beta max_0y_0(ax_0^2 + by_0^2 - 1)^{m-1} - \mu, \\ J_{22} = \beta(ax_0^2 + by_0^2 - 1)^{m-1}(ax_0^2 + b(2m+1)y_0^2 - 1). \end{cases}$$

Therefore, the first-order linear approximation of the system is

$$\dot{X} = J_F X, \tag{10.2.1}$$

where $X = [x, y]^{\mathrm{T}}$. Matrix X describes how small changes in the initial value $X_0 = [x_0, y_0]^{\mathrm{T}}$ propagate to the end $X(t) = [x(t), y(t)]^{\mathrm{T}}$. The limit is

$$\Lambda = \lim_{t\to\infty} \frac{1}{2t} \log(X(t)X^{\mathrm{T}}(t)). \tag{10.2.2}$$

The conditions for the existence of the limit are given by Oseledets theorem [433, 434]. The Lyapunov exponent is defined by the eigenvalues of matrix Λ. For almost all initial values of the components of the dynamic system, the Lyapunov exponent set is the same.

According to the above conclusions, we first need to find solution X of system (10.2.1). Note that J_F is a second-order real constant-coefficient matrix. Set its eigenvalues be λ_1 and λ_2 corresponding to the eigenvectors $\mathbf{v}_1$ and $\mathbf{v}_2$, respectively. Then the basic solution matrix of the system is

$$\Phi(t) = \exp J_F t = [e^{\lambda_1 t}\mathbf{v}_1, e^{\lambda_2 t}\mathbf{v}_2].$$

Characteristic values are

$$\begin{aligned} \lambda_1 &= \tfrac{J_{11}+J_{22}-w}{2} \\ &= \tfrac{(\alpha+\beta)(ax^2+by^2-1)^m}{2} + (a\alpha x^2 - b\beta y^2)(ax^2+by^2-1) - \tfrac{\sqrt{P_1+Q_1}}{2}, \end{aligned}$$

where

$$\begin{aligned} P_1 &= (ax^2+by^2-1)^{2m-2}\left(\alpha - \beta - a(\alpha-\beta+2\alpha m)x^2 - b(\alpha-\beta-2\beta m)y^2\right)^2, \\ Q_1 &= 2m\left(\mu - 2a\beta xy(ax^2+by^2-1)\right)\left(\lambda + 2\alpha bmxy(ax^2+by^2-1)\right) \end{aligned}$$

and

$$\lambda_2 = \frac{J_{11}+J_{22}+w}{2}$$
$$= \frac{(\alpha+\beta)(ax^2+by^2-1)^m}{2} + (a\alpha x^2 - b\beta y^2)(ax^2+by^2-1) + \frac{\sqrt{P_2+Q_2}}{2},$$

where

$$P_2 = (ax^2+by^2-1)^{2m-2}\left(\alpha-\beta-a(\alpha-\beta+2\alpha m)x^2-b(\alpha-\beta-2\beta m)y^2\right)^2,$$
$$Q_2 = 2\left(\mu-2a\beta mxy(ax^2+by^2-1)\right)\left(\lambda+2\alpha bmxy(ax^2+by^2-1)\right).$$

The corresponding feature vectors are

$$\mathbf{v}_1 = \begin{bmatrix} \frac{J_{11}-J_{22}-w}{2J_{21}} \\ 1 \end{bmatrix} \quad \text{and} \quad \mathbf{v}_2 = \begin{bmatrix} \frac{J_{11}-J_{22}+w}{2J_{21}} \\ 1 \end{bmatrix}.$$

Denote

$$P_3 = (ax^2+by^2-1)^{2m-2}\left(\alpha-\beta-a(\alpha-\beta+2\alpha m)x^2-b(\alpha-\beta-2\beta m)y^2\right)^2,$$
$$Q_3 = 2\left(\mu-2a\beta mxy(ax^2+by^2-1)\right)\left(\lambda+2\alpha bmxy(ax^2+by^2-1)\right),$$
$$R_3 = 2\left(\mu-2a\beta mxy(ax^2+by^2-1)\right),$$
$$P_4 = (ax^2+by^2-1)^{2m-2}\left(\alpha-\beta-a(\alpha-\beta+2\alpha m)x^2-b(\alpha-\beta-2\beta m)y^2\right)^2,$$
$$Q_4 = 2\left(\mu-2a\beta mxy(ax^2+by^2-1)\right)\left(\lambda+2\alpha bmxy(ax^2+by^2-1)\right),$$
$$R_4 = 2\left(\mu-2a\beta mxy(ax^2+by^2-1)\right),$$
$$P_5 = (ax^2+by^2-1)^{2m-2}\left(\alpha-\beta-a(\alpha-\beta+2\alpha m)x^2-b(\alpha-\beta-2\beta m)y^2\right)^2,$$
$$Q_5 = 2\left(\mu-2a\beta mxy(ax^2+by^2-1)\right)\left(\lambda+2\alpha bmxy(ax^2+by^2-1)\right).$$

Then

$$\frac{J_{11}-J_{22}-w}{2J_{21}} = (\beta-\alpha)(ax^2+by^2-1)^m$$
$$- 2m(a\alpha x^2 - b\beta y^2)(ax^2+by^2-1) + \frac{\sqrt{P_2+Q_3}}{R_3},$$
$$\frac{J_{11}-J_{22}+w}{2J_{21}} = (\beta-\alpha)(ax^2+by^2-1)^m$$
$$- 2m(a\alpha x^2 - b\beta y^2)(ax^2+by^2-1)^{m-1} - \frac{\sqrt{P_4+Q_4}}{R_4},$$

where $w = \sqrt{(J_{11}-J_{22})^2 + 4J_{12}J_{21}} = \sqrt{P_5+Q_5}$. Therefore, system (10.2.1)'s solution satisfying the initial condition $X_0 = [x_0, y_0]^{\mathrm{T}}$ is

$$X(t) = \Phi(t)X_0 = [x_0 e^{\lambda_1 t}\mathbf{v}_1, y_0 e^{\lambda_2 t}\mathbf{v}_2].$$

Substituting it into formula (10.2.2), we can get

Table 10.1 The chaos of the model

(λ_1, λ_2)	$(+,+)$	$(+,0)$	$(0,0)$	$(0,-)$	$(-,-)$
State	Hyperchaos	Chaos	Critical point	Limit cycle	Fixed point

$$\Lambda = \lim_{t\to\infty}\frac{1}{2t}\log(x_0^2e^{2\lambda_1 t}\mathbf{v}_1\mathbf{v}_1^{\mathrm{T}} + y_0^2e^{2\lambda_2 t}\mathbf{v}_2\mathbf{v}_2^{\mathrm{T}}).$$

Then find the eigenvalues of matrix Λ, one gets Lyapunov exponents λ_1 and λ_2. Then the chaos of the model can be discussed, seeing Table 10.1.

If $\alpha = 1$, $\beta = 1$, $a = 2$, $b = 1$, $\lambda\omega = \frac{1}{\sqrt{2}}$, $\mu\omega = \sqrt{2}$ and $m = 2$, then the model degenerates to

$$\begin{cases} \dot{x} = \frac{y}{\sqrt{2}} + x\left(2x^2 + y^2 - 1\right)^2, \\ \dot{y} = -\sqrt{2}x + y\left(2x^2 + y^2 - 1\right)^2. \end{cases} \tag{10.2.3}$$

The Jacobian matrix of the model at $X_0 = [0.1, 0.1]^{\mathrm{T}}$ is

$$J_F = \begin{bmatrix} (2x_0^2 + y_0^2 - 1)(10x_0^2 + y_0^2 - 1) & 4x_0y_0(2x_0^2 + y_0^2 - 1) + \frac{1}{\sqrt{2}} \\ 8x_0y_0(2x_0^2 + y_0^2 - 1) + \sqrt{2} & (2x_0^2 + y_0^2 - 1)(2x_0^2 + 5y_0^2 - 1) \end{bmatrix} = \begin{bmatrix} 0.8633 & 0.6683 \\ 1.3370 & 0.9021 \end{bmatrix}.$$

Its characteristic values are

$$\lambda_1 = -0.0626 \quad \text{and} \quad \lambda_2 = 1.8280.$$

The corresponding feature vectors are

$$\mathbf{v}_1 = \begin{bmatrix} -0.7218 \\ 1.0000 \end{bmatrix} \quad \text{and} \quad \mathbf{v}_2 = \begin{bmatrix} 0.6927 \\ 0.0000 \end{bmatrix}.$$

Substituting it into formula (10.2.2), one gets

$$\Lambda = \lim_{t\to\infty}\frac{1}{2t}\log(x_0^2e^{2\lambda_1 t}\mathbf{v}_1\mathbf{v}_1^{\mathrm{T}} + y_0^2e^{2\lambda_2 t}\mathbf{v}_2\mathbf{v}_2^{\mathrm{T}}) = [3.656, 0]^{\mathrm{T}}.$$

Lyapunov exponents $\lambda_1 = 3.656$ and $\lambda_2 = 0$, so the model is chaotic and the attractor is chaotic one. Noticing that it is a semi-stable limit cycle, so it is a semi-strange attractor. See Fig. 10.1.

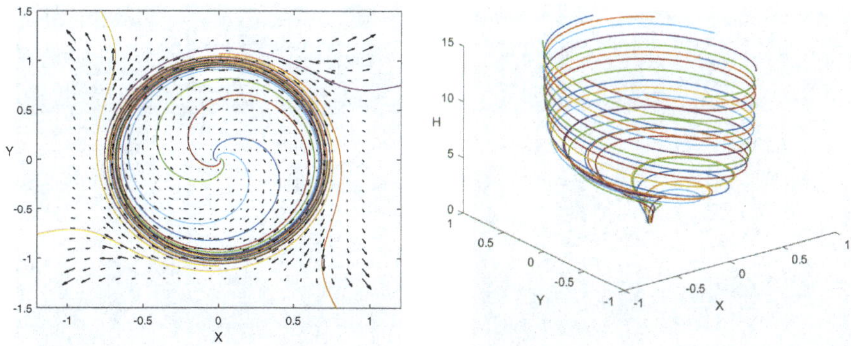

Fig. 10.1 Planar phase diagram of the attractor and three-dimensional structure of nonlinear system (10.2.3)

10.3 Singularity of Mesoscale Eddy and its Physical Meaning

For the nonlinear system (3.3.16) of ocean mesoscale eddies, the plane system under the condition of geostrophic equilibrium is considered as

$$\begin{cases} \frac{\mathrm{d}x}{\mathrm{d}t} = \alpha x(ax^2 + by^2 - R^2)^m + \omega\lambda y, \\ \frac{\mathrm{d}y}{\mathrm{d}t} = \beta y(ax^2 + by^2 - R^2)^m - \omega\mu x, \end{cases} \tag{10.3.1}$$

whose equilibrium equation is

$$\begin{cases} \alpha x(ax^2 + by^2 - R^2)^m + \omega\lambda y = 0, \\ \beta y(ax^2 + by^2 - R^2)^m - \omega\mu x = 0. \end{cases} \tag{10.3.2}$$

1. Trivial singularity

It is easy to see from Eq. (10.3.2) that $(x, y) = (0, 0)$ is an equilibrium point, which is called the **trivial/ordinary singularity** of system (10.3.1).

2. Nontrivial singularity

It is easy to see from Eq. (10.3.2), let $\xi = \frac{x}{y}$, because x and y are the two directions of seawater movement, so ξ is called the fusion space of seawater movement in the area, and there are

$$(ax^2 + by^2 - R^2)^m = -\frac{\omega\lambda}{\alpha} \cdot \frac{y}{x} = \frac{\omega\mu}{\beta} \cdot \frac{x}{y}.$$

Because of the mesoscale properties of mesoscale eddies, without loss of generality, take $\alpha = \beta = \lambda = \mu$, then $\xi^2 = -1$, that is $\xi = \pm\mathrm{i}$

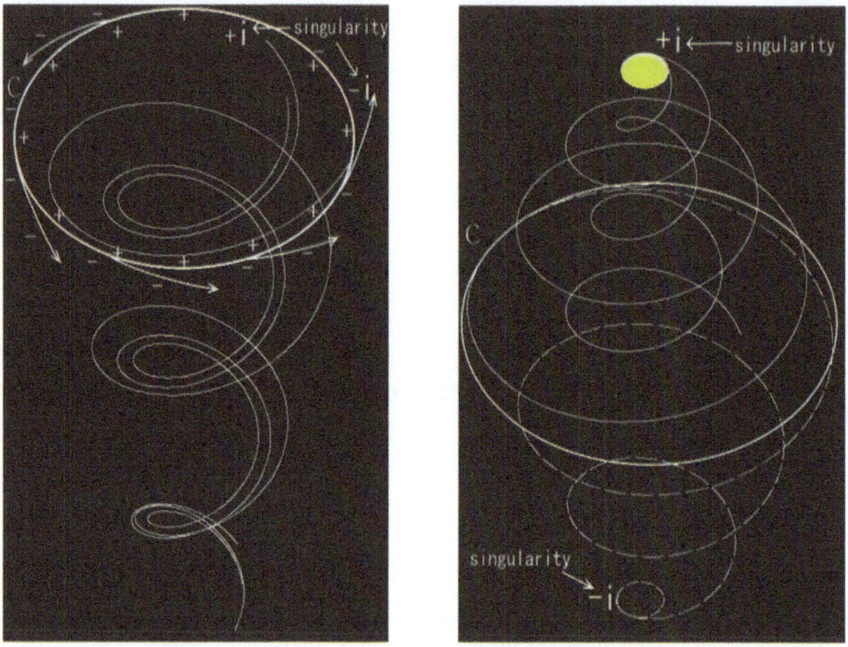

(a) Mesoscale cold eddy (b) Mesoscale warm eddy

Fig. 10.2 Physical meaning of nontrivial singularity

$$\xi = \pm \mathrm{i}, \tag{10.3.3}$$

which is called the **nontrivial singularity** of the mesoscale eddy.

3. The physical meaning of nontrivial singularity

The singularity of seawater flow behavior is $\pm$i, which is in the semi-stable limit cycle, corresponding to cold eddy and warm eddy. In either case, $\pm$i shows the motion behavior of instantaneous convergence and divergence of seawater fluid: if the singularity exists, it is continuous convergence; if the singularity does not exist, it is continuous divergence, but it becomes a certain state. The existence and nonexistence are the physical meaning of singularity $\pm$i. See Fig. 10.2.

4. Two examples of cold eddy and warm eddy on singularity

Using systems (4.5.1) and (4.6.1), the motion of cold eddy and warm eddy about singularity $\pm$i is studied. See Fig. 10.3.

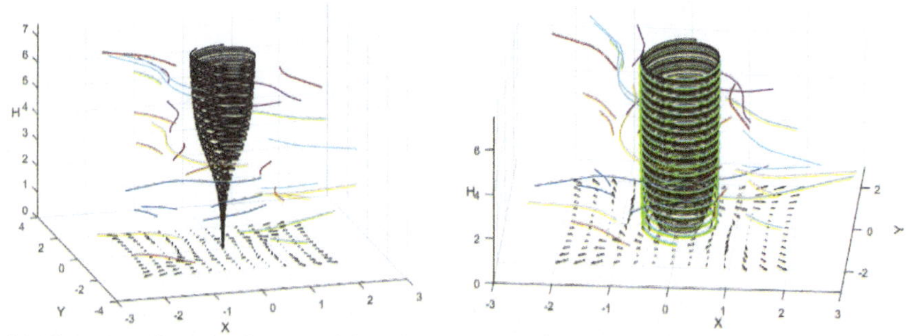

(a) The case of internally unstable motion in a cold eddy to singularity

(b) Points within and near C tends to its singularity

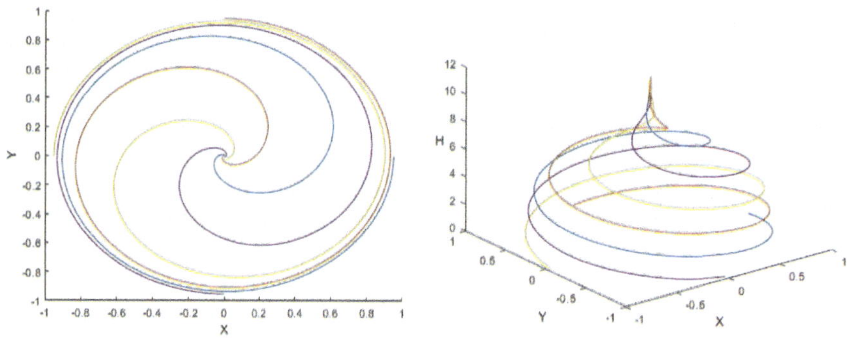

(c) Singularity plane trend of warm eddy

(d) Singularity space trend of warm eddy

Fig. 10.3 Motion of cold eddy and warm eddy about singularity

Chapter 11
Same Solution Between Momentum Balance Equations and Mesoscale Eddies

11.1 Navier-Stokes Equation

For incompressible marine viscous fluid, any motion is in a definite coordinate reference system, and any motion of seawater is related to the rotation of the earth. Therefore, with the earth as coordinate reference system, the Navier-Stokes equation in the motion of the viscous fluid of the unit mass of seawater is

$$\frac{\partial \mathbf{V}}{\partial t} + (\mathbf{V} \cdot \nabla)\mathbf{V} = -\frac{1}{\rho}\nabla P + \mu \nabla^2 \mathbf{V}, \tag{11.1.1}$$

where $\mathbf{V} = \mathbf{V}(u, v, w)$ is the velocity vector field, P is the pressure, ρ is the seawater density, ∇ is Hamilton operator, μ is the viscosity coefficient, and $\nabla^2 = \dfrac{\partial^2}{\partial x^2} + \dfrac{\partial^2}{\partial y^2} + \dfrac{\partial^2}{\partial z^2}$ is Laplace operator. For the right end of Eq. (11.1.1), there is

$$\begin{cases} \frac{du}{dt} = \frac{\partial u}{\partial t} + u\frac{\partial u}{\partial x} + v\frac{\partial u}{\partial y} + w\frac{\partial u}{\partial z}, \\ \frac{dv}{dt} = \frac{\partial v}{\partial t} + u\frac{\partial v}{\partial x} + v\frac{\partial v}{\partial y} + w\frac{\partial v}{\partial z}, \\ \frac{dw}{dt} = \frac{\partial w}{\partial t} + u\frac{\partial w}{\partial x} + v\frac{\partial w}{\partial y} + w\frac{\partial w}{\partial z}. \end{cases}$$

In addition, if the gravitational acceleration is $\mathbf{g}$ in Eq. (11.1.1), and the direction of $\mathbf{g}$ is the horizontal direction pointing to the vertical direction of the center of the sphere, then $\mathbf{g} = -g\mathbf{k}$. Take Coriolis force as

$$-2\omega \times \mathbf{V} = -2\begin{vmatrix} \mathbf{i} & \mathbf{j} & \mathbf{k} \\ 0 & \omega\cos\varphi & \omega\sin\varphi \\ u & v & w \end{vmatrix} = (fv - f_1\omega)\mathbf{i} - fv\mathbf{j} + f_1u\mathbf{k}.$$

Since $f_1\omega$ is very small, it can be ignored. Similarly, $f_1\mu$ can also be ignored. Reorganizing Eq. (11.1.1), we get

S.-T. Liu et al., *Mathematical Principle and Fractal Analysis of Mesoscale Eddy*,
https://doi.org/10.1007/978-981-16-1839-0_11

$$\frac{\partial \mathbf{V}}{\partial t} + (\mathbf{V} \cdot \nabla)\mathbf{V} = -\frac{1}{\rho}\nabla P - \mathbf{g} - 2\omega \times \mathbf{V} + \mu \nabla^2 \mathbf{V}. \tag{11.1.2}$$

And the component form of Eq. (11.1.2) is

$$\begin{cases} \frac{du}{dt} = \frac{\partial u}{\partial t} + u\frac{\partial u}{\partial x} + v\frac{\partial u}{\partial y} + w\frac{\partial u}{\partial z} = -\frac{1}{\rho}\nabla P + fv + \mu\nabla^2 u, \\ \frac{dv}{dt} = \frac{\partial v}{\partial t} + u\frac{\partial v}{\partial x} + v\frac{\partial v}{\partial y} + w\frac{\partial v}{\partial z} = -\frac{1}{\rho}\nabla P - fu + \mu\nabla^2 v, \\ \frac{dw}{dt} = \frac{\partial w}{\partial t} + u\frac{\partial w}{\partial x} + v\frac{\partial w}{\partial y} + w\frac{\partial w}{\partial z} = -\frac{1}{\rho}\nabla P - g + \mu\nabla^2 w. \end{cases} \tag{11.1.3}$$

Especially for the characteristics of the mesoscale eddy, we first focus on his most prominent personality feature, which is its planar behavior. From Eq. (11.1.3), under the condition of floor tiles, two-dimensional momentum balance equation of seawater motion is

$$\begin{cases} \frac{du}{dt} = \frac{\partial u}{\partial t} + u\frac{\partial u}{\partial x} + v\frac{\partial u}{\partial y} + w\frac{\partial u}{\partial z} = -\frac{1}{\rho}\nabla P + fv + \mu\nabla^2 u, \\ \frac{dv}{dt} = \frac{\partial v}{\partial t} + u\frac{\partial v}{\partial x} + v\frac{\partial v}{\partial y} + w\frac{\partial v}{\partial z} = -\frac{1}{\rho}\nabla P - fu + \mu\nabla^2 v \end{cases}$$

or

$$\begin{cases} \frac{du}{dt} = -\frac{1}{\rho}\nabla P + fv + \mu\nabla^2 u, \\ \frac{dv}{dt} = -\frac{1}{\rho}\nabla P - fu + \mu\nabla^2 v. \end{cases} \tag{11.1.4}$$

11.2 Same Solution Between the Mathematical Model of Mesoscale Eddy and the Momentum Balance Equation of Seawater Motion

From system (11.1.4), there are

1. In objective reality, the disk-shaped eddy E on the sea surface is first observed, which is an ellipse in random motion.
2. In the rotating coordinate system, the Taylor number [516] theorem shows that the motion of the entire water tends to be two dimensional, that is, in the process of its ocean depth, its motion behavior is consistent up and down.
3. According to Sect. 3.2, in the mesoscale eddy E, any point p is asymptotic and has a uniform motion behavior that tends to rotate.

Based on the above (1)–(3), it can be seen that in the flow field represented by two-dimensional momentum balance Eq. (11.1.4) of seawater movement, the motion behavior of water particles is consistent with the Lagrange particle motion behavior given by system (3.3.16). Since the water mass represented by Eq. (11.1.2) is a mesoscale eddy, it has an asymptotic uniform motion tendency. So the particle motion behavior of system (3.3.16) is included in the flow field of system (11.1.4). The particle motion behavior of system (3.3.16) is the motion behavior of system (11.1.4). The difference is that the solution of system (3.3.16) is only the initial value

(x_{i0}, y_{i0}), $i = 1, 2, 3, \ldots$. It does not care about the initial value of particle motion. **System** (11.1.4) **and system** (3.3.16) **are mathematical systems with the same solution without considering the initial value of particle motion.**

The particle motion described by system (3.3.16) is in the Lagrangian form, while the two-dimensional equilibrium equation of seawater motion under geostrophic conditions described by system (11.1.4) is the Euler form. Despite the essential difference, we get the following conclusions from the above (1)–(3).

conclusion 1 As long as E is a mesoscale eddy, all the particles in the flow field at any time position in system (11.1.4) move in the form of Lagrangian particles.

conclusion 2 The movement of the particle described in Lagrangian form and the movement of the flow field at the coordinate position described in Euler form can be processed together, that is, they are mathematically the same system.

11.3 Necessary Conditions for Existence of Mesoscale Eddies in Special Model

The special model of the mesoscale eddy refers to nonlinear system (3.3.16), and if the system parameters satisfy Tables 4.5 and 4.6, then the system has a mesoscale cold ocean eddy or a mesoscale warm ocean eddy. From Sect. 10.1, we can see that nonlinear systems (11.1.4) and (3.3.16) have the same solution. Note that system (3.3.16) is a specific one, so we can use its specific form to give a necessary condition for the existence of mesoscale eddies:

Theorem 11.1 *For two-dimensional momentum balance Eq.* (11.1.4) *of seawater motion with friction under geostrophic conditions, if there is a mesoscale eddy, then its pressure gradient must satisfy*

$$\left\{\begin{aligned}
\frac{\partial P}{\partial x} &= -\alpha^2 \rho x(ax^2+by^2-R^2)^{2m} \\
&\quad - \alpha\rho m x[2(\alpha a x^2+\beta b y^2)(ax^2+by^2-R^2)^m + 2(\lambda a-\gamma b)\omega xy]^{m-1} \\
&\quad - [(\alpha+\beta)\lambda\omega - f\beta]\rho y(ax^2+by^2-R^2)^m \\
&\quad + 2\rho\mu\alpha m y(ax^2+by^2-R^2)^{m-2} \\
&\quad \times [a(3b+2am-a)x^2 + b(a+b+2bm)y^2 - (a+3b)R^2] \\
&\quad + \rho\gamma\omega(\lambda\omega x - f)x, \\
\frac{\partial P}{\partial y} &= -\beta^2 \rho y(ax^2+by^2-R^2)^{2m} \\
&\quad - \beta\rho m y[2(\alpha a x^2+\beta b y^2)(ax^2+by^2-R^2)^m + 2(\lambda a-\gamma b)\omega xy]^{m-1} \\
&\quad + [(\alpha+\beta)\gamma\omega - f\alpha]\rho x(ax^2+by^2-R^2)^m \\
&\quad + 2\rho\mu\beta m x(ax^2+by^2-R^2)^{m-2} \\
&\quad \times [a(a+b+2am)x^2 + b(3a+2bm-b)y^2 - (3a+b)R^2] \\
&\quad + \rho\lambda\omega(\gamma\omega - f)y.
\end{aligned}\right. \tag{11.3.1}$$

Especially when $\mu = 0$ in (11.3.1), that is, the system does not have friction terms, then we have the following corollary.

Corollary 11.1 *For two-dimensional momentum balance Eq.* (11.1.4) *of seawater motion without friction under geostrophic conditions, if there is a mesoscale eddy, then its pressure gradient must satisfy*

$$\begin{cases} \frac{\partial P}{\partial x} = -\alpha^2 \rho x(ax^2 + by^2 - R^2)^{2m} \\ \quad - \alpha\rho m x[2(\alpha a x^2 + \beta b y^2)(ax^2 + by^2 - R^2)^m + 2(\lambda a - \gamma b)\omega xy]^{m-1} \\ \quad - [(\alpha + \beta)\lambda\omega - f\beta]\rho y(ax^2 + by^2 - R^2)^m + \rho\gamma\omega(\lambda\omega x - f)x, \\ \frac{\partial P}{\partial y} = -\beta^2 \rho y(ax^2 + by^2 - R^2)^{2m} \\ \quad - \beta\rho m y[2(\alpha a x^2 + \beta b y^2)(ax^2 + by^2 - R^2)^m + 2(\lambda a - \gamma b)\omega xy]^{m-1} \\ \quad + [(\alpha + \beta)\gamma\omega - f\alpha]\rho x(ax^2 + by^2 - R^2)^m + \rho\lambda\omega(\gamma\omega - f)y. \end{cases}$$

Proof For system (3.3.16), we have

$$\begin{cases} u = \frac{dx}{dt} = \alpha x(ax^2 + by^2 - R^2)^m + \omega\lambda y, \\ v = \frac{dy}{dt} = \beta y(ax^2 + by^2 - R^2)^m - \omega\gamma x. \end{cases} \tag{11.3.2}$$

Substituting (11.3.2) into system (11.1.4), then

$$\begin{aligned} \frac{du}{dt} &= -\frac{1}{\rho} \cdot \frac{\partial P}{\partial x} + fv + \mu\nabla^2 u \\ &= -\frac{1}{\rho} \cdot \frac{\partial P}{\partial x} + f[\beta y(ax^2 + by^2 - R^2)^m - \omega\gamma x] \\ &+ \mu\nabla^2[\alpha x(ax^2 + by^2 - R^2)^m + \omega\lambda y] \\ &= -\frac{1}{\rho} \cdot \frac{\partial P}{\partial x} + f[\beta y(ax^2 + by^2 - R^2)^m - \omega\gamma x] \\ &+ 2\mu\alpha m y(ax^2 + by^2 - R^2)^{m-2} \\ &\times [a(3b + 2am - a)x^2 + b(a + b + 2bm)y^2 - (a + 3b)R^2] \end{aligned} \tag{11.3.3}$$

and

$$
\begin{aligned}
\frac{\mathrm{d}v}{\mathrm{d}t} &= -\frac{1}{\rho}\cdot\frac{\partial P}{\partial y} - fu + \mu\nabla^2 v \\
&= -\frac{1}{\rho}\cdot\frac{\partial P}{\partial y} - f[\alpha x(ax^2+by^2-R^2)^m + \omega\lambda y] \\
&\quad + \mu\nabla^2[\beta y(ax^2+by^2-R^2)^m - \omega\gamma x] \\
&= -\frac{1}{\rho}\cdot\frac{\partial P}{\partial y} - f[\alpha x(ax^2+by^2-R^2)^m + \omega\lambda y] \\
&\quad + 2\mu\beta m x(ax^2+by^2-R^2)^{m-2} \\
&\quad \times [a(a+b+2am)x^2 + b(3a+2bm-b)y^2 - (3a+b)R^2].
\end{aligned}
\tag{11.3.4}
$$

Then take the derivative of the time t on both sides of system (3.3.16), and then get

$$
\begin{cases}
\frac{\mathrm{d}u}{\mathrm{d}t} = \frac{\mathrm{d}^2x}{\mathrm{d}t^2} = \alpha^2 x(ax^2+by^2-R^2)^{2m} + \alpha m x[2(\alpha a x^2+\beta b y^2)(ax^2+by^2-R^2)^m \\
\quad + 2(\lambda a - \gamma b)\omega x y]^{m-1} + (\alpha+\beta)\lambda\omega y(ax^2+by^2-R^2)^m - \lambda\gamma\omega^2 x; \\
\frac{\mathrm{d}v}{\mathrm{d}t} = \frac{\mathrm{d}^2y}{\mathrm{d}t^2} = \beta^2 y(ax^2+by^2-R^2)^{2m} + \beta m y[2(\alpha a x^2+\beta b y^2)(ax^2+by^2-R^2)^m \\
\quad + 2(\lambda a - \gamma b)\omega x y]^{m-1} - (\alpha+\beta)\gamma\omega x(ax^2+by^2-R^2)^m - \lambda\gamma\omega^2 y.
\end{cases}
\tag{11.3.5}
$$

Substituting (11.3.5) into (11.3.3) and (11.3.4) canceling $\frac{\mathrm{d}u}{\mathrm{d}t}$ and $\frac{\mathrm{d}v}{\mathrm{d}t}$, then (11.3.1) is met.

11.4 Sufficient and Necessary Conditions for the Existence of Mesoscale Eddies in the General Model

11.4.1 No Stickiness

Take $\mu = 0$ in system (11.1.4). From $\mu\nabla^2 u = \mu\nabla^2 v = 0$, suppose the pressure field is constant, that is, pressure gradient is a known function about the spatial position (x, y), i.e.,

$$
\begin{cases}
\frac{\partial P}{\partial x} = R(x, y), \\
\frac{\partial P}{\partial y} = S(x, y).
\end{cases}
\tag{11.4.1}
$$

Substituting (11.4.1) into the first two equations in system (11.1.4), we get

$$
\begin{cases}
\frac{\mathrm{d}u}{\mathrm{d}t} = fv - \frac{1}{\rho}R(x, y), \\
\frac{\mathrm{d}v}{\mathrm{d}t} = -fu - \frac{1}{\rho}S(x, y).
\end{cases}
\tag{11.4.2}
$$

We obtain sufficient conditions for the existence of mesoscale eddies.

Theorem 11.2 *Set*

$$\begin{cases} \tilde{R}(x,y) = fv - \frac{1}{\rho}R(x,y), \\ \tilde{S}(x,y) = -fu - \frac{1}{\rho}S(x,y). \end{cases} \tag{11.4.3}$$

Under the action of the Coriolis force, in the absence of viscosity, if Conclusions 1 and 2 are established, $\widetilde{R}(x,y)$ *and* $\widetilde{S}(x,y)$ *satisfy*

$$\int_0^T \left(\frac{\partial \tilde{R}}{\partial x} + \frac{\partial \tilde{S}}{\partial y} \right) dt < 0,$$

then differential dynamic system (11.4.2) *has a limit cycle.*

Proof It is directly obtained from Theorem 2.9 in Sect. 2.2.

Example 11.1 If the pressure change rate $R(x,y)$ and $S(x,y)$ of differential dynamic system (11.3.5) are taken as

$$\begin{cases} R(x,y) = 0, \\ S(x,y) = -\rho\sigma u^2 + \rho\lambda v(u^3 - u^2 + v^2) - \rho\gamma v, \end{cases}$$

then differential dynamic system (11.4.3) is transformed into

$$\begin{cases} \frac{du}{dt} = fv, \\ \frac{dv}{dt} = -fu - \sigma u^2 + \lambda v(u^3 - u^2 + v^2) - \rho\gamma v, \end{cases} \tag{11.4.4}$$

where

$$\begin{cases} \tilde{R}(x,y) = fv, \\ \tilde{S}(x,y) = -fu - \sigma u^2 + \lambda v(u^3 - u^2 + v^2) - \rho\gamma v. \end{cases}$$

Set $f = 2, \sigma = 3, \lambda = 0.8$, and $\gamma = 0$. Then

$$\begin{cases} \tilde{R}(x,y) = 2v, \\ \tilde{S}(x,y) = 2u - 4u^2 + 0.7v(u^3 - u^2 + v^2). \end{cases}$$

So

$$\int_0^T \left(\frac{\partial \tilde{R}}{\partial x} + \frac{\partial \tilde{S}}{\partial y} \right) dt < 0.$$

According to Theorem 2.9 in Sect. 2.2, differential dynamical system (11.4.4) has a limit cycle. Furthermore, according to the semi-stability of the differential dynamical system, it can be known that differential dynamical system (11.4.3) is internally stable but externally unstable. Therefore, there exists a mesoscale warm eddy and it is an ideal mesoscale warm eddy. The simulation of differential dynamic system (11.4.3) is shown in Fig. 11.1.

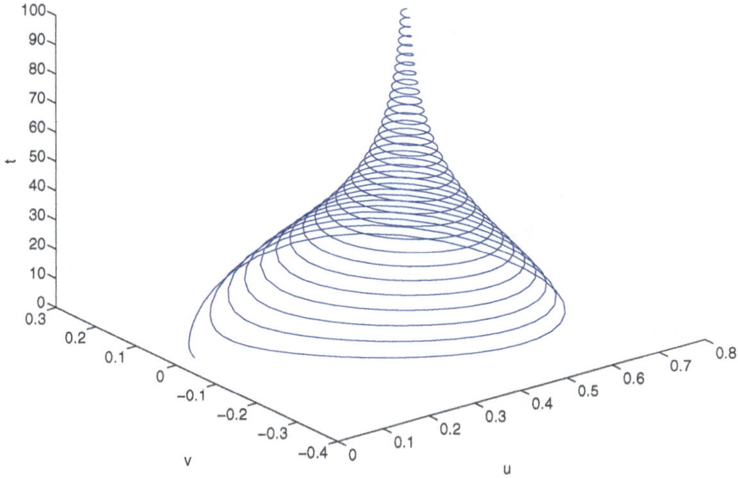

Fig. 11.1 Mesoscale warm eddy with $f = 2, \sigma = 4, \lambda = 0.7$, and $\gamma = 0$

11.4.2 Stickiness

The viscosity of the fluid is one of the important factors that produce the mesoscale eddy. Now, we will discuss in detail in two parts.

Approximation Treatment of Sticky Terms

First of all, in order to simplify the discussion, we express the viscous term as a Newtonian form with the opposite speed, namely,

$$\begin{cases} \mu\nabla^2 u \approx -\mu u, \\ \mu\nabla^2 v \approx -\mu v. \end{cases} \tag{11.4.5}$$

Then nonlinear system (11.1.4) becomes

$$\begin{cases} \frac{du}{dt} = -\frac{1}{\rho}\frac{\partial P}{\partial x} + fv - \mu u, \\ \frac{dv}{dt} = -\frac{1}{\rho}\frac{\partial P}{\partial y} - fu - \mu v. \end{cases} \tag{11.4.6}$$

We still assume that the pressure gradient satisfies Eq. (11.4.1).

(1) Sticky items only consider the situation in x direction

Set

$$\begin{cases} \frac{du}{dt} = -\frac{1}{\rho}\frac{\partial P}{\partial x} + fv - \mu u, \\ \frac{dv}{dt} = -\frac{1}{\rho}\frac{\partial P}{\partial y} - fu. \end{cases} \tag{11.4.7}$$

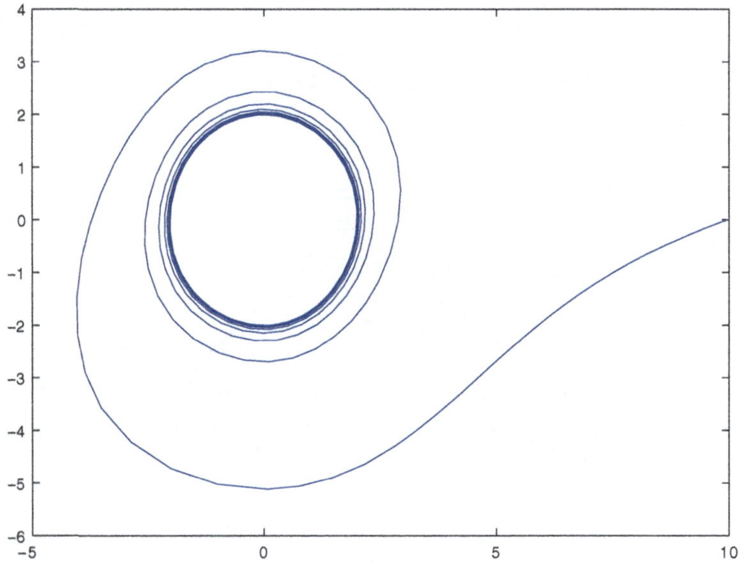

Fig. 11.2 Newtonian approximation with only one axis

Substitute formula (11.4.1) into system (11.1.4), there is

$$\begin{cases} \frac{du}{dt} = -\mu u + fv - \frac{1}{\rho}R(x, y), \\ \frac{dv}{dt} = -fu - \frac{1}{\rho}S(x, y). \end{cases} \tag{11.4.8}$$

Substituting the first expression in Eq. (11.4.8) for the time derivative once and then substituting the second expression into it, we get

$$\frac{d^2u}{dt^2} + \mu\frac{du}{dt} + f^2u = 0.$$

In this way, a dissipation term $\mu\frac{du}{dt}$ appears in Eq. (11.4.8). Under the action of Coriolis force, it is easy to get the solution of system (11.4.8) mathematically as

$$u = e^{-\frac{\mu}{2}t}\sin\sqrt{f^2 - \left(\frac{\mu}{2}\right)^2} \cdot t. \tag{11.4.9}$$

When $\mu = 0$, it becomes a non-viscous solution. Equation (11.4.9) shows that with the increase of time t, solution u will decay in oscillation, which means that the viscosity turns the closed non-rotation into a spiral vortex. The simulation of (11.4.9) is shown in Fig. 11.2. From this we obtain the following sufficient conditions for the existence of mesoscale eddies.

Theorem 11.3 *In the case of viscosity in x-direction, differential dynamic system* (11.4.7) *has a limit cycle. Thus, there is a mesoscale eddy.*

(2) Consider the problem of stickiness on both sides

Substituting (11.4.1) into (11.4.6), we have

$$\begin{cases} \frac{du}{dt} = -\mu u + fv - \frac{1}{\rho}R(x, y), \\ \frac{dv}{dt} = -fu - \mu v - \frac{1}{\rho}S(x, y). \end{cases} \tag{11.4.10}$$

Theorem 11.4 *Set*

$$\begin{cases} \tilde{R}(x, y) = -\mu u + fv - \frac{1}{\rho}R(x, y), \\ \tilde{S}(x, y) = -fu - \mu v - \frac{1}{\rho}S(x, y). \end{cases}$$

Under the action of Coriolis force, if both squares are sticky, assumptions (1)–(3) in Sect. 10.1 are satisfied, as well as $\widetilde{R}(x, y)$ *and* $\widetilde{S}(x, y)$ *satisfy*

$$\int_0^T \left(\frac{\partial \tilde{R}}{\partial x} + \frac{\partial \tilde{S}}{\partial y} \right) dt < 0,$$

then differential dynamic system (11.4.10) *has a limit cycle.*

Example 11.2 If the pressure change rate $R(x, y)$ and $S(x, y)$ of differential dynamic system (11.3.5) are taken as

$$\begin{cases} R(x, y) = -\rho\sigma \left(\frac{u^3}{3} - u \right), \\ S(x, y) = 0, \end{cases}$$

then differential dynamic system (11.4.2) is transformed into

$$\begin{cases} \frac{du}{dt} = -\mu u + fv - \sigma \left(\frac{u^3}{3} - u \right), \\ \frac{dv}{dt} = -fu - \mu v, \end{cases} \tag{11.4.11}$$

where

$$\begin{cases} \tilde{R}(x, y) = -\mu u + fv - \sigma \left(\frac{u^3}{3} - u \right) 0. \\ \tilde{S}(x, y) = -fu - \mu v. \end{cases}$$

Set $f = 1, \sigma = 0.1$ and $\mu = 0.01$. Then

$$\begin{cases} \tilde{R}(x, y) = -0.01u + v - 0.1 \left(\frac{u^3}{3} - u \right), \\ \tilde{S}(x, y) = -u - 0.01v. \end{cases}$$

So

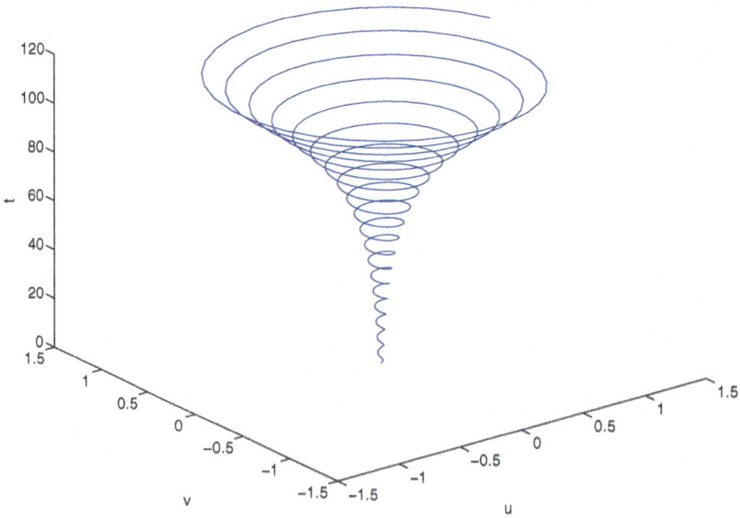

Fig. 11.3 Mesoscale cold eddy with $f = 1, \sigma = 0.1, \mu = 0.01$

$$\int_0^T \left(\frac{\partial \tilde{R}}{\partial x} + \frac{\partial \tilde{S}}{\partial y} \right) dt < 0.$$

Therefore, according to Theorem 11.2, differential dynamical system (11.4.11) has a limit cycle. Further, according to the semi-stability of the differential dynamical system, it can be known that differential dynamical system (11.4.11) is internally unstable and externally stable. Therefore, there exists a mesoscale cold eddy and it is an ideal mesoscale cold eddy. The simulation of differential dynamic system (11.4.11) is shown in Fig. 11.3.

No Approximate

(1) Numerical approximation form of mesoscale eddy

To facilitate processing, we assume that the pressure gradient is

$$\begin{cases} \frac{\partial P}{\partial x} = \rho(fS + \mu\nabla^2 R - R'), \\ \frac{\partial P}{\partial y} = \rho(-fR + \mu\nabla^2 S - S'), \end{cases} \tag{11.4.12}$$

where $R = R(x, y, t)$, $S = S(x, y, t)$, $R' = R'(x, y, t)$, and $S' = S'(x, y, t)$ are known functions. Substituting formula (11.4.12) into the first two equations in system (11.1.4), we get

$$\begin{aligned} \frac{\mathrm{d}u}{\mathrm{d}t} &= -fS - \mu\nabla^2 R + R' + fv + \mu\nabla^2 u \\ &= f(v - S) + \mu\nabla^2(u - R) + R', \end{aligned}$$

that is,

$$\frac{\mathrm{d}u}{\mathrm{d}t} - R' = f(v - S) + \mu\nabla^2(u - R),$$

namely,

$$\frac{\mathrm{d}(u - R)}{\mathrm{d}t} = f(v - S) + \mu\nabla^2(u - R); \tag{11.4.13}$$

and

$$\begin{aligned}\frac{\mathrm{d}v}{\mathrm{d}t} &= fR - \mu\nabla^2 S + S' - fu + \mu\nabla^2 v\\ &= -f(u - R) + \mu\nabla^2(v - S) + S',\end{aligned}$$

that is,

$$\frac{\mathrm{d}v}{\mathrm{d}t} - S' = -f(u - R) + \mu\nabla^2(v - S),$$

namely,

$$\frac{\mathrm{d}(v - S)}{\mathrm{d}t} = -f(u - R) + \mu\nabla^2(v - S). \tag{11.4.14}$$

Comprehensive (11.4.12) and (11.4.13), there is

$$\begin{cases}\frac{\mathrm{d}(u-R)}{\mathrm{d}t} = f(v - S) + \mu\nabla^2(u - R),\\ \frac{\mathrm{d}(v-S)}{\mathrm{d}t} = -f(u - R) + \mu\nabla^2(v - S).\end{cases} \tag{11.4.15}$$

Denote

$$\begin{cases}\tilde{u} = u - R,\\ \tilde{v} = v - S.\end{cases}$$

Then system (11.4.15) becomes

$$\begin{cases}\frac{\mathrm{d}\tilde{u}}{\mathrm{d}t} = f\tilde{v} + \mu\nabla^2\tilde{u},\\ \frac{\mathrm{d}\tilde{v}}{\mathrm{d}t} = -f\tilde{u} + \mu\nabla^2\tilde{v}.\end{cases} \tag{11.4.16}$$

At this point, we can consider using numerical methods to deal with system (11.4.16).

(2) Approximate treatment of pressure change rate (11.4.11)
(i) Viscous force is expressed in Newtonian form with opposite speed. In order to facilitate further discussion, we still express the viscous force as a Newtonian form with opposite speed, namely,

$$\begin{cases}\mu\nabla^2\tilde{u} \approx -\mu\tilde{u},\\ \mu\nabla^2\tilde{v} \approx -\mu\tilde{v}.\end{cases} \tag{11.4.17}$$

Substituting (11.4.17) into (11.4.15), there is

$$\begin{cases} \frac{d\tilde{u}}{dt} = f\tilde{v} - \mu\tilde{u}, \\ \frac{d\tilde{v}}{dt} = -f\tilde{u} - \mu\tilde{v}. \end{cases} \tag{11.4.18}$$

The coefficient matrix of system (11.4.18) is

$$\mathbf{A} = \begin{bmatrix} -\mu & f \\ -f & -\mu \end{bmatrix},$$

whose characteristic equation is

$$(\lambda + \mu)^2 + f^2 = 0.$$

Characteristic roots are

$$\lambda_1 = -\mu + \mathrm{i}f \quad \text{and} \quad \lambda_2 = -\mu - \mathrm{i}f,$$

where i is the imaginary unit. The two linearly independent eigenvectors corresponding to λ_1 and λ_2 are

$$\mathbf{V}_1 = \begin{bmatrix} 1 \\ \mathrm{i} \end{bmatrix} \quad \text{and} \quad \mathbf{V}_2 = \begin{bmatrix} \mathrm{i} \\ 1 \end{bmatrix},$$

respectively. Then matrix

$$\Phi(t) = \begin{bmatrix} e^{-(\mu-\mathrm{i}f)t} & \mathrm{i}e^{-(\mu+\mathrm{i}f)t} \\ \mathrm{i}e^{-(\mu-\mathrm{i}f)t} & e^{-(\mu+\mathrm{i}f)t} \end{bmatrix}$$

is a basis solution matrix. Its standard basis solution matrix is

$$\exp \mathbf{A}t = \Phi(t)\Phi(0)^{-1} = \begin{bmatrix} e^{-(\mu-\mathrm{i}f)t} & \mathrm{i}e^{-(\mu+\mathrm{i}f)t} \\ \mathrm{i}e^{-(\mu-\mathrm{i}f)t} & e^{-(\mu+\mathrm{i}f)t} \end{bmatrix} \begin{bmatrix} 1 & \mathrm{i} \\ \mathrm{i} & 1 \end{bmatrix}^{-1} = e^{-\mu t} \begin{bmatrix} \cos ft & \sin ft \\ -\sin ft & \cos ft \end{bmatrix}.$$

Set $\mathbf{C} = [c_1, c_2]^{\mathrm{T}}$ be an arbitrary constant vector. Then the general solution of Eq. (11.4.18) is

$$\begin{bmatrix} \tilde{u} \\ \tilde{v} \end{bmatrix} = \exp \mathbf{A}t \cdot \mathbf{C} = e^{-\mu t} \begin{bmatrix} \cos ft & \sin ft \\ -\sin ft & \cos ft \end{bmatrix} \begin{bmatrix} c_1 \\ c_2 \end{bmatrix} = e^{-\mu t} \begin{bmatrix} c_1 \cos ft + c_2 \sin ft \\ c_2 \cos ft - c_1 \sin ft \end{bmatrix},$$

that is,

$$\begin{cases} \tilde{u} = (c_1 \cos ft + c_2 \sin ft)e^{-\mu t}, \\ \tilde{v} = (c_2 \cos ft - c_1 \sin ft)e^{-\mu t}, \end{cases} \tag{11.4.19}$$

is the Coriolis force. As for the value of c_1 and c_2, solution (11.4.19) is in an eddy form, which we will discuss further in future work.

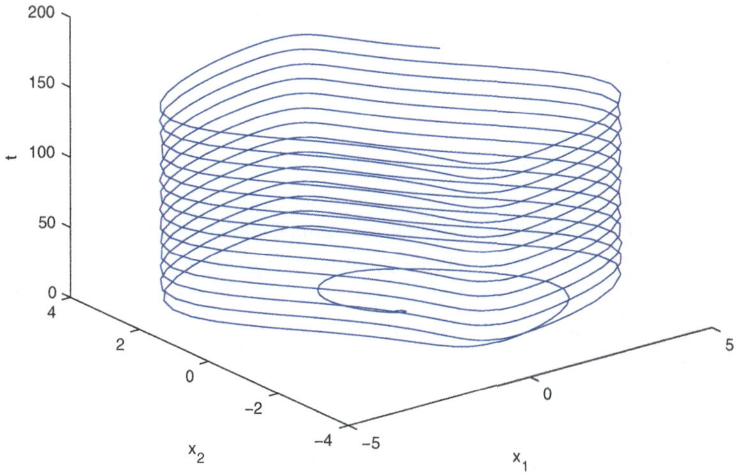

Fig. 11.4 Mesoscale eddy with $f = 1, \mu = 1$

Remark 11.1 The above can also be used as a separate question for interested researchers.

(ii) Viscous force is expressed as other forms of opposite speed. Now we put the viscous force as

$$\begin{cases} \mu\nabla^2\tilde{u} \approx -\mu\left(\frac{\tilde{u}^3}{3} - \tilde{u}\right), \\ \mu\nabla^2\tilde{v} \approx 0. \end{cases} \tag{11.4.20}$$

Substituting (11.4.20) into (11.4.15), we get

$$\begin{cases} \frac{\mathrm{d}\tilde{u}}{\mathrm{d}t} = f\tilde{v} - \mu\left(\frac{\tilde{u}}{3} - \tilde{u}\right), \\ \frac{\mathrm{d}\tilde{v}}{\mathrm{d}t} = -f\tilde{u}. \end{cases} \tag{11.4.21}$$

The simulation of differential dynamic system (11.4.21) is shown in Fig. 11.4.

11.4.3 Perturbation Terms of Parameters with Pressure Change Rate

Without considering the viscosity, suppose that the pressure field is constant and has certain perturbation, that is, the pressure gradient is a known function of spatial position (x, y) and interference term (ξ, η).

$$\begin{cases} \frac{\partial P}{\partial x} = R(x, y) + \xi, \\ \frac{\partial P}{\partial y} = S(x, y) + \eta, \end{cases} \tag{11.4.22}$$

where ξ and η are perturbation terms. Substituting (11.4.22) into the first two equations in system (11.1.4), there is

$$\begin{cases} \frac{du}{dt} = fv - \frac{1}{\rho}R(x, y) + \xi, \\ \frac{dv}{dt} = -fu - \frac{1}{\rho}S(x, y) + \eta. \end{cases} \tag{11.4.23}$$

If $R(x, y)$ and $S(x, y)$ and the perturbation items ξ and η meet certain conditions, limit cycles may appear in differential dynamic system (11.4.23).

Furthermore, considering the influence of the viscous term, and expressing the viscous force as the Newtonian form with the opposite speed as (1) in this section, Eq. (11.4.10) is corrected to

$$\begin{cases} \frac{du}{dt} = -\mu u + fv - \frac{1}{\rho}R(x, y) + \xi, \\ \frac{dv}{dt} = -fu - \mu v - \frac{1}{\rho}S(x, y) + \eta. \end{cases} \tag{11.4.24}$$

If $R(x, y)$ and $S(x, y)$ and the perturbation items ξ and η meet certain conditions, limit cycles may appear in differential dynamic system (11.4.24).

Finally, when the approximation is not used, Eq. (11.4.12) can be modified as

$$\begin{cases} \frac{\partial P}{\partial x} = \rho(fS + \mu\nabla^2 R - R') + \xi, \\ \frac{\partial P}{\partial y} = \rho(-fR + \mu\nabla^2 S - S') + \eta, \end{cases}$$

where $R = R(x, y, t)$, $S = S(x, y, t)$, $R' = R'(x, y, t)$, and $S' = S'(x, y, t)$ are known functions, ξ and η are perturbation items. At this time, formula (11.4.16) is corrected as

$$\begin{cases} \frac{d\tilde{u}}{dt} = f\tilde{v} + \mu\nabla^2\tilde{u} + \xi, \\ \frac{d\tilde{v}}{dt} = -f\tilde{u} + \mu\nabla^2\tilde{v} + \eta. \end{cases} \tag{11.4.25}$$

So far, we can consider using numerical methods to deal with Eq. (11.4.25).

11.4.4 Necessary Conditions

Suppose a nonlinear dynamic system with a semi-stable limit cycle:

$$\begin{cases} \frac{dx}{dt} = \tilde{P}(x(t), y(t)), \\ \frac{dy}{dt} = \tilde{Q}(x(t), y(t)), \end{cases} \tag{11.4.26}$$

within the changing time t, for the Navier-Stokes equation in Euler form

$$\begin{cases} \frac{du}{dt} = -\frac{1}{\rho}\frac{\partial P}{\partial x} + fv + \mu\nabla^2 u, \\ \frac{dv}{dt} = -\frac{1}{\rho}\frac{\partial P}{\partial y} - fu + \mu\nabla^2 v. \end{cases} \tag{11.4.27}$$

In statistical sense, the behavior of all particles in the infinitesimal volume has a unified particle motion. Then from system (11.4.26), we have

$$\begin{cases} x' = u, \\ y' = v, \end{cases}$$

then

$$\begin{cases} \frac{du}{dt} = u' = x'' = \tilde{P}'(x, y), \\ \frac{dv}{dt} = v' = y'' = \tilde{Q}'(x, y). \end{cases} \tag{11.4.28}$$

Substituting system (11.4.26) into system (11.4.27), we get

$$\begin{cases} \tilde{P}'(x, y) = -\frac{1}{\rho}\frac{\partial P}{\partial x} + f\tilde{Q}(x, y) + \mu\nabla^2\tilde{P}(x, y) \\ \qquad = -\frac{1}{\rho}\frac{\partial P}{\partial x} + f\tilde{Q}(x, y) + \mu\left(\frac{\partial^2\tilde{P}(x, y)}{\partial x^2} + \frac{\partial^2\tilde{P}(x, y)}{\partial y^2}\right), \\ \tilde{Q}'(x, y) = -\frac{1}{\rho}\frac{\partial P}{\partial y} - f\tilde{P}(x, y) + \mu\nabla^2\tilde{Q}(x, y) \\ \qquad = -\frac{1}{\rho}\frac{\partial P}{\partial y} - f\tilde{P}(x, y) + \mu\left(\frac{\partial^2\tilde{Q}(x, y)}{\partial x^2} + \frac{\partial^2\tilde{Q}(x, y)}{\partial y^2}\right), \end{cases}$$

then

$$\begin{cases} \frac{\partial P}{\partial x} = \rho\left[f\tilde{Q}(x, y) + \mu\left(\frac{\partial^2\tilde{P}(x,y)}{\partial x^2} + \frac{\partial^2\tilde{P}(x,y)}{\partial y^2}\right) - \tilde{P}'(x, y)\right], \\ \frac{\partial Q}{\partial y} = \rho\left[-f\tilde{P}(x, y) + \mu\left(\frac{\partial^2\tilde{Q}(x,y)}{\partial x^2} + \frac{\partial^2\tilde{Q}(x,y)}{\partial y^2}\right) - \tilde{Q}'(x, y)\right], \end{cases} \tag{11.4.29}$$

from which we get the necessary conditions for the existence of mesoscale eddies as follows.

Theorem 11.5 *If the geostrophic balance Eq.* (11.4.27) *has mesoscale eddies, the pressure change rate must satisfy*

$$\begin{cases} \frac{\partial P}{\partial x} = \rho\left[f\tilde{Q}(x, y) + \mu\left(\frac{\partial^2\tilde{P}(x,y)}{\partial x^2} + \frac{\partial^2\tilde{P}(x,y)}{\partial y^2}\right) - \tilde{P}'(x, y)\right], \\ \frac{\partial Q}{\partial y} = \rho\left[-f\tilde{P}(x, y) + \mu\left(\frac{\partial^2\tilde{Q}(x,y)}{\partial x^2} + \frac{\partial^2\tilde{Q}(x,y)}{\partial y^2}\right) - \tilde{Q}'(x, y)\right]. \end{cases}$$

Example 11.3 In differential dynamic system (11.4.26), take $\widetilde{P}'(x, y)$ and $\widetilde{Q}'(x, y)$ as

$$\begin{cases} \tilde{P}(x,y) = -0.01x + y - 0.1\left(\frac{x^3}{3} - x\right), \\ \tilde{Q}(x,y) = -x - 0.01y, \end{cases} \tag{11.4.30}$$

respectively. Then

$$\begin{cases} \tilde{P}'(x,y) = -0.01x' + y' - 0.1[x^2 - 1]x' \\ \qquad\qquad = -0.01u + v - 0.1[x^2 - 1]u, \\ \tilde{Q}'(x,y) = -x' - 0.01y' \\ \qquad\qquad = -u - 0.01v, \end{cases} \tag{11.4.31}$$

and

$$\begin{cases} \nabla^2 \tilde{P}(x,y) = \left(\frac{\partial^2 \tilde{P}(x,y)}{\partial x^2} + \frac{\partial^2 \tilde{P}(x,y)}{\partial y^2}\right) = -0.2, \\ \nabla^2 \tilde{Q}(x,y) = \left(\frac{\partial^2 \tilde{Q}(x,y)}{\partial x^2} + \frac{\partial^2 \tilde{Q}(x,y)}{\partial y^2}\right) = 0. \end{cases} \tag{11.4.32}$$

Substituting (11.4.30)–(11.4.32) into formula (11.4.29), we get

$$\begin{aligned} \frac{\partial P}{\partial x} &= \rho\left[f\tilde{Q}(x,y) + \mu\left(\frac{\partial^2 \tilde{P}(x,y)}{\partial x^2} + \frac{\partial^2 \tilde{P}(x,y)}{\partial y^2}\right) - \tilde{P}'(x,y)\right] \\ &= \rho\left[f(-x - 0.01y) - 0.2\mu + 0.01u - v + 0.1(x^2 - 1)u\right] \\ &= -\rho f x - 0.01\rho f y + 0.1(x^2 - 0.9)\rho u - \rho v - 0.2\rho\mu \end{aligned}$$

and

$$\begin{aligned} \frac{\partial Q}{\partial y} &= \rho\left[-f\tilde{P}(x,y) + \mu\left(\frac{\partial^2 \tilde{Q}(x,y)}{\partial x^2} + \frac{\partial^2 \tilde{Q}(x,y)}{\partial y^2}\right) - \tilde{Q}'(x,y)\right] \\ &= \rho\left[-f\left(-0.01x + y - 0.1\left(\frac{x^3}{3} - x\right)\right) - (-u - 0.01v)\right] \\ &= 0.1\rho f\left(\frac{x^2}{3} - 0.9\right)x - \rho f y + u + 0.01v. \end{aligned}$$

Therefore, a necessary condition for the existence of a limit cycle in two-dimensional geostrophic balance Eq. (11.4.28) is that the pressure change rate needs to satisfy

$$\begin{cases} \frac{\partial P}{\partial x} = -\rho f x - 0.01\rho f y + 0.1(x^2 - 0.9)\rho u - \rho v - 0.2\rho\mu, \\ \frac{\partial Q}{\partial y} = 0.1\rho f\left(\frac{x^2}{3} - 0.9\right)x - \rho f y + u + 0.01v, \end{cases}$$

where f is the Coriolis force. It can be seen that the Coriolis force plays an important role in the formation of mesoscale eddies!

Chapter 12
Momentum Balance Equation Based on Truncation Function and Mathematical Model of Mesoscale Eddies

Generally, the Navier-Stokes equation of Euler flow field describes the ocean flow field in the way of nonlinear partial differential dynamic system. However, the strong nonlinear partial differential dynamic system brings almost insurmountable barrier to solve practical problems. Therefore, the key of this chapter is to interpret the highly difficult nonlinear partial differential dynamic system as a nonlinear ordinary differential dynamic system to deal with the related problems of mesoscale eddies.

12.1 Sufficient Conditions of Mesoscale Eddies for the Two-Dimensional Momentum Balance Equation in Euler Form under the Truncated Function

From system (11.1.4), consider the approximation of β-plane inviscidically rotating the equilibrium equation to the nonlinear dynamic system. Note that the Coriolis-force at this time is

$$f = f_0 + \beta x_2.$$

System (11.1.4) considers the two-dimensional fluid motion on the sea surface and tries to obtain the corresponding dynamic model.

First, set rectangular coordinate system $O - x_1x_2$, the direction velocity of x_1 as u and the direction velocity of x_2 as v. Then acceleration a_1 in x_1 direction and acceleration a_2 in x_2 direction satisfy

$$\begin{cases} a_1 = \frac{\mathrm{d}u}{\mathrm{d}t} = \frac{\partial u}{\partial t} + u\frac{\partial u}{\partial x_1} + v\frac{\partial u}{\partial x_2} = \frac{\partial u}{\partial t} + (\mathbf{u} \cdot \nabla)u, \\ a_2 = \frac{\mathrm{d}v}{\mathrm{d}t} = \frac{\partial v}{\partial t} + u\frac{\partial v}{\partial x_1} + v\frac{\partial v}{\partial x_2} = \frac{\partial v}{\partial t} + (\mathbf{u} \cdot \nabla)v. \end{cases}$$

S.-T. Liu et al., *Mathematical Principle and Fractal Analysis of Mesoscale Eddy*,
https://doi.org/10.1007/978-981-16-1839-0_12

So in the directions x_1 and x_2, system (11.1.4) has

$$\frac{\mathrm{d}u}{\mathrm{d}t} = \frac{\partial u}{\partial t} + (\mathbf{u} \cdot \nabla)u = -\frac{1}{\rho} \cdot \frac{\partial P}{\partial x_1} + fv \tag{12.1.1}$$

and

$$\frac{\mathrm{d}v}{\mathrm{d}t} = \frac{\partial v}{\partial t} + (\mathbf{u} \cdot \nabla)v = -\frac{1}{\rho} \cdot \frac{\partial P}{\partial x_2} + fu. \tag{12.1.2}$$

Combining Eqs. (12.1.1)–(12.1.2) and continuity equation

$$\frac{\partial u}{\partial x_1} + \frac{\partial v}{\partial x_2} = 0,$$

we obtain the following two-dimensional momentum balance equation of seawater motion under geostrophic conditions:

$$\begin{cases} \frac{\mathrm{d}u}{\mathrm{d}t} = \frac{\partial u}{\partial t} + (\mathbf{u} \cdot \nabla)u = -\frac{1}{\rho} \cdot \frac{\partial P}{\partial x_1} + fv, \\ \frac{\mathrm{d}v}{\mathrm{d}t} = \frac{\partial v}{\partial t} + (\mathbf{u} \cdot \nabla)v = -\frac{1}{\rho} \cdot \frac{\partial P}{\partial x_2} + fu, \\ \frac{\partial u}{\partial x_1} + \frac{\partial v}{\partial x_2} = 0. \end{cases} \tag{12.1.3}$$

In order to facilitate problem handling, we introduce the stream function $\varphi(x_1, x_2, t)$, which satisfies

$$\mathbf{u} = (u, v)^{\mathrm{T}} = \left(-\frac{\partial \varphi}{\partial x_2}, \frac{\partial \varphi}{\partial x_1}\right)^{\mathrm{T}}.$$

The flow function φ is a scalar function related to the continuity equation in fluid mechanics. It has important applications in fluid plane motion and axis-symmetric motion. The contour of the flow function is the streamline. For two-dimensional incompressible flow, whether it is swirling or non-swirling, or whether the fluid is viscous or not, there must be a flow function. The flow function also satisfies

$$\Delta\varphi = \frac{\partial u}{\partial x_2} - \frac{\partial v}{\partial x_1}. \tag{12.1.4}$$

Derivation of the two sides of the first formula in Eq. (12.1.3) with $\frac{\partial}{\partial x_2}$, there is

$$\begin{aligned} &\frac{\partial}{\partial t}\left(\frac{\partial u}{\partial x_2}\right) + \frac{\partial u}{\partial x_2} \cdot \frac{\partial u}{\partial x_1} + u \cdot \frac{\partial^2 u}{\partial x_1 \partial x_2} + \frac{\partial v}{\partial x_2} \cdot \frac{\partial u}{\partial x_2} + u_2 \cdot \frac{\partial^2 u}{\partial x_2^2} \\ &= -\frac{1}{\rho} \cdot \frac{\partial^2 P}{\partial x_1 \partial x_2} + \left(f \cdot \frac{\partial v}{\partial x_2} + v \cdot \frac{\partial f}{\partial x_2}\right) + \mu\left(\frac{\partial^3 u}{\partial {x_1}^2 \partial x_2} + \frac{\partial^3 u}{\partial x_2^3}\right); \end{aligned}$$

derivation of the two sides of the second formula in Eq. (12.1.3) with $\frac{\partial}{\partial x_1}$, there is

$$\frac{\partial}{\partial t}\left(\frac{\partial v}{\partial x_1}\right)+\frac{\partial u_1}{\partial x_1}\cdot\frac{\partial v}{\partial x_1}+u\cdot\frac{\partial^2 v}{\partial {x_1}^2}+\frac{\partial v}{\partial x_1}\cdot\frac{\partial v}{\partial x_2}+u_2\cdot\frac{\partial^2 v}{\partial x_1\partial x_2}$$
$$=-\frac{1}{\rho}\cdot\frac{\partial^2 P}{\partial x_1\partial x_2}-\left(f\cdot\frac{\partial u}{\partial x_1}+u\cdot\frac{\partial f}{\partial x_1}\right)+\mu\left(\frac{\partial^3 v}{\partial {x_1}^3}+\frac{\partial^3 v}{\partial x_1\partial {x_2}^2}\right).$$

So we have

$$\frac{\partial}{\partial t}\left(\frac{\partial u}{\partial x_2}-\frac{\partial v}{\partial x_1}\right)+\frac{\partial u}{\partial x_1}\left(\frac{\partial u}{\partial x_2}-\frac{\partial v}{\partial x_1}\right)+\frac{\partial u_2}{\partial x_2}\left(\frac{\partial u}{\partial x_2}-\frac{\partial v}{\partial x_1}\right)$$
$$+u\frac{\partial}{\partial x_1}\left(\frac{\partial u}{\partial x_2}-\frac{\partial v}{\partial x_1}\right)+v\frac{\partial}{\partial x_2}\left(\frac{\partial u}{\partial x_2}-\frac{\partial v}{\partial x_1}\right)$$
$$=\mu\left[\frac{\partial^2}{\partial {x_1}^2}\left(\frac{\partial u}{\partial x_2}-\frac{\partial v}{\partial x_1}\right)+\frac{\partial^2}{\partial {x_2}^2}\left(\frac{\partial u}{\partial x_2}-\frac{\partial v}{\partial x_1}\right)\right]+f\left(\frac{\partial u}{\partial x_1}+\frac{\partial v}{\partial x_2}\right)$$
$$+\left(u\frac{\partial f}{\partial x_1}+u_2\frac{\partial f}{\partial x_2}\right).$$

Notice system (12.1.4). Then we have

$$\frac{\partial\Delta\varphi}{\partial t}+\frac{\partial u}{\partial x_1}\cdot\Delta\varphi+u\cdot\frac{\partial\Delta\varphi}{\partial x_1}+v\cdot\frac{\partial\Delta\varphi}{\partial x_2}+\frac{\partial v}{\partial x_2}\cdot\Delta\varphi$$
$$=\mu\left(\frac{\partial^2\Delta\varphi}{\partial {x_1}^2}+\frac{\partial^2\Delta\varphi}{\partial {x_2}^2}\right)+\left(u\frac{\partial f}{\partial x_1}+v\frac{\partial f}{\partial x_2}\right).$$

Noticing the continuous equation again, we have

$$\frac{\partial\Delta\varphi}{\partial t}+u\cdot\frac{\partial\Delta\varphi}{\partial x_1}+v\cdot\frac{\partial\Delta\varphi}{\partial x_2}=\mu\left(\frac{\partial^2\Delta\varphi}{\partial {x_1}^2}+\frac{\partial^2\Delta\varphi}{\partial {x_2}^2}\right)+\left(u\frac{\partial f}{\partial x_1}+v\frac{\partial f}{\partial x_2}\right).$$

According to the definition of flow function, the above equation is equivalent to

$$\frac{\partial\Delta\varphi}{\partial t}-\frac{\partial\varphi}{\partial x_2}\frac{\partial\Delta\varphi}{\partial x_1}+\frac{\partial\varphi}{\partial x_1}\frac{\partial\Delta\varphi}{\partial x_2}=\mu\left[\frac{\partial^2\Delta\varphi}{\partial {x_1}^2}+\frac{\partial^2\Delta\varphi}{\partial {x_2}^2}\right]+\left(-\frac{\partial\varphi}{\partial x_2}\frac{\partial f}{\partial x_1}+\frac{\partial\varphi}{\partial x_1}\frac{\partial f}{\partial x_2}\right),$$

where f is the Coriolis-force. In β-plane, $f=\beta x_2$ is a constant. So there is

$$[\varphi,f]=\frac{\partial\varphi}{\partial x_1}\cdot\frac{\partial f}{\partial x_2}-\frac{\partial\varphi}{\partial x_2}\cdot\frac{\partial f}{\partial x_1}=\beta v\neq 0.$$

Then system (12.1.3) turns into

$$\frac{\partial\Delta\varphi}{\partial t}+[\varphi,\Delta\varphi]=\mu\Delta^2\varphi+[\varphi,f]=\mu\Delta^2\varphi+\beta v, \tag{12.1.5}$$

where $[f_1,f_2]:=\frac{\partial f_1}{\partial x_1}\cdot\frac{\partial f_2}{\partial x_2}-\frac{\partial f_1}{\partial x_2}\cdot\frac{\partial f_2}{\partial x_1}$ is Poisson bracket.

According to [486, 487], the truncated function is used to select the seawater motion under geostrophic conditions to study the asymptotic geometric behavior of the plane flow of mesoscale eddies caused by the two-dimensional momentum balance equation. The approximate stochastic ellipse form as the subject is

$$(ax_1^2 + bx_2^2 - R^2)^2,$$

where a, b, and R are positive real numbers. We obtain the following sufficient conditions for the existence of mesoscale eddies.

Theorem 12.1 *Suppose there is a flow function $\varphi(x_1, x_2, t)$ that satisfies*

$$\begin{aligned}
&\frac{\partial \Delta\varphi}{\partial t} - \frac{\partial \varphi}{\partial x_2}\frac{\partial \Delta\varphi}{\partial x_1} + \frac{\partial \varphi}{\partial x_1}\frac{\partial \Delta\varphi}{\partial x_2} \\
&= \dot{x}F(x_1, x_2, c_1) + \dot{y}G(x_1, x_2, c_2)\, u\left(\frac{\partial^2 \Delta\varphi}{\partial {x_1}^2} + \frac{\partial^2 \Delta\varphi}{\partial {x_2}^2}\right) + \beta\left(\frac{\partial \varphi}{\partial x_1}\right) \\
&= [\alpha x(ax^2 + by^2 - R^2)^2 + \omega\lambda y]F(x_1, x_2, c_1) \\
&\quad + [\beta y(ax^2 + by^2 - R^2)^2 + \omega\mu x]G(x_1, x_2, c_2),
\end{aligned} \tag{12.1.6}$$

where $F(x_1, x_2, c_1)$ and $G(x_1, x_2, c_2)$ are known functions. When parameters a, b, α, β, λ, μ, and ω

1. *in the range of Table 4.4, the Euler flow field system has mesoscale eddies. In particular:*
2. *in the range of Table 4.5, the system is internally stable but externally unstable, producing mesoscale cold eddies.*
3. *in the range of Table 4.6, the system is internally unstable but externally stable, producing mesoscale warm eddies.*

Proof: From Poisson brackets

$$[\varphi, \Delta\varphi] = -\frac{\partial \varphi}{\partial x_2}\frac{\partial \Delta\varphi}{\partial x_1} + \frac{\partial \varphi}{\partial x_1}\frac{\partial \Delta\varphi}{\partial x_2},$$

we have

$$\frac{\partial \Delta\varphi}{\partial t} + [\varphi, \Delta\varphi] = \frac{\partial \Delta\varphi}{\partial t} - \frac{\partial \varphi}{\partial x_2}\frac{\partial \Delta\varphi}{\partial x_1} + \frac{\partial \varphi}{\partial x_1}\frac{\partial \Delta\varphi}{\partial x_2}. \tag{12.1.7}$$

From the left side of the second formula in (12.1.6) and (12.1.7), we get

$$\begin{aligned}
\frac{\partial \Delta\varphi}{\partial t} + [\varphi, \Delta\varphi] &= \frac{\partial \Delta\varphi}{\partial t} - \frac{\partial \varphi}{\partial x_2}\frac{\partial \Delta\varphi}{\partial x_1} + \frac{\partial \varphi}{\partial x_1}\frac{\partial \Delta\varphi}{\partial x_2} \\
&= \dot{x}F(x_1, x_2, c_1) + \dot{y}G(x_1, x_2, c_2).
\end{aligned} \tag{12.1.8}$$

For $\mu\Delta^2\varphi + \beta v$, from the second formula in (12.1.6), we have

$$\begin{aligned}
\mu\Delta^2\varphi + \beta v &= \mu\Delta\left(\frac{\partial^2\varphi}{\partial x_1^2} + \frac{\partial^2\varphi}{\partial x_2^2}\right) + \beta\left(\frac{\partial\varphi}{\partial x_1}\right) \\
&= \mu\left(\frac{\partial^2\Delta\varphi}{\partial x_1^2} + \frac{\partial^2\Delta\varphi}{x_2^2}\right) + \beta\left(\frac{\partial\varphi}{\partial x_1}\right) \\
&= \left[\alpha x\left(ax^2 + by^2 - R^2\right)^2 + \omega\lambda y\right] F\left(x_1, x_2, c_1\right) \\
&\quad + \left[\beta y\left(ax^2 + by^2 - R^2\right)^2 - \omega\mu x\right] G\left(x_1, x_2, c_2\right).
\end{aligned} \tag{12.1.9}$$

From system (12.1.5), and combining (12.1.8) with (12.1.9), comparing the coefficients of similar terms, we get

$$\begin{cases} \dot{x} = \alpha x\left(ax^2 + by^2 - R^2\right)^2 + \omega\lambda y \\ \dot{y} = \beta y\left(ax^2 + by^2 - R^2\right)^2 - \omega\mu x. \end{cases}$$

This is system (3.3.16). When the various parameters of system (3.3.16) are in the range of Tables 4.4–4.6, the conclusion in Chap. 4 is that there are mesoscale eddies.

12.2 Existence of Mesoscale Eddies in Two-Dimensional Momentum Balance Equation Based on the Compound Periodic Truncation Function

12.2.1 *β-Plane Approximation and Viscosity*

Since there is no sticky term, from (12.1.5), that is, $\mu = 0$, it becomes

$$\frac{\partial\Delta\varphi}{\partial t} + [\varphi, \Delta\varphi] = [\varphi, f] = \beta v. \tag{12.2.1}$$

In view of the behavior of the almost periodic characteristics of mesoscale eddies, from [486, 487], we set the flow function φ to be a double Fourier series

$$
\begin{aligned}
&\varphi(x_1, x_2, t) \\
&= \frac{a_{00}(t)}{4} + \frac{1}{2}\sum_{m=1}^{\infty}\left[a_{m0}(t)\cos\left(\frac{n\pi x_1}{K}\right) + b_{m0}(t)\sin\left(\frac{n\pi x_1}{K}\right)\right] \\
&\quad + \frac{1}{2}\sum_{n=1}^{\infty}\left[a_{0n}(t)\cos\left(\frac{n\pi x_2}{L}\right) + b_{0n}(t)\sin\left(\frac{n\pi x_2}{L}\right)\right] \\
&+ \sum_{m=1}^{\infty}\sum_{n=1}^{\infty}\left[a_{mn}(t)\cos\left(\frac{n\pi x_1}{K}\right)\cos\left(\frac{n\pi x_2}{L}\right)\right. \\
&\quad + b_{mn}(t)\cos\left(\frac{n\pi x_1}{K}\right)\sin\left(\frac{n\pi x_2}{L}\right) \\
&\quad + c_{mn}(t)\sin\left(\frac{n\pi x_1}{K}\right)\cos\left(\frac{n\pi x_2}{L}\right) \\
&\quad \left.+ d_{mn}(t)\sin\left(\frac{n\pi x_1}{K}\right)\sin\left(\frac{n\pi x_2}{L}\right)\right],
\end{aligned} \tag{12.2.2}
$$

where K represents the wavelength in the direction of x_1, L represents the wavelength in the direction of x_2, the coefficient a_{ij} is only related to t and t is the time parameter. Formula (12.2.2) is the general representation of the flow function φ by a two-parameter Fourier series. In a specific problem, in order to deal with the problem conveniently and without loss of generality, we take $K = L = 1$. Series (12.2.2) takes the following truncation function:

$$
\varphi = x(t)\sin(\pi a x_1) + x(t)\cos(\pi a x_1) + y(t)\sin(2\pi b x_2), \tag{12.2.3}
$$

where $a = \frac{\pi}{K} \neq 0$ and $b = \frac{\pi}{L} \neq 0$. Note the definition of the flow function φ, we can see that formula (12.2.3) also indirectly gives an approximate characterization of the velocity field $\mathbf{u} = (u_1, u_2)^{\mathrm{T}}$. From formula (12.2.3), there is

$$
\Delta\varphi' = -a^2\pi^2 x'(t)[\sin(\pi a x_1) + \cos(\pi a x_1) - 4b^2\pi^2 y'(t)\sin(2\pi b x_2)], \tag{12.2.4}
$$

$$
\Delta\varphi = -a^2\pi^2 x(t)[\sin(\pi a x_1) + \cos(\pi a x_1)] - 4b^2\pi^2 y(t)\sin(2\pi b x_2), \tag{12.2.5}
$$

$$
\begin{aligned}
[\varphi, \Delta\varphi] &= 2\sqrt{2}ab(a^2 - 4b^2)\pi^4 x(t)y(t) \\
&\quad \times \left[\cos(\pi a x_1)\cos\frac{\pi}{4} - \sin(\pi a x_1)\sin\frac{\pi}{4}\right][1 - 2\sin^2(\pi b x_2)] \\
&= 2ab(a^2 - 4b^2)\pi^4 x(t)y(t)\cdot[\cos(\pi a x_1) - \sin(\pi a x_1)] \\
&\quad \times [1 - 2\sin^2(\pi b x_2)]
\end{aligned} \tag{12.2.6}
$$

and

$$\begin{aligned}\beta v &= \beta\frac{\partial\varphi}{\partial x_1}\\ &= \pi a\beta x(t)[\cos(\pi a x_1) - \sin(\pi a x_1)]\\ &= \pi a\beta x(t)[\cos(\pi a x_1) + \sin(\pi a x_1) - 2\sin(\pi a x_1)]\\ &= \pi a\beta x(t)[\cos(\pi a x_1) + \sin(\pi a x_1)] - 2\pi a\beta x(t) sin(\pi a x_1).\end{aligned} \tag{12.2.7}$$

Substituting (12.2.4)–(12.2.7) into (12.2.1), we get

$$\begin{aligned}&- a^2\pi^2 x'(t)[\sin(\pi a x_1) + \cos(\pi a x_1)] - 4b^2\pi^2 y'(t)\sin(2\pi b x_2)\\ &+ 2ab(a^2 - 4b^2)\pi^4 x(t)y(t)[\sin(\pi a x_1) + \cos(\pi a x_1)]\\ &= \pi a\beta x(t)[\cos(\pi a x_1) + \sin(\pi a x_1)].\end{aligned} \tag{12.2.8}$$

Comparing the coefficients of similar terms in Eq. (12.2.8), we get

$$\begin{cases} -a^2\pi^2 x'(t) + 2ab(a^2 - 4b^2)\pi^4 x(t)y(t) = \pi a\beta x(t),\\ -4b^2\pi^2 y'(t) = 0.\end{cases}$$

Since $ab \neq 0$, one has

$$\begin{cases} x' = \pi a\beta x + \frac{2b(a^2-4b^2)\pi^2}{a} xy,\\ y' = 0.\end{cases} \tag{12.2.9}$$

where x and y are proportional to the intensity of convective motion. The increase or decrease of the value of x and y indicates the rise of warm fluid and the decline of cold fluid.

From the partial differential system of Euler flow field, a nonlinear ordinary differential system is obtained, which is transformed into the purpose of portraying the smooth ocean with a simple ordinary differential system. This fully reveals and theoretically proves that the difficult nonlinear partial differential system of Euler flow field can be attributed to the new ideas and new methods of solving problems in nonlinear ordinary differential systems.

From formula (12.2.3) to system (12.2.9), the truncation of Fourier series is convenient and has practical meaning. In fact, if we start from double Fourier series (12.2.2) directly, we will theoretically get infinitely many ordinary differential equations, which will lead to great difficulties and even be impossible to deal with in the following calculations. This is also unnecessary in practical situations. By using (12.2.3), the infinitely many ordinary differential equations will be reduced to the finite element ordinary differential equations.

Since we use the truncated Fourier series, there is an error between system (12.2.9) and the actual measurement. In fact, any system will have noise perturbation in its behavior. When analyzing and studying certain fluid mechanics problems, due to random disturbances, the original velocity field is changed, especially the circulation factor of the circulation motion behavior. For this, we naturally introduce nonlinear disturbance terms $f(\sigma, \rho, x, y)$ and $g(\sigma, \rho, x, y)$ in system (12.2.9), where ρ and σ

is a constant parameter. So, there is

$$\begin{cases} x' = \pi a\beta x + \frac{2b(a^2-4b^2)\pi^2}{a}xy + f(\sigma,\rho,x,y), \\ y' = g(\sigma,\rho,x,y). \end{cases}$$

Under certain conditions, there is an attraction cycle in the phase space, which is also the basic parameter equation of a mesoscale eddy under certain parameter conditions plane structure.

Remark 12.1 The above method can be built on f-plane, and the derivation process is similar.

12.2.2 β-Plane Approximation and Nonviscosity

Due to the viscous term, the Euler flow field system is (12.1.5). Without loss of generality, here is arbitrarily selected $K = L = 1$ and truncated (12.2.2) to obtain

$$\varphi = x(t)\sin(\pi a x_1) + x(t)\cos(\pi a x_1) + y(t)\sin(2\pi b x_2), \tag{12.2.10}$$

where $a = \frac{\pi}{K} \neq 0,\ b = \frac{\pi}{L} \neq 0$. According to formula (12.1.9), we have

$$\Delta' = -a^2\pi^2 x'(t)[\sin(\pi a x_1) + \cos(\pi a x_1) - 4b^2\pi^2 y'(t)\sin(2\pi b x_2)], \tag{12.2.11}$$

$$\Delta\varphi = -a^2\pi^2 x(t)[\sin(\pi a x_1) + \cos(\pi a x_1)] - 4b^2\pi^2 y(t)\sin(2\pi b x_2), \tag{12.2.12}$$

$$\Delta^2\varphi = a^4\pi^4 x(t)[\sin(\pi a x_1) + \cos(\pi a x_1)] + 16b^4\pi^4 y(t)\sin(2\pi b x_2), \tag{12.2.13}$$

$$\begin{aligned}[\varphi, \Delta\varphi] &= 2\sqrt{2}ab(a^2-4b^2)\pi^4 x(t)y(t) \\ &\quad\times\left[\cos(\pi a x_1)\cos\frac{\pi}{4} - \sin(\pi a x_1)\sin\frac{\pi}{4}\right][1-2\sin^2(\pi b x_2)] \\ &= 2ab(a^2-4b^2)\pi^4 x(t)y(t)\cdot[\cos(\pi a x_1) - \sin(\pi a x_1)] \\ &\quad\times[1-2\sin^2(\pi b x_2)]\end{aligned} \tag{12.2.14}$$

and

$$\begin{aligned}\beta v &= \beta\frac{\partial\varphi}{\partial x_1} \\ &= \pi a\beta x(t)[\cos(\pi a x_1) - \sin(\pi a x_1)] \\ &= \pi a\beta x(t)[\cos(\pi a x_1) + \sin(\pi a x_1) - 2\sin(\pi a x_1)] \\ &= \pi a\beta x(t)[\cos(\pi a x_1) + \sin(\pi a x_1)] - 2\pi a\beta x(t) sin(\pi a x_1).\end{aligned} \tag{12.2.15}$$

Substituting (12.2.11)–(12.2.15) into (12.1.5), we get

$$\begin{aligned}&-a^2\pi^2x'(t)[\sin(\pi ax_1)+\cos(\pi ax_1)]-4b^2\pi^2y'(t)\sin(2\pi bx_2)\\&+2ab(a^2-4b^2)\pi^4x(t)y(t)[\sin(\pi ax_1)+\cos(\pi ax_1)]\\&=\mu a^4\pi^4x(t)[\sin(\pi ax_1)+\cos(\pi ax_1)]+16\mu b^4\pi^4y(t)\sin(2\pi bx_2)\\&+\pi a\beta x(t)[\cos(\pi ax_1)+\sin(\pi ax_1)].\end{aligned} \tag{12.2.16}$$

Comparing the coefficients of similar terms in Eq. (12.2.16), we get

$$\begin{cases}-a^2\pi^2x'(t)+2ab(a^2-4b^2)\pi^4x(t)y(t)=\mu a^4\pi^4x(t)+\pi a\beta x(t),\\-4b^2\pi^2y'(t)=16vb^4\pi^4y(t).\end{cases}$$

Since $ab\neq 0$, one has

$$\begin{cases}x'=(\beta-\mu a\pi)\pi ax+\frac{2b(a^2-4b^2)\pi^2}{a}xy,\\y'=-4\mu b^2\pi^2y,\end{cases}$$

where x and y are proportional to the intensity of convective motion. The increase or decrease of the value of x and y indicates the rise of warm fluid and the decline of cold fluid. β is the rate of change of Coriolis-force.

Due to the random disturbance, the original velocity field is changed. For this reason, we naturally introduce the nonlinear disturbance term $f(\sigma,\rho,x,y)$ and $g(\sigma,\rho,x,y)$, where ρ and σ are constant parameters. Then, there is

$$\begin{cases}x'=(\pi a\beta-\mu a^2\pi^2)x+\frac{2b(a^2-4b^2)\pi^2}{a}xy+f(\sigma,\rho,x,y),\\y'=-4\mu b^2\pi^2y+g(\sigma,\rho,x,y).\end{cases} \tag{12.2.17}$$

Under certain conditions, there is an attraction cycle in the phase space, which is also the basic parameter equation for generating a mesoscale eddy under certain parameter conditions plane structure.

Remark 12.2 The above method can be built on f-plane, and the derivation process is similar.

12.3 Mesoscale Cold and Warm Eddies Produced by Truncation Function and Circulation Factor

Example 12.1 For nonlinear noise items $f(\sigma,\rho,x,y)$ and $g(\sigma,\rho,x,y)$ in system (12.2.17), we take

$$\begin{cases}f(\sigma,\rho,x,y)=y-\sigma x\left(\frac{1}{\rho}x^2-1\right)-(\alpha x+\beta xy),\\g(\sigma,\rho,x,y)=\sigma x.\end{cases}$$

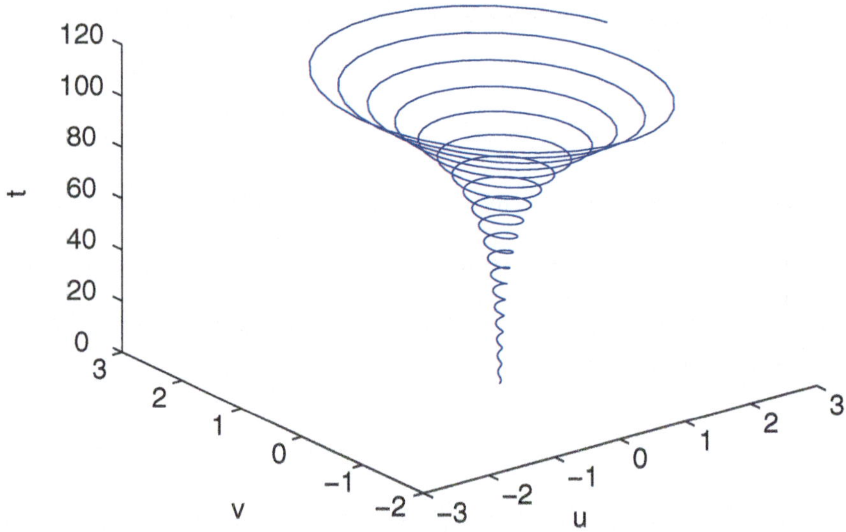

Fig. 12.1 Mesoscale cold eddy with $\sigma = 0.09$, $\rho = -1$

Then Eq. (12.2.17) is transformed into

$$\begin{cases} \frac{dx}{dt} = y - \sigma\left(\frac{x^3}{3} - x\right), \\ \frac{dy}{dt} = \rho x. \end{cases} \tag{12.3.1}$$

Computing the divergence of (12.3.1), we have

$$\nabla \cdot \mathbf{V} = \frac{\partial u}{\partial x} + \frac{\partial v}{\partial y} = \frac{\partial x'}{\partial x} + \frac{\partial y'}{\partial y} = -\sigma(x^2(t) - 1).$$

When $|x| > 1$, then $\nabla \cdot \mathbf{V} < 0$, that is, system (12.3.1) is a dissipative system. And the dissipative system must have an attraction cycle. According to the semi-stability theorem of differential dynamical systems, system (12.3.1) is internally stable and externally unstable. So there is a mesoscale cold eddy and it is an ideal mesoscale cold one. See Fig. 12.1.

Example 12.2 For the nonlinear perturbation items $f(\sigma, \rho, x, y)$ and $g(\sigma, \rho, x, y)$ in (12.2.17), we take

$$\begin{cases} f(\sigma, \rho, x, y) = \rho y - (\alpha x + \beta x y), \\ g(\sigma, \rho, x, y) = x(\sigma - \rho x) + \rho y\left(x^3 - x^2 + y^2\right) - \gamma y. \end{cases}$$

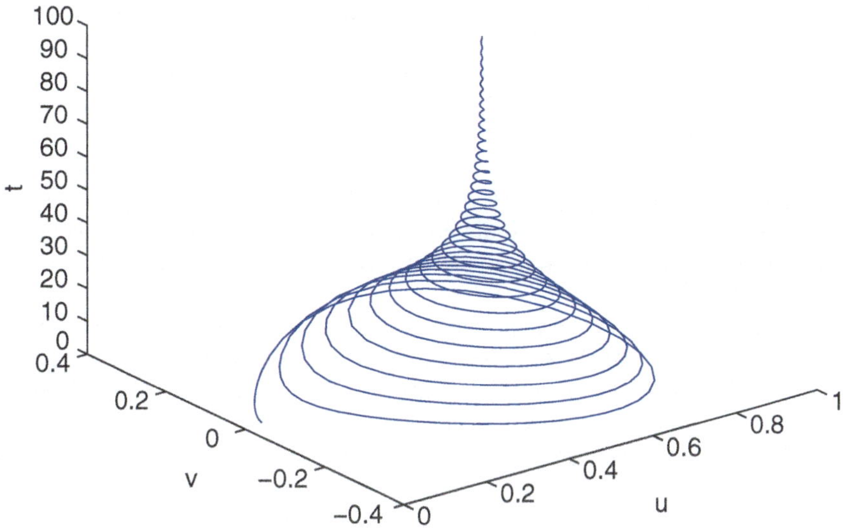

Fig. 12.2 Mesoscale warm eddy with $\sigma = 2$, $\rho = 3.5$ and $\lambda = 1.0$

Then Eq. (12.2.17) is transformed into the following nonlinear dynamic system:

$$\begin{cases} \frac{\mathrm{d}x}{\mathrm{d}t} = \sigma y, \\ \frac{\mathrm{d}y}{\mathrm{d}t} = \sigma x - \rho x^2 + \lambda y(x^3 - x^2 + y^2). \end{cases} \tag{12.3.2}$$

Computing the divergence of (12.3.2), we have

$$\begin{aligned} \nabla \cdot \mathbf{V} &= \frac{\partial u}{\partial x} + \frac{\partial v}{\partial x} = \frac{\partial x'}{\partial x} + \frac{\partial y'}{\partial y} \\ &= 0 + \lambda[(x^3 - x^2 + y^2) + y(2y)] \\ &= \lambda[x^3 - x^2 + 2y^2] \\ &= \lambda[x^2(x-1) + 2y^2]. \end{aligned}$$

When $| x |> 1, \lambda < 0$, then $\nabla \cdot \mathbf{V} < 0$, that is, system (12.3.2) is a dissipative system. And the dissipative system must have an attraction cycle. According to the semi-stability theorem of differential dynamical systems, system (12.3.2) is internally stable and externally unstable. So there is a mesoscale warm eddy and it is an ideal mesoscale warm one. See Fig. 12.2.

The above examples show that in the space of $\mathbb{R}^3$, from the continuous change of time t, the formation process of the mesoscale eddy in the undisturbed state. In summary, according to the stability of the limit cycle theory, we have the following conclusions:

1. Using the truncation function, the nonlinear partial differential equation representing the Euler flow field system can be simplified to a nonlinear ordinary differential system, and the circulation factor is introduced to solve the problem, which fully reveals the partial differential system of the Euler flow field theoretically. This can be attributed to new ideas and methods to solve problems in linear ordinary differential systems.
2. This is also a new method that we use the Navier-Stokes equation of a simple flow field on the plane (two-dimensional momentum balance equation under geostrophic conditions) to directly use mathematical analytical methods to study mesoscale eddies.
3. Although this method has some shortcomings, for example, how to choose the appropriate flow function $\varphi(x_1, x_2, t)$ to cause the nonlinear differential system without adding the circulation factor and then directly generate the mesoscale eddy. After all, we have pioneered such a new way, and the precise results below can be used as an open question to continue research!

Chapter 13
Interpolation Prediction of Mesoscale Eddies

From the perspective of describing the data changes of the radius of mesoscale eddies, their movement behavior is studied from the beginning to the extinction.

1. Using the observed radius database of mesoscale eddies, we give the trajectory function of the continuous motion of the mesoscale eddy from the beginning to the extinction.
2. Using the mesoscale eddy in the interval $D = [a, b]$, the continuous and more detailed motion characteristics of a certain interval included in the interval D are predicted through the whole trajectory function from the beginning to the extinction, that is, assuming $[c, d] \subseteq [a, b]$, the continuous variation behavior of point vortex motion at any value in $[c, d]$ is predicted from the motion of mesoscale eddy on $[a, b]$.

From the radius data of a mesoscale eddy in Table 3.1, we use an interpolation method [488–494] to give the change of the radius of mesoscale eddy and then describe the mesoscale eddy from one side, which describes the evolution of mesoscale eddies from the beginning to the extinction.

Noting that according to the interpolation method and the theory of linear equations, consider polynomial

$$f(x) = a_{27}x^{27} + a_{26}x^{26} + \cdots + a_1 x + a_0,$$

where the coefficient matrix can be obtained by solving with MATLAB. But it is generally difficult to realize in practice because of the high number of times.

From Table 3.1, we use a genetic algorithm and piecewise cubic spline interpolation to get the following numerical results shown in Fig. 13.1. The coefficients of its cubic spline function $C_i(x) = a_i + b_i x + c_i x^2 + d_i x^3$ in the interval $[i, i + 1]$ are shown as in Table 13.1.

S.-T. Liu et al., *Mathematical Principle and Fractal Analysis of Mesoscale Eddy*,
https://doi.org/10.1007/978-981-16-1839-0_13

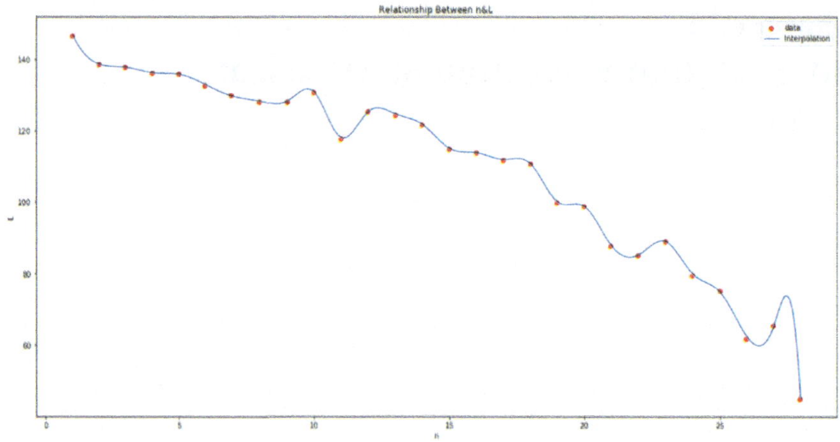

Fig. 13.1 Interpolation results of cubic spline

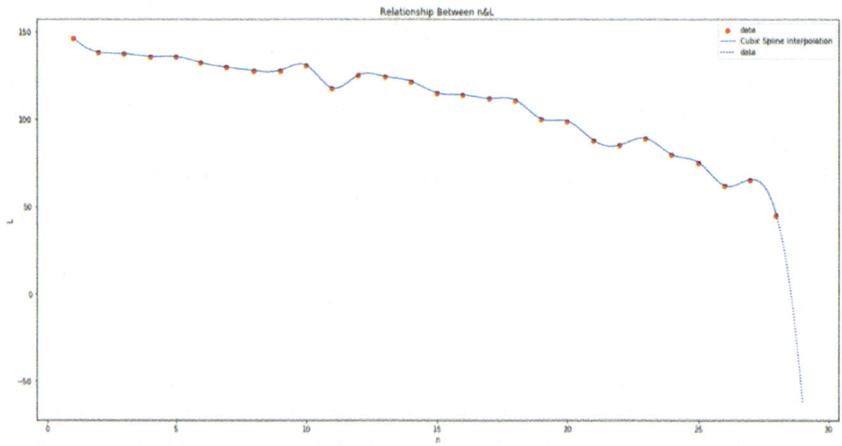

Fig. 13.2 Extrapolation prediction results of cubic spline

For the points in the data range, the accurate numerical value of the radius can be accurately predicted. See Table 13.2. Using extrapolation prediction, the numerical correspondence shown in Fig. 13.2 can be obtained.

It can be seen that its radius gradually decays with the increase of time, and it is expected to decrease greatly at the next moment, which reflects the overall situation

Table 13.1 Interpolation result of cubic spline

Interval	*a*	*b*	*c*	*d*
[1, 2]	−1.9462	9.4542	−15.5729	146.739
[2, 3]	−1.9462	3.6155	−2.5033	138.674
[3, 4]	1.6401	−2.2232	−1.1109	137.84
[4, 5]	−2.0862	2.6971	−0.637	136.146
[5, 6]	1.6466	−3.5614	−1.5012	136.12
[6, 7]	−0.3224	1.3785	−3.6841	132.704
[7, 8]	−0.3012	0.4114	−1.8942	130.076
[8, 9]	2.4653	−0.4923	−1.975	128.292
[9, 10]	−8.702	6.9036	4.4363	128.29
[10, 11]	14.1775	−19.2022	−7.8623	130.928
[11, 12]	−12.0921	23.3303	−3.7342	118.041
[12, 13]	5.3759	−12.946	6.6501	125.545
[13, 14]	−2.8675	3.1817	−3.1142	124.625
[14, 15]	3.9992	−5.4209	−5.3534	121.825
[15, 16]	−3.3835	6.5769	−4.1974	115.05
[16, 17]	2.7276	−3.5736	−1.1941	114.046
[17, 18]	−5.4821	4.6094	−0.1583	112.006
[18, 19]	8.4047	−11.8369	−7.3858	110.975
[19, 20]	−8.6865	13.3772	−5.8455	100.157
[20, 21]	6.8316	−12.6822	−5.1504	99.0023
[21, 22]	−0.4246	7.8126	−10.02	88.0013
[22, 23]	−6.9644	6.5389	4.3315	85.3693
[23, 24]	8.4106	−14.3542	−3.4838	89.2753
[24, 25]	−8.3346	10.8777	−6.9603	79.8479
[25, 26]	11.0926	−14.1261	−10.2087	75.4307
[26, 27]	−10.3993	19.1518	−5.1829	62.1886
[27, 28]	−10.3993	−12.0461	1.9228	65.7582

Table 13.2 Precise prediction results

Time	Radius
6.1583	132.1541
10.6821	121.1304
15.4688	114.1791
24.7258	77.3397

of mesoscale eddy from the beginning to the extinction. So we can get the radius data of mesoscale eddies within this range. This is what we call more perfect and meticulous work, which is also the main meaning of this chapter.

Chapter 14
Random Elliptic Curve and Brownian Motion Trajectory of Mesoscale Eddy

The motion trajectory and prediction of mesoscale eddies is an important problem in the field of physical oceanography research. At present, satellite observation and altimeter can track from the beginning to the extinction of a mesoscale eddy very well, but it is still ongoing research to better prediction by using limited data.

14.1 Trajectory of Elliptical Arc

The motion behavior of a mesoscale eddy is very complex due to noise perturbation from outside, and the natural plane shape is also random and changeable. However, no matter what the transformation, coordinate reference system is constantly rotating because the earth is constantly autobiography around the axis from west to east at an average angular velocity $\omega = 7.292 \times 10^{-5}$ rad/s. Therefore, it is a non-inertial system. It is necessary to consider the earth's rotation effect (also called Coriolis-force or Coriolis effect). Therefore, the basic motion is a random arc process, which is usually dominated by the following arcs.

14.1.1 Mesoscale Cold Eddy

Consider nonlinear system

$$\begin{cases} \frac{\mathrm{d}u}{\mathrm{d}t} = -0.01u + v - 0.1\left(\frac{u^3}{3} - u\right), \\ \frac{\mathrm{d}v}{\mathrm{d}t} = -u - 0.01v, \\ \frac{\mathrm{d}w}{\mathrm{d}t} = 1. \end{cases} \tag{14.1.1}$$

S.-T. Liu et al., *Mathematical Principle and Fractal Analysis of Mesoscale Eddy*,
https://doi.org/10.1007/978-981-16-1839-0_14

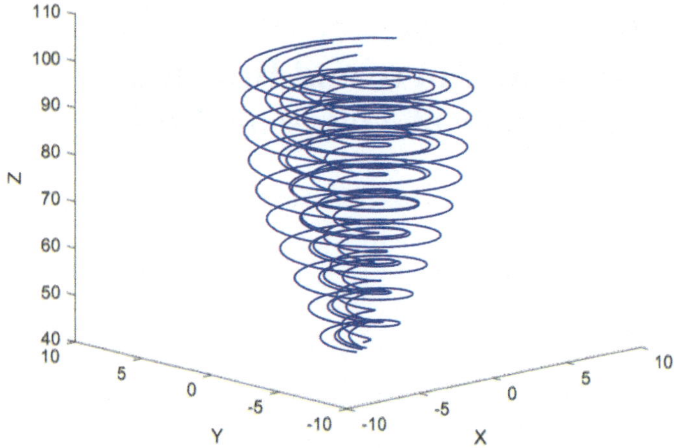

Fig. 14.1 Three-dimensional spatial behavior of the mesoscale cold eddy

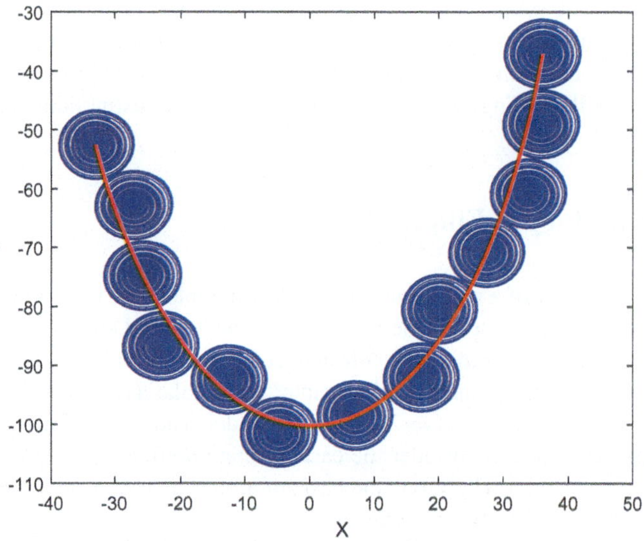

Fig. 14.2 Plane motion trajectory (the red line in the figure is $\alpha x^2 + \beta y^2 = r^2$, where $\alpha = 0.6$, $\beta = 0.3$ and $r^2 = 900$

As system (14.1.1), it can be seen from semi-stability results in Chap. 2 that the system is internally stable but externally unstable. Thus it is in the form of a mesoscale cold eddy, seeing Fig. 14.1.

Along ellipse $\alpha x^2 + \beta y^2 = r^2$, where $\alpha = 0.6$, $\beta = 0.3$, and $r^2 = 900$, segment motion trajectory shown in Figs. 14.2 and 14.3.

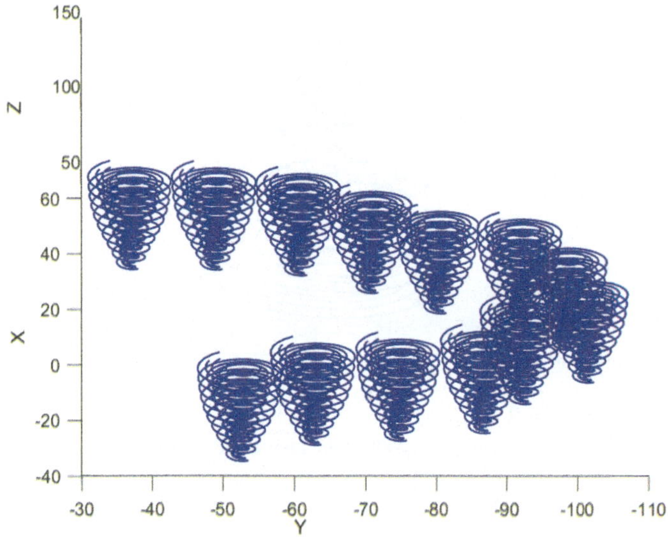

Fig. 14.3 Spatio-temporal motion behavior

14.1.2 *Mesoscale Warm Eddy*

Consider nonlinear system

$$\begin{cases} \frac{\mathrm{d}u}{\mathrm{d}t} = v, \\ \frac{\mathrm{d}v}{\mathrm{d}t} = -u - 3u^2 + 0.8v\left(u^3 - u^2 + v^2\right), \\ \frac{\mathrm{d}w}{\mathrm{d}t} = 1. \end{cases} \tag{14.1.2}$$

As system (14.1.2), it can be seen from semi-stability results in Chap. 2 that the system is internally unstable and externally stable. Thus it is in the form of a mesoscale warm eddy, seeing Fig. 14.4.

Along ellipse $\alpha x^2 + \beta y^2 = r^2$, where $\alpha = 0.3$, $\beta = 0.4$ and $r^2 = 800$, segment motion trajectory shown in Figs. 14.5 and 14.6.

14.2 Trajectory of Brownian Curve

According to the numerical law of mesoscale eddies motion in SSH database, the trajectory of mesoscale eddy conforms to the following fractional Brownian motion [472, 484, 495–497]:

$$B_H(t) - B_H(0) = \frac{1}{\Gamma\left(H + \frac{1}{2}\right)} \int_{-\infty}^{t} K\left(t - t'\right) dB\left(t'\right)$$

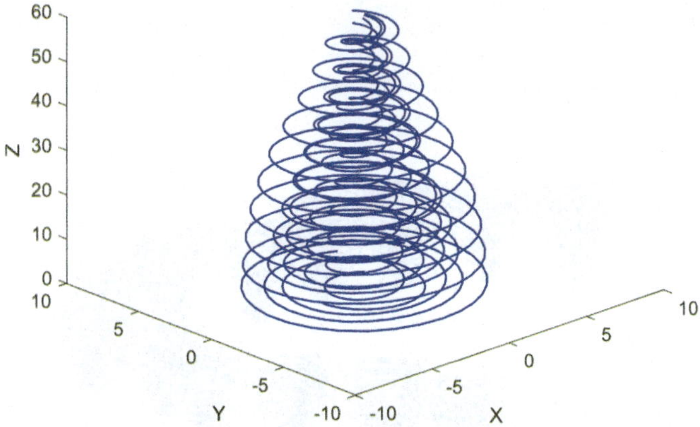

Fig. 14.4 Three-dimensional spatial behavior of the mesoscale warm eddy

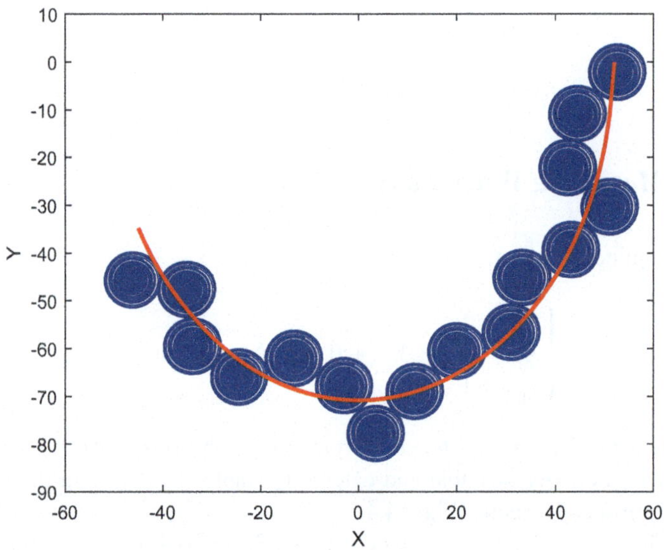

Fig. 14.5 Plane motion trajectory (the red line in the figure is $\alpha x^2 + \beta y^2 = r^2$, where $\alpha = 0.3$, $\beta = 0.4$ and $r^2 = 800$

and

$$K(t-t') = \begin{cases} (t-t')^{H-\frac{1}{2}}, & 0 \le t' \le t, \\ (t-t')^{H-\frac{1}{2}} - (-t')^{H-\frac{1}{2}}, & t' \le 0, \end{cases}$$

where $t > 0$, $K(t-t')$ is a kernel function, parameter $0 < H < 1$, $B(t)$ is an ordinary Gaussian random process with a mean value of zero and $\langle B^2(t) \rangle = t$.

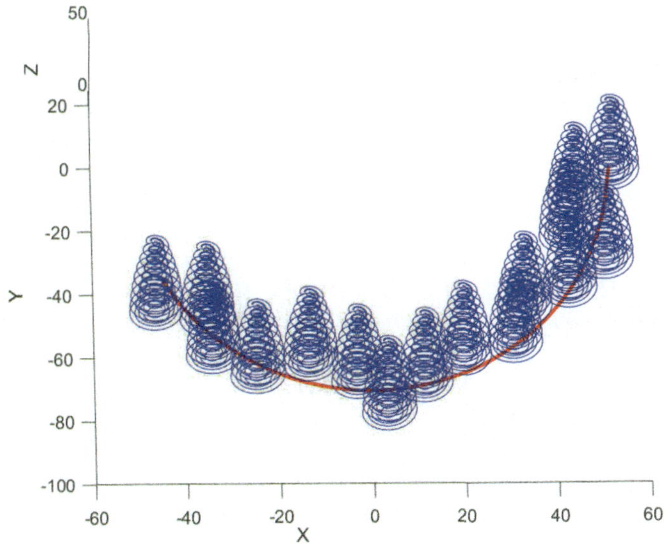

Fig. 14.6 Spatio-temporal motion behavior

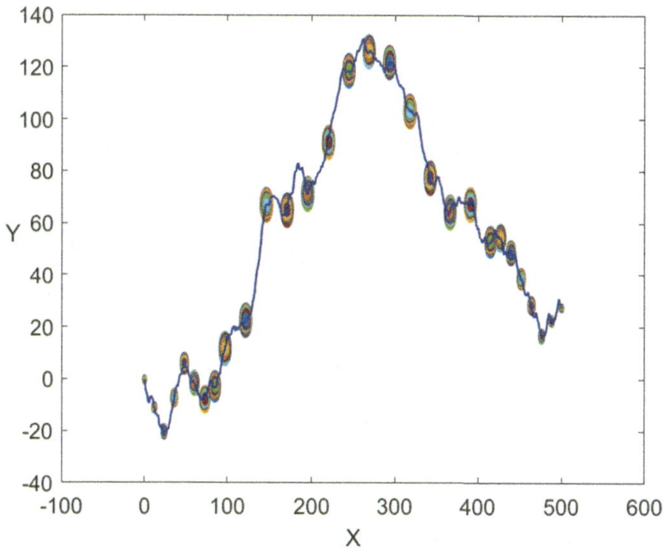

Fig. 14.7 Behavior of Brownian curve in mesoscale cold eddy plane

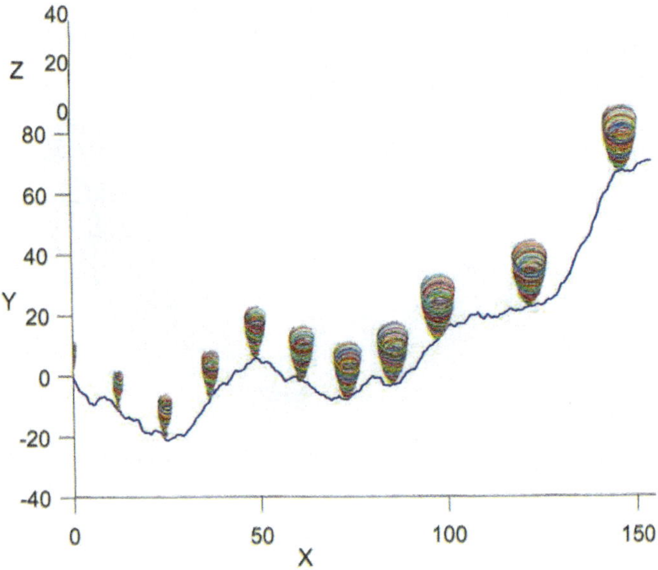

Fig. 14.8 Growth process of mesoscale cold eddy moving along Brownian curve

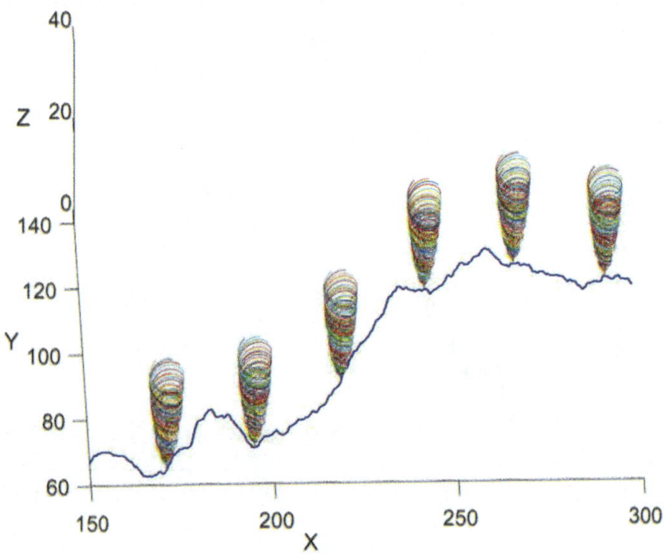

Fig. 14.9 Growth process of mesoscale cold eddy

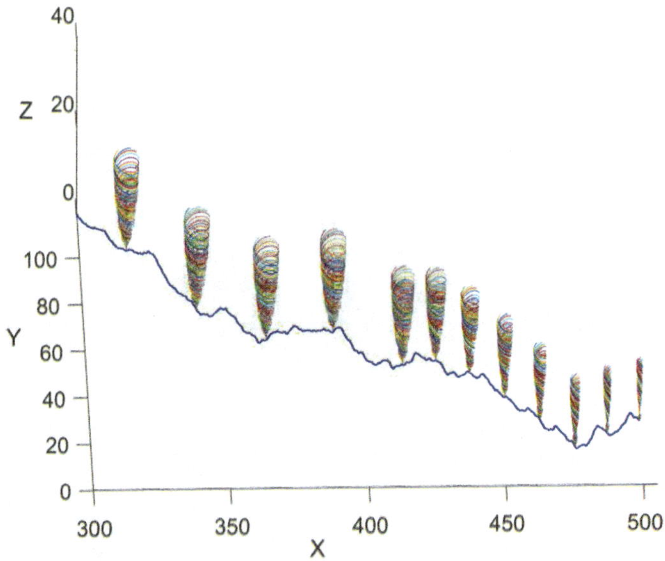

Fig. 14.10 Extinction process of mesoscale cold eddy

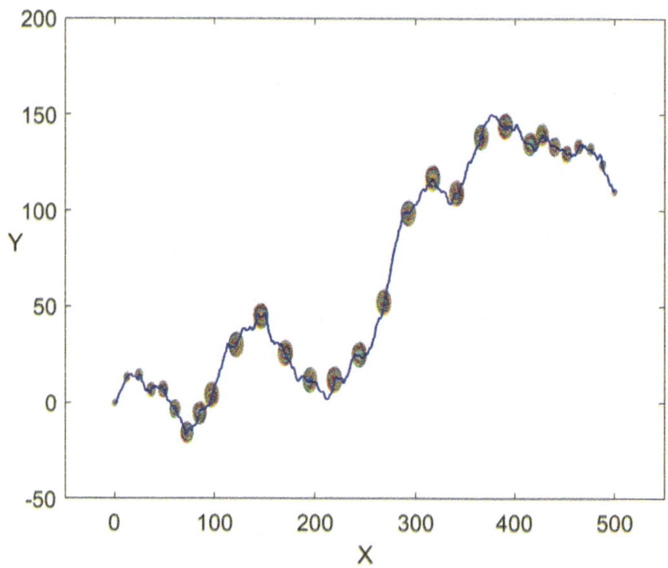

Fig. 14.11 Behavior of Brownian curve in mesoscale warm eddy plane

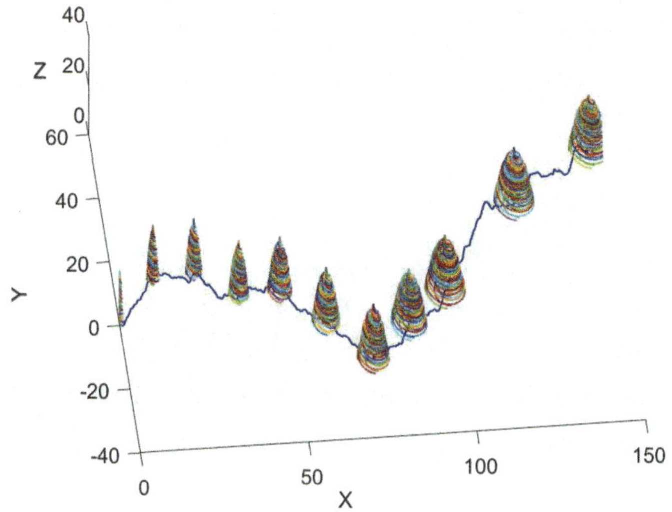

Fig. 14.12 Growth process of mesoscale warm eddy moving along Brownian curve

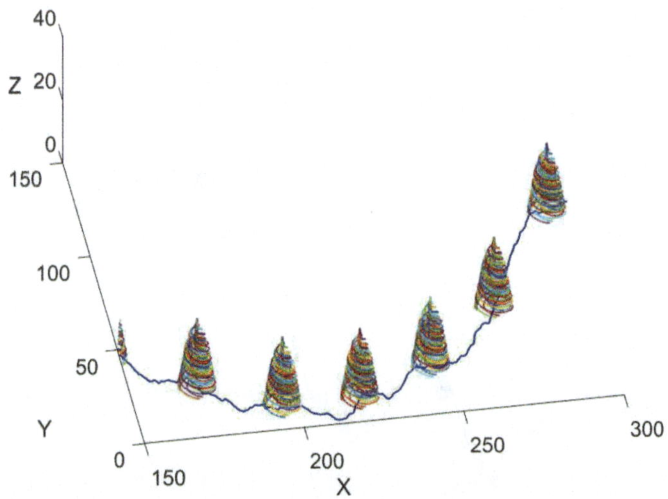

Fig. 14.13 Growth process of mesoscale warm eddy

1. The mesoscale cold eddy (see Fig. 14.1) is the motion behavior of stochastic ellipse conforming to the Brownian curve on the sea surface, that is, the fractal Brownian motion curve with the blue line ($H = 0.9$) in Fig. 14.7. Then the growth process, the continuous growth process, and the disappearance process can be seen in Figs. 14.8, 14.9, and 14.10.
2. The mesoscale warm eddy (see Fig. 14.4) is the motion behavior of stochastic ellipse conforming to Brownian curve on the sea surface, that is, the fractal Brow-

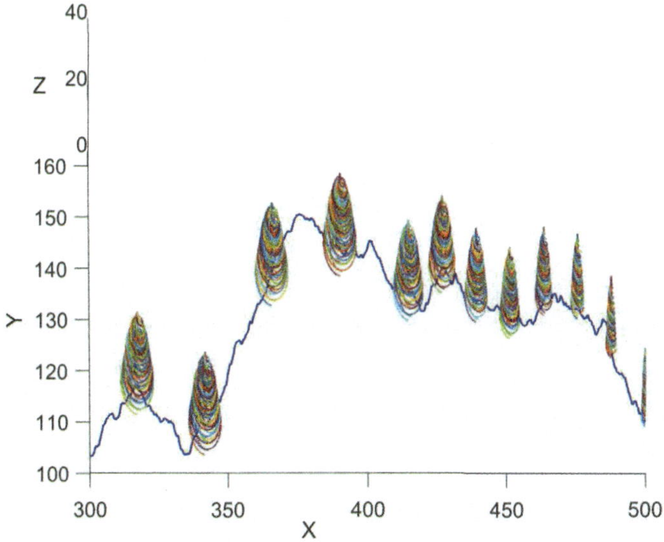

Fig. 14.14 Extinction process of mesoscale warm eddy

nian motion curve with the blue line ($H = 0.9$) in Fig. 14.11. Then the growth process, the continuous growth process, and the disappearance process can be seen in Figs. 14.12, 14.13, and 14.14.

Chapter 15
Mathematical Model for Edge Waves of Mesoscale Eddies and Its Spatio-Temporal Fractal Structures

As a physical oceanography phenomenon, edge waves of a mesoscale eddy (called edge waves for short) are typically nonlinear behavior. See Fig. 15.1. For mesoscale eddies, it is concerned that the maximum kinetic energy is not in the center, but in the region with the maximum linear velocity of rotation of the seawater, many parts of which are on the eddy edge. Mesoscale eddies are aroused by the internal and external inverse problems of the semi-stable limit cycle for stability. Edge waves have complex random characteristics and nonlinear behaviors. They are also one of the important research contents of the physical oceanography.

Fig. 15.1 Edge wave

S.-T. Liu et al., *Mathematical Principle and Fractal Analysis of Mesoscale Eddy*,
https://doi.org/10.1007/978-981-16-1839-0_15

In this chapter, using a dynamic nonlinear differential system, Poincaré cross-section, self-similarity theory of iterative function and random fractal theory, the motion model of the edge wave is fundamentally established. Then the basic fractal characteristic information under this model is given, which makes up for the vacancy in the research of motion behavior of the edge wave.

15.1 Mathematical Model of Edge Waves

Traditionally, the study of edge waves is carried out by using observation statistics and using sine waves, cosine waves, and Stokes waves in fluid dynamics [498]. Edge waves are regarded as sine wave superposition with different amplitudes and phases and then studied by using spectral analysis method [499, 500]. For example, the following edge wave models are given in [501]:

$$\xi(x, y, t) = \sum_{i=1}^{m}\sum_{j=1}^{n} A_n(\omega_i, \theta_j)\cos(k_i x\cos\theta_j + k_i y\sin\theta_j - \omega_i t + \varepsilon_{ij}),$$

where $A_n = \sqrt{2S(\omega, \theta)\Delta\omega\Delta\theta}$ is the amplitude of simple sine wave with different frequencies, k_i is the wave number, $\varepsilon_{ij} \in (0, 2\pi)$ is a random initial value phase with uniform distributions, $S(\omega, \theta) = S_n(\omega) \times G(\omega, \theta)$ is called directional spectrum function, $S_n(\omega)$ is the spectrum function and $G(\omega, \theta)$ is called direction function.

However, edge waves have essentially different physical forms and characteristics from various types of waves in the ocean. To study edge waves only by using the method of wave motion, it is bound to bring great errors and even contradictory problems with the basic characteristics of edge waves. Therefore, in order to reveal the essential law of the edge wave, it is necessary to start from the basic law of seawater particle motion and give the mathematical model of the overall change of the edge waves and the corresponding law properties.

Generally, for nonlinear systems, their complex characteristics usually make it impossible to observe the basic laws and morphological characteristics. At the end of the nineteenth century, the French mathematician Poincaré proposed the "cross-section method" to study the macroscopic motion of nonlinear behavior. It has gradually become an effective method to study intricate nonlinear behavior. The motion of an edge wave is a complex nonlinear behavior in the ocean. Therefore, it will objectively show the realistic behavior of edge wave by using Poincaré map to study the random complex motion of the edge wave.

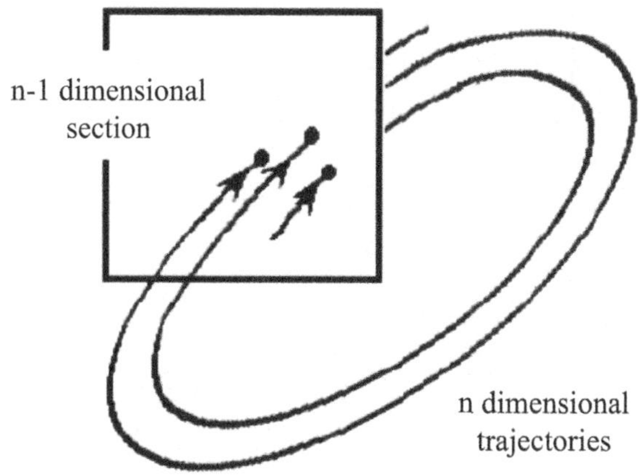

Fig. 15.2 Poincaré cross-section

15.1.1 Poincaré Cross-Section

As for edge waves, their fixed physical forms will appear in people's minds. This shows that the shape of edge waves has a certain inherent form and is an approximately fixed thing. Edge waves jump along with eddies, and the trajectory of the jump is the key to describe them. In order to study this complex trajectory, we introduce the Poincaré cross-section [502–510] in the phase space and take a constant coordinate section to study the intersection point of the phase trajectory and the section and analyze the complex behavior of the system. See Fig. 15.2.

Definition 15.1 Set (R, M, φ) as a global dynamic system, R is real number, M as phase space and φ as evolution function. Take γ as the periodic orbit pass through point p, S as the local differentiable section of φ to p, which called Poincaré cross-section cross over p.

When the phase trajectories of M pass through the space W, the cross point trajectories reflecting the trajectory of complex operation behavior.

The Poincaré cross-section can be used to describe the nonlinear and complex behavior of edge waves. It is no doubt that the trajectory through the cross-section points can be used to describe the motion of edge waves. Moreover, how to find the nonlinear dynamic system to describe edge waves is the key to solve the problem.

15.1.2 Edge Wave Motion and Its Duffing Dynamical System

The Coriolis-force can be neglected in the small-scale behavior of the edge wave [498], so that the governing equation is

$$\begin{cases} \frac{\partial u}{\partial t} + u\frac{\partial u}{\partial x} + v\frac{\partial u}{\partial y} + w\frac{\partial u}{\partial z} = -\frac{1}{\rho}\frac{\partial p}{\partial x}, \\ \frac{\partial v}{\partial t} + u\frac{\partial v}{\partial x} + v\frac{\partial v}{\partial y} + w\frac{\partial v}{\partial z} = -\frac{1}{\rho}\frac{\partial p}{\partial y}, \\ \frac{\partial w}{\partial t} + u\frac{\partial w}{\partial x} + v\frac{\partial w}{\partial y} + w\frac{\partial w}{\partial z} = -\frac{1}{\rho}\frac{\partial p}{\partial z} - g \end{cases}$$

and continuity equation

$$\frac{\partial u}{\partial x} + \frac{\partial v}{\partial y} + \frac{\partial w}{\partial z} = 0$$

with kinematic boundary conditions

$$z = 0, \quad \frac{\partial \xi}{\partial t} + \frac{\partial \xi}{\partial x} + \frac{\partial \xi}{\partial y} = \omega,$$

solid boundary condition

$$V_n = 0$$

and dynamic boundary conditions

$$z = 0, p = p_0(x, y, t).$$

And the linear wave, standing wave, elliptical cosine wave and Stokes wave are the main factors leading to the wave. Set the velocity potential as $\varphi(x, t) = \varphi_0(z)\cos(kx - \omega t)$, the depth as d and $a = -\omega/gD\cosh(kd)$, where k, ω, D are constants. Then the trajectories of the linear wave can described as

$$\begin{cases} \frac{dx}{dt} = u = \frac{gka}{\omega}\frac{\mathrm{ch}(d+z_0)}{\mathrm{ch}(kd)}\sin(kx_0 - \omega t), \\ \frac{dz}{dt} = w = -\frac{gka}{\omega}\frac{\mathrm{sh}(k(d+z_0))}{\mathrm{ch}(kd)}\cos(kx_0 - \omega t). \end{cases} \tag{15.1.1}$$

The standing wave and wave group are also closely related to (15.1.1).

In addition, set s as the wave direction line, l as the distance between the two wave direction lines and l_0 as distance between the two wave direction lines. Take $\beta = l/l_0$. Then the wave direction distribution equation related to edge waves is

$$\frac{d^2\beta}{ds^2} + p\frac{d\beta}{ds} + q\beta = f(\beta, s). \tag{15.1.2}$$

Moreover, the phase behavior of Stokes wave and elliptic cosine wave is also caused by the nonlinear dynamical systems (15.1.1) and (15.1.2).

It can be seen from the above that the motion behavior of edge waves can always be attributed to the motion behavior of a second-order nonlinear dynamic system. The fundamental reason is that edge waves are a kind of vibration behavior of water particles, and the system describing the vibration behavior of matter is the Duffing dynamic system. In particular, the complexity of the Duffing dynamic system is described by the Poincaré cross-section. In order

to reveal the law of edge wave motion fundamentally, the Duffing dynamic system, which satisfied the vibration behavior of water particles, is necessary for the complexity, variability, and self-similar fractal behavior of edge wave structure.

To facilitate research and generalize the problem to the nonlinear behavior in the ocean, we introduce Duffing dynamic

$$x'' + p(x)x' + q(x) = 0, \tag{15.1.3}$$

which is in fact a homogeneous form of Eq. (15.1.2) without perturbation. If we take $f(x) = \int_0^x p(\xi)\mathrm{d}\xi$, then

$$\begin{cases} \frac{\mathrm{d}x}{\mathrm{d}t} = y - f(x), \\ \frac{\mathrm{d}y}{\mathrm{d}t} = -q(x). \end{cases}$$

This is the form of Duffing dynamic system similar to system (15.1.1). From Eq. (15.1.3), further considering the noise perturbation, then there is

$$x'' + p(x)x' + q(x) = f(t, x(t)), \tag{15.1.4}$$

where $\forall t \in \mathbb{R}$ and $p(x) > 0$. The nonlinear function $q(x)$ and the nonlinear disturbance term $f(t, x(t))$ are the functions with physical characteristics given by the practical needs of the problem. System (15.1.4) is a more general Duffing dynamical system.

It is noticed that there are very complex and interfering water bodies in the process of the formation of edge waves. In mathematics, the wave is usually expressed as a function of periodic vibration. In order to solve the problem easily and not lose the general characteristics of the edge wave, the disturbance term $f(t, x(t))$ is a periodic function composed of sine and cosine functions, that is,

$$f(t, x(t)) = f(t, x(t), \sin \omega t, \cos \omega t).$$

Similarly, for simplification of the problem, the nonlinear function can be taken as a polynomial function as

$$q(x) = a_0 + a_1x + a_2x^2 + \cdots + a_nx^n, \quad n \geq 2.$$

In equation system (15.1.4), let

$$\begin{cases} p(x) = \text{constant} = \delta > 0, \\ q(x) = \sum\limits_{i=0}^{N} a_ix^i, \\ f(t, x(t)) = \text{compound periodic function}, \ f(t, x(t), \sin \omega t, \cos \omega t). \end{cases}$$

Then equation system (15.1.4) can be rewritten as

$$x'' + p(x)x' + \sum_{i=0}^{N} a_i x^i = f(t, x(t), \sin \omega t, \cos \omega t). \tag{15.1.5}$$

Thus, in general understanding, Eq. (15.1.5) is a Duffing dynamic system.

It is noteworthy that many mathematical models of nonlinear vibration problems in engineering can be transformed into Eq. (15.1.5) to study, such as ship rolling motion, structural vibration, destruction of chemical bonds, fault detection of sharp rubbing rotor, weak periodic signal detection, periodic oscillation analysis of dynamic system, simulation, and control of periodic circuit system. In addition, the axial tension perturbation model of the transverse wave equation and the dynamic equation of the rotor bearing is similar to the Duffing dynamic system. Of course, there are still many problems with the Duffing dynamic system that has not been thoroughly studied, such as fractional harmonic vibration, superharmonic vibration, combined vibration, and so on. And the research findings can be extended to other similar systems. Therefore, the study of nonlinear Duffing dynamic system is the foundation for many complex dynamic systems.

15.2 Mathematical Model of Edge Waves Based on Poincaré Cross-Section and Random Fractal

15.2.1 Generators of Edge Waves

In this section, we describe the nonlinear characteristics of edge waves by using the Poincaré cross-section of the Duffing dynamic system. It is known from the previous section that $q(x)$ is a polynomial function and $f(t, x)$ is a compound periodic function. **It is noted that different Duffing dynamic systems and different Poincaré cross-sections can be obtained by taking different parameters and noise functions in Eq.** (15.1.5). **So different wave generators can be obtained at** $\omega_0, \omega_1, \omega_2, \ldots, \omega_n, \ldots$. **The key is that waves are generated by a finite iteration and superposition of generators.** For example, we can choose randomly

$$\begin{cases} p(x) = 0.19(1 - 8.11x^2), \\ q(x) = x^3, \\ f(t, x) = 0.34 \cos t. \end{cases}$$

Thus Eq. (15.1.5) is transformed into

$$x'' - 0.19(1 - 8.11x^2)x' + x^3 = 0.34 \cos t. \tag{15.2.1}$$

Fig. 15.3 Poincaré cross-section of system (15.2.1)

In fact, for functions $p(x)$, $q(x)$ and $f(t, x)$, the method can be taken in other forms. Here is just an illustration. According to Eq. (15.2.1), the Poincaré cross-section in Fig. 15.3 can be obtained.

Let the trajectory of Fig. 15.3 be ω_0. As mentioned above, the Poincaré cross-section of the Duffing dynamical system is specially generated for characterizing and recognizing extremely complex nonlinear motion behavior. Therefore, for highly complex edge waves, we use this section to recognize and describe it appropriately following the reality, and thus ω_0 is generated as an edge waves. A basic initial element is the independent variable to edge waves mathematical model. It can be observed that the independent variable here is a geometric curve in random form, rather than an independent variable in unknown or parametric form, which is different from the past.

In the following, ω_0 is used as an independent variable to construct a mathematical model of edge waves.

15.2.2 A Mathematical Model for Random Fractal of Edge Waves

Now we use the random fractal theory to mathematically model edge waves. As for the ocean, what we usually call "wave-linking" and "wave-surfing" contain profound mathematical ideas and connotations. The mathematical essence of the so-called "edge-wave-surfing" is an iterative relationship we introduced above, that is,

Edge wave-surfing of mesoscale eddies = Iterative behavior of functions.

Strictly speaking, this is a superposition of random iterations and random behaviors or a recursive relationship of random behaviors.

It is noted that f is a definite and regular transformation, while the edge wave is a physical form with strong ups and downs and jumping behavior, which are irregular thing but an inherent physical form on the whole. Therefore, the transformation f of the generated edge waves must be a random behavior and thus a random function. Denote the wave as ζ. As can be seen from Sect. 3.1, ω_0 is a geometric shape and is also a geometric behavior made by the Poincaré cross-section, i.e.,

$$\begin{aligned} w_0 &= \text{set of all points in Poincaré cross-section of nonlinear system (15.2.1)} \\ &= \{c|c \in \text{ Poincaré cross-section of nonlinear system (15.2.1)}\}. \end{aligned}$$

Taking ω_0 as the basic element of edge waves, that is, ω_0 is the generator of edge waves ζ, we establish the following random fractal model of edge waves:

$$\begin{aligned} &f_U(t)(\omega_0) \text{ is } t \text{ values surging upward for } \omega_0, \;\; t \in [0, 1], \\ &f_D(t)(\omega_0) \text{ is } t \text{ values downward upward for } \omega_0, \;\; t \in [0, 1], \\ &f_L(t)(\omega_0) \text{ is } t \text{ values surging to the left for } \omega_0, \;\; t \in [0, 1], \\ &f_R(t)(\omega_0) \text{ is } t \text{ values surging to the right for } \omega_0, \;\; t \in [0, 1], \\ &h(\theta_s)(\omega_0) \text{ is the horizontal rotation angle } \theta_s, \;\; \theta_s \in [0, 2\pi], \\ &\tilde{h}(\alpha_s)(\omega_0) \text{ is the vertical tilt angle } \alpha_s, \;\; \alpha_s \in [0, 2\pi] \end{aligned}$$

and their composite functions as

$$h(\theta) f_U(t)(\omega_0), \quad h(\theta) f_D(t)(\omega_0), \quad h(\theta) g_L(t)(\omega_0), \quad h(\theta) g_R(t)(\omega_0).$$

It is worth noting that the commutative law of the above composite functions is established, that is,

$$\begin{aligned} h(\theta) f_U(t) &= f_U(t) h(\theta), \qquad & h(\theta) f_D(t) &= f_D(t) h(\theta), \\ h(\theta) g_L(t) &= g_L(t) h(\theta), \qquad & h(\theta) g_R(t) &= g_R(t) h(\theta). \end{aligned}$$

Furthermore, when $\theta = 0$, we have

$$\begin{aligned} &h(\theta)(\omega_0) = h(0)(\omega_0) = \omega_0, \quad \text{and} \\ &f_U(t)(\omega_0) = f_D(t)(\omega_0) = g_L(t)(\omega_0) = g_R(t)(\omega_0)|_{t=0} \equiv \omega_0. \end{aligned}$$

Meanwhile, there are finite operations of the following five sets of functions:

$$\eta_1(\zeta)(\omega_0) := \sum_{s=1}^{p_1}\sum_{i=1}^{q_1} h_s(\theta_s) f_{Ui}(t)(\zeta)(\omega_0),$$

$$\eta_2(\zeta)(\omega_0) := \sum_{s=1}^{p_2}\sum_{j=1}^{q_2} h_s(\theta_s) f_{Dj}(t)(\zeta)(\omega_0),$$

$$\eta_3(\zeta)(\omega_0) := \sum_{s=1}^{p_3}\sum_{m=1}^{q_3} h_s(\theta_s) g_{Lm}(t)(\zeta)(\omega_0),$$

$$\eta_4(\zeta)(\omega_0) := \sum_{s=1}^{p_4}\sum_{n=1}^{q_4} h_s(\theta_s) g_{Rn}(t)(\zeta)(\omega_0),$$

$$\eta_5(\zeta)(\omega_0) := \sum_{s=1}^{p_5} \tilde{h}_s(\alpha_s)(\zeta)(\omega_0),$$

where $p_i, q_i \in \mathbb{N}$. Thus, the mathematical model of edge waves generated by ω_0 is

$$\zeta(\omega_0) = \sum_{k=1}^{n} \zeta_k(\omega_0) = \zeta_1(\omega_0) + \zeta_2(\omega_0) + \cdots + \zeta_n(\omega_0), \tag{15.2.2}$$

where ζ_k is the k-th iteration according to the following rules:

$$\begin{aligned}
\zeta_k(\omega_0) &= \mathscr{R}(\zeta_{k-1})(\omega_0)\\
&:= \sum_{s=1}^{p_1}\sum_{i=1}^{q_1} h_s(\theta_s) f_{Ui}(t)(\zeta_{k-1})(\omega_0)\\
&+ \sum_{s=1}^{p_2}\sum_{j=1}^{q_2} h_s(\theta_s) f_{Dj}(t)(\zeta_{k-1})(\omega_0)\\
&+ \sum_{s=1}^{p_3}\sum_{m=1}^{q_3} h_s(\theta_s) g_{Lm}(t)(\zeta_{k-1})(\omega_0)\\
&+ \sum_{s=1}^{p_4}\sum_{n=1}^{q_4} h_s(\theta_s) g_{Rn}(t)(\zeta_{k-1})(\omega_0)\\
&+ \sum_{s=1}^{p_5} \tilde{h}_s(\alpha_s)(\zeta_{k-1})(\omega_0)\\
&= \eta_1(\zeta_{k-1})(\omega_0) + \eta_2(\zeta_{k-1})(\omega_0) + \eta_3(\zeta_{k-1})(\omega_0)\\
&+ \eta_4(\zeta_{k-1})(\omega_0) + \eta_5(\zeta_{k-1})(\omega_0)\\
&:= \sum_{i=1}^{5} \eta_i(\zeta_{k-1})(\omega_0),
\end{aligned} \tag{15.2.3}$$

where $p_i, q_i \in \mathbb{N}$ and $i = 1, 2, 3, 4, 5$. The random iteration $\mathscr{R}$ represents the horizontal and vertical transformation of the finite number of upward, downward, left, right, and angular movements of the previous $(k-1)$-th element. It is noted that formula (15.2.3) is a recursive relation and model (15.2.2) is another superposition relationship, thus revealing the natural law of random fractal of self-similarity of the wave pushing wave and wave striking wave. From model (15.2.2) with formula (15.2.3), we can conclude

Theorem 15.1 *Edge waves $\zeta(\omega_0)$ is a finite order iteration and finite recursive superposition of the self-similarity of the Poincaré cross-section of the Duffing dynamic system. And the different forms of edge waves depend on the number of iterations and superpositions.*

In particular, in model (15.2.2), we take $n = 10$ and in formula (15.2.3), we get

$$\begin{aligned}\zeta_k(\omega_0) &= \mathscr{R}(\zeta_{k-1})(\omega_0)\\ &= f_U(t)(\zeta_{k-1})(\omega_0) + g_L(t)(\zeta_{k-1})(\omega_0) + \tilde{h}_U(\alpha_s)(\zeta_{k-1})(\omega_0),\end{aligned}$$

where $k = 1, 2, \ldots, 10$. Random iteration $\mathscr{R}$ moves **up** $t_1 \in [0, 0.01]$, **left** $t_2 \in [0, 0.1]$ and rotate **clockwise** $\theta \in [0, 5°]$ randomly, i.e.,

$$\mathscr{R}(\cdot) = f_U(t_1)(\cdot)|_{t_1 \in [0,0.01]} + g_L(t_2)(\cdot)|_{t_2 \in [0,0.1]} + h(\theta)(\cdot)|_{\theta \in [0,5°]}. \qquad (15.2.4)$$

Take ω_0 as the Poincaré cross-section of system (15.2.1). Then from model (15.2.2) and formula (15.2.4), we have

$$\zeta(\omega_0) = \zeta_1(\omega_0) + \zeta_2(\omega_0) + \cdots + \zeta_{10}(\omega_0),$$

from which the waves generated in shown in Fig. 15.4.

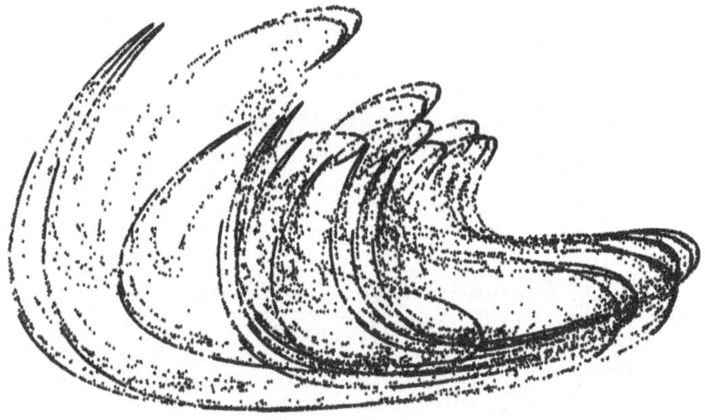

Fig. 15.4 Edge wave behavior of self-similar fractal at $n = 10$

Notably, the result of each summation in formula (15.2.3) is that each diagram is similar to each other, and the whole is similar to the parts. Thus, formula (15.2.3) is a random fractal form of edge waves. It can be viewed on the above that the Poincaré cross-section of the Duffing dynamic system, which reflects the form of edge waves of vibration can be used to describe the edge waves. It actually interprets the dynamic behavior of the waves in the local part of the ocean. It can be observed that the wave is only a finite iteration and a finite superposition of this dynamic behavior in the sense of random fractal. In summary, we give the following definitions.

Definition 15.2 Edge waves are a nonlinear dynamic system behavior under external noise perturbation, and a finite number of iterations of a random fractal structure and a finite number of superimposed synthetic ocean fluids.

Take ω_0 as the independent variable for generating edge waves. In system (15.1.4), we take

$$\begin{cases} p(x) = -\varepsilon\left[a + 2\cos x - (x')^2\right], \\ q(x) = \sin x, \\ f(t, x) = F\cos vt. \end{cases}$$

Then system (15.1.4) becomes

$$x'' - \varepsilon\left[a + 2\cos x - (x')^2\right]x' + \sin x = F\cos vt.$$

At this time, the independent variable ω_0 of edge waves in formula (15.2.3) can be taken as $\overline{\omega}_0$ and $\overline{\overline{\omega}}_0$ according to different parameters, seeing Fig. 15.5.

Different independent variables ω_0, $\overline{\omega}_0$, $\overline{\overline{\omega}}_0, \ldots$ can generate different edge waves. Moreover, for different random function relations, different edge waves can also be generated for the same independent variable.

Edge waves always have a specific form, so there is always a suitable random function to generate corresponding edge waves. A database of edge waves can be

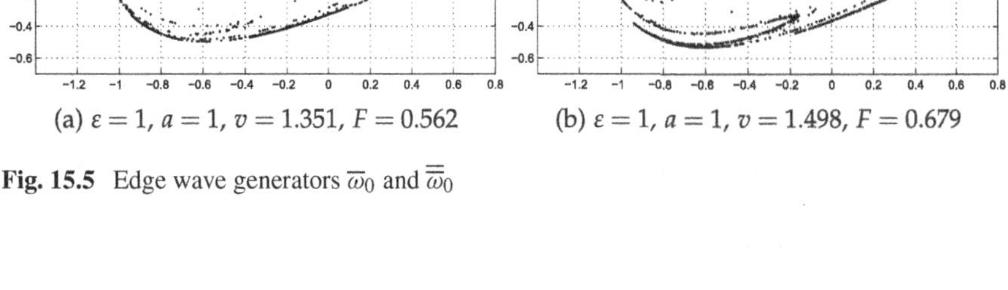

(a) $\varepsilon = 1$, $a = 1$, $v = 1.351$, $F = 0.562$ (b) $\varepsilon = 1$, $a = 1$, $v = 1.498$, $F = 0.679$

Fig. 15.5 Edge wave generators $\overline{\omega}_0$ and $\overline{\overline{\omega}}_0$

Table 15.1 Database of edge waves

Parameter	T_0: without transformation	T_1: 15° clockwise rotation	T_2: 30° clockwise rotation	T_3: superposition of T_0, T_1, T_2 with 1 unit left translation
$v = 1.152$ $F = 0.397$		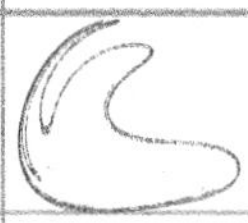		
$v = 1.249$ $F = 0.478$				
$v = 1.351$ $F = 0.561$				
$v = 1.500$ $F = 0.682$		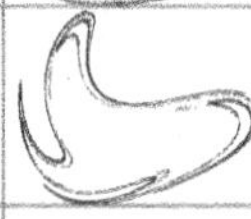		
$v = 0.397$ $F = 0.279$				

generated by a finite number of different random functions to form a set of edge waves as

$$\Omega = \{\xi(\omega_0), \xi(\omega_1), \ldots, \xi(\omega_n)\}.$$

In the ocean, there is always a the edge wave whose behavior belongs to Ω, which can be shown in Table 15.1.

15.3 Fractal Analysis of Internal Structure Complexity of Edge Waves

The edge wave is a self-similar finite iteration and finite superposition of the Poincaré cross-section of the Duffing dynamic system. Therefore, the spatial structure of the edge wave is a self-similar fractal structure. The complexity of the spatial structure of the edge wave is studied below. Its complexity can be described by a fractal dimension. For this reason, the vertical cross-section of the edge wave is first taken, and a series of sections are obtained by finite vertical interception. When the interval between them is infinitesimal, it can be seen that the edge wave is composed of a

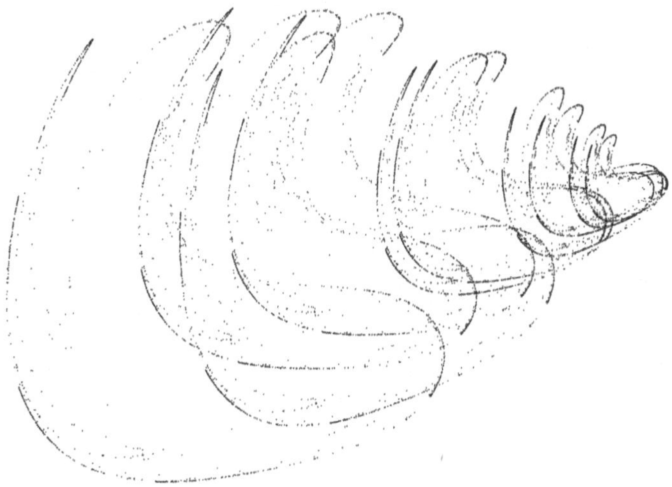

Fig. 15.6 v= 13.5 and F = 0.56 to generate edge wave map

Table 15.2 Fractal dimension D and number of boxes $\ln(N)$ of edge waves

	1	2	3	4	5	6	7	8	9	10
D	1.083	1.188	1.284	1.334	1.378	1.393	1.406	1.406	1.469	1.484
$\ln(N)$	7.301	7.901	8.394	8.709	9.053	9.360	9.509	9.756	9.882	9.975

plurality of sections with infinitesimal intervals, thus the complexity of the three-dimensional structure of the edge wave can be revealed according to the fractal dimension of the sections. The edge wave generated when $v = 13.5$ and $F = 0.56$ is shown in Fig. 15.6. Take 10 groups of vertical cross-sections from inside to outside of the generated edge waves. The result is shown in Fig. 15.7.

From the fractal point of view, the fractal dimension can depict the variation degree of the cross-section, that is, change of the section figure will cause the change of the fractal dimension, and the fractal dimension is corresponding to the figure of sections. Through adjusting parameters, fractal graphs of different cross-sections can be obtained. According to the box dimension method of a fractal, 10 cross-sections are taken to calculate their fractal dimension and box number values, as shown in Table 15.2.

In Table 15.2, $\ln(N)$ denotes the number of small squares with a length of 0.001 covering the edge of the cross-section of edge waves, and fractal dimension D describes the complexity of edge waves. Figure 15.7 corresponds to the first, third, sixth, and tenth groups of data in Table 15.2. These four cross-section graphs describe the increasing of fractal dimension and box number one by one. They also reflect the trend that the fractal dimension and box number of these 10 groups of edge waves gradually increase from edge section to deep section, and the complexity of edge waves increases from weak to strong.

(a) Cross-section 1 of the edge wave

(b) Cross-section 3 of the edge wave

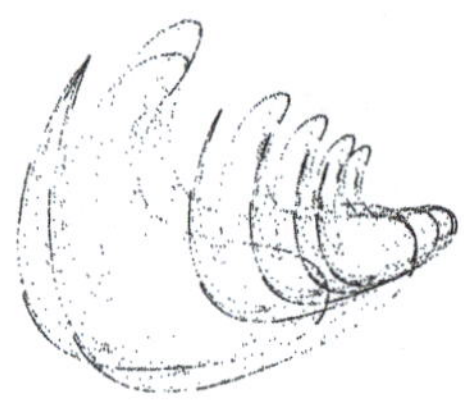

(c) Cross-section 6 of the edge wave

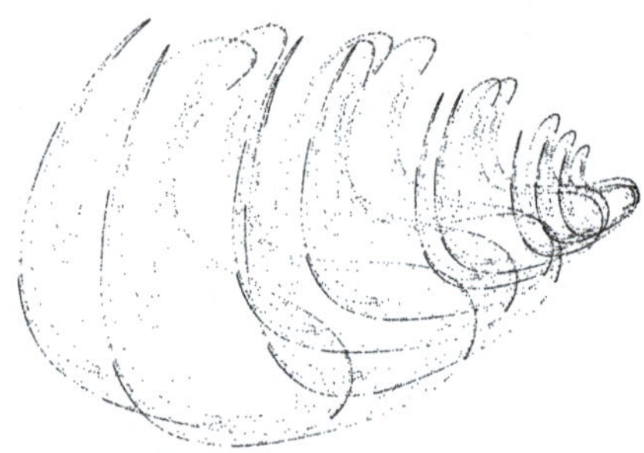

(d) Cross-section 10 of the edge wave

Fig. 15.7 Cross-section of the edge wave

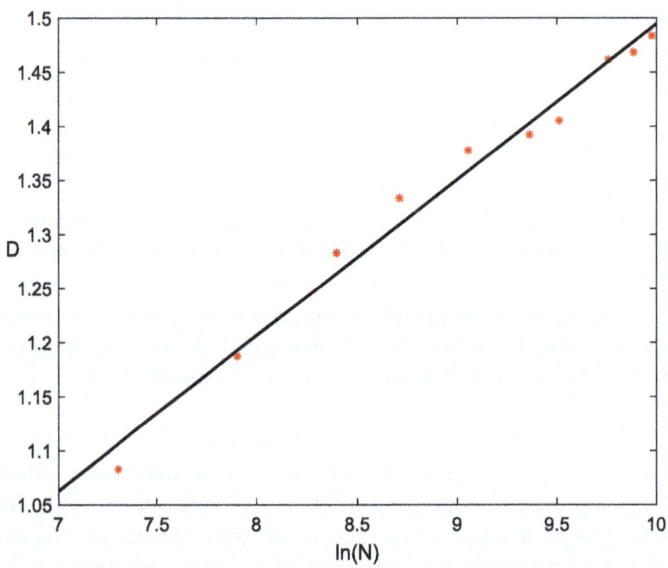

Fig. 15.8 The relation curve between the fractal dimension and the number of boxes of edge waves ("*": data points; "-": fitting curve)

The data in Table 15.2 are fitted by MATLAB and the numerical curve shown in Fig. 15.8 is obtained. From Fig. 15.8, it can be seen that the 10 cross-section data points of an edge wave in Table 15.2 are evenly distributed on the fitting curve. The fractal dimension of the edge wave is linearly and positively correlated with the number of boxes. The fractal dimension increases with the increase of the number of boxes and the spatial structure of edge waves. The relationship between the fractal dimension and the number of boxes of edge waves is expressed as follows:

$$D = \alpha \ln N + \beta,$$

where $\alpha = 0.144$ and $\beta = 0.0547$, it depicts the complexity of the internal structure of the edge waves.

15.4 New Problems Arising from Random Fractal Models of Edge Waves

In this chapter, the self-similarity theory, random fractal theory of nonlinear differential dynamic system, Poincaré cross-section, and iteration function are used to reveal the complex, random, and nonlinear behavior of seawater motion. The mathematical model of the edge wave motion is established fundamentally for the first time, and the basic characteristic information under this model is given. The behavior of edge waves is essentially a self-similar fractal structure, so we use fractal dimension to analyze the mathematical relationship of the complexity of the internal structure of edge waves, which makes up for the long-term vacancy in the study of the behavior law of edge waves. Five interesting open questions are given later, which is also one of our next research contents.

1. Calculating the fractal dimension of edge waves.
2. Prediction of edge waves from the fractal dimension of edge waves: (b) Kinetic energy; (c) Potential energy; (d) Relationship between basic the edge wave model and the edge wave spectrum; (e) other problems.
3. The relationship between the variation of the generated element ω_0 and the kinetic energy and potential energy.
4. Other marine information is contained in the generator ω_0.
5. Generator ω_0 and analysis of primary spectral characteristics.

At present, the study of edge waves is based on the combination of different amplitude and the superposition of different phases. Although formula (15.2.3) is also the superposition part, it is not the superposition of elements in edge waves. However, the finite iteration and finite superposition of random fractal curves are generated by generators ω_0, which are basically fractal structures.

References

1. Sverdrup H U. Wind-driven currents in a baroclinic ocean; with application to the equatorial currents of the eastern pacific. Proceedings of the National Academy of Sciences of the United States of America, 1947, 33(11): 318–326.
2. Mao H L. Progress of marine science in the past 20 years. Marine Sciences, 1979, 1: 1–9. (In Chinese)
3. Zheng Q A. Application of remote sensing in the study of marine mesoscale vortices and interpretation of satellite photos. Acta Oceanologica Sinica, 1979, 1(2): 342–357. (In Chinese)
4. Flierl G R. Isolated eddy models in geophysics. Annual Review of Fluid Mechanics, 1987, 19(1): 493–530.
5. Flierl G R. A simple model for the structure of warm and cold core rings. Journal of Geophysical Research: Oceans, 1979, 84(C2): 781–785.
6. Aref H, Siggia E D. Vortex dynamics of the two-dimensional turbulent shear layer. Journal of Fluid Mechanics, 1980, 100(4): 705–737.
7. Aref H. Integrable, chaotic, and turbulent vortex motion in two-dimensional flows. Annual Review of Fluid Mechanics, 1983, 15(1): 345–389.
8. Larichev V D. Two dimensional Rossby soliton: An exact solution. Rep. USSR Acad. Sci., 1976, 231: 1007–1079.
9. Berestov A L. Some new solutions for the Rossby solitons. Izvestiya Akademii Nauk Sssr Fizika Atmosfery I Okeana, 1981, 17(1): 82–87.
10. Berestov A L. Solitary Rossby waves. Izvestiya, Academy of Sciences, USSR, Atmospheric Oceanic Physics, 1979, 15(6): 648–654.
11. Boyd J P. Equatorial solitary waves. Part 3: westward-traveling modons. Journal of Physical Oceanography, 1985, 15(1): 46–54.
12. Killworth P D. On the motion of isolated lenses on a beta-plane. Journal of Physical Oceanography, 1983, 13(3): 368–376.
13. Killworth P D. On the propagation of isolated multilayer and continuously stratified eddies. Journal of Physical Oceanography, 1986, 16(4): 709–716.
14. Chelton D B, Schlax M G. Global observations of oceanic Rossby waves. Science, 1996, 272(5259): 234–238.
15. Fu L L, Chelton D B. Chapter 2 large-scale ocean circulation// International Geophysics. Elsevier Science & Technology, 2001: 133–169.

S.-T. Liu et al., *Mathematical Principle and Fractal Analysis of Mesoscale Eddy*,
https://doi.org/10.1007/978-981-16-1839-0

16. Zhang Z, Tian J, Qiu B, Zhao W, Chang P, Wu D, Wan X. Observed 3D structure, generation, and dissipation of oceanic mesoscale eddies in the south china sea. Scientific Reports, 2016, 6(1): 24349
17. Zhang Z, Zhang Y, Wang W. Three-compartment structure of subsurface-intensified mesoscale eddies in the ocean. Journal of Geophysical Research, 2017, 122(3): 1653–1664.
18. Pearson B, Foxkemper B, Bachman S, Bryan F. Evaluation of scale-aware subgrid mesoscale eddy models in a global eddy-rich model. Ocean Modelling, 2017: 42–58.
19. Roemmich D, Gilson J. Eddy transport of heat and thermocline waters in the north pacific: a key to interannual/decadal climate variability? Journal of Physical Oceanography, 2001, 31(3): 675–688.
20. Rhines P B. Mesoscale eddies. Encyclopedia of Ocean Sciences, 2001, 23: 755–767.
21. Chaigneau A, Gizolme A, Grados C. Mesoscale eddies off Peru in altimeter records: Identification algorithms and eddy spatio-temporal patterns. Progress in Oceanography, 2008, 79(2): 106–119.
22. Chelton D B, Gaube P, Schlax M, Early J J, Samelson, R M. The influence of nonlinear mesoscale eddies on near-surface oceanic chlorophyll. Science, 2011, 334(6054): 328–332.
23. Li C, Zhang Z, Zhao W Tian J, A statistical study on the subthermocline submesoscale eddies in the northwestern pacific ocean based on argo data. Journal of Geophysical Research: Oceans, 2017, 122(5): 3586–3598.
24. Zhang D X, Lee T N, Johns W E, Liu C T, Zantopp R. The kuroshio east of taiwan: modes of variability and relationship to interior ocean mesoscale eddies. Journal of Physical Oceanography, 2001, 31(4): 1054–1074.
25. Cheng X H, Qi Y Q. Distribution and propagation of mesoscale eddies in the global oceans learnt from altimetric data. Advances in Marine Science, 2008, 26(4): 447–453. (In Chinese)
26. Gao L, Liu Y G, Rong Z R. Sea level anomaly and mesoscale eddiesin the kuroshio extension region. Transactions of Oceanology and Limnology, 2007(1): 14–23. (In Chinese)
27. Chow C H, Liu Q. Eddy effects on sea surface temperature and sea surface wind in the continental slope region of the northern South China Sea. Geophysical Research Letters, 2012, 39(2): L02601.
28. Cheng X H, Qi Y Q, Wang W Q. Seasonal and interannual variabilities of mesoscale eddies in South China Sea. Journal of Tropical Oceanography, 2005, 24(4): 51–59. (In Chinese)
29. Wang G H, Su J L, Qi Y Q. Advances in studying mesoscale eddies in South China Sea . Advance in Earth Sciences, 2005, 20(8): 882–886. (In Chinese)
30. Lin P F, Wang F, Chen Y L, Tang X H. Temporal and spatial variation characteristics on eddies in the South China Sea I. Statistical analyses. Acta Oceanologica Sinica, 2007, 29(3): 14–22. (In Chinese)
31. Chen G X. Spatial and Temporal Characteristics of Mesoscale Eddies in the South China Sea. Qingdao: The Institute of Oceanology, Chinese Academy of Sciences, 2010. (In Chinese)
32. Chen G, Hou Y, Chu X. Mesoscale eddies in the South China Sea: Mean properties, spatiotemporal variability, and impact on thermohaline structure. Journal of Geophysical Research: Oceans, 2011, 116(C6): 102–108.
33. Kuo Y C, Chern C S, Zheng Z W. Numerical study on the interactions between the Kuroshio current in the Luzon Strait and a mesoscale eddy. Ocean Dynamics, 2017: 1–13.
34. Zhang Z G, Zhang Y, Wang W, Huang R X. Universal structure of mesoscale eddies in the ocean. Geophysical Research Letters, 2013, 40(14): 3677–3681.
35. Early J J, Samelson R M, Chelton D B. The evolution and propagation of quasigeostrophic ocean eddies. Journal of Physical Oceanography, 2011, 41(8): 1535–1555.
36. Benitez-Nelson C R, Bidigare R R, Dickey T D, Landry M R, Leonard C L, Brown S L, Yang E J. Mesoscale eddies drive increased silica export in the subtropical pacific ocean. Science, 2007, 316(5827): 1017–21.
37. Mcneil J D, Jannasch H W, Dickey T, Mcgillicuddy D J, Brzezinski M, Sakamoto C M. New chemical, bio-optical and physical observations of upper ocean response to the passage of a mesoscale eddy off Bermuda. Journal of Geophysical Research: Oceans, 1999, 104(7): 15537–15548.

38. Martin A P, Richards K J. Mechanisms for vertical nutrient transport within a North Atlantic mesoscale eddy. Deep-Sea Research Part II, 2001, 48(4): 757–773.
39. Chelton D B, Samelson R M, Schlax M G. The influence of nonlinear mesoscale eddies on near-surface oceanic chlorophyll. Science, 2011, 334(6054): 328–332.
40. Liu F, Tang S, Chen C. Impact of nonlinear mesoscale eddy on phytoplankton distribution in the northern South China Sea. Journal of Marine Systems, 2013, 123–124(4): 33–40.
41. Amores A, Monserrat S, Melnichenko O, V, Maximenko N. On the shape of sea level anomaly signal on periphery of mesoscale ocean eddies. Geophysical Research Letters, 2017, 44(13): 6926–6932.
42. Gao W, Wang Z, Zhang K. Controlling effects of mesoscale eddies on thermohaline structure and in situ, chlorophyll distribution in the western North Pacific. Journal of Marine Systems, 2017: 24–35.
43. Zhang Z G. Mesoscale Eddies. Qingdao: Ocean University of China, 2014. (In Chinese)
44. Zheng Q A, Xie L L, Zheng Z W, Hu J Y. Progress in research of mesoscale eddies in the South China Sea. Advances in Marine Science, 2017, 35(2): 131–158.
45. Anstey J A, Zanna L. A deformation-based parametrization of ocean mesoscale eddy reynolds stresses. Ocean Modelling, 2017, 112: 99–111.
46. Huang X, Zhang Z, Zhang X, Qian H, Zhao W, Tian J. Impacts of a mesoscale eddy pair on internal solitary waves in the northern South China Sea revealed by mooring array observations. Journal of Physical Oceanography, 2017, 47(7): 1539–1554.
47. Prants S V, Budyansky M V, Uleysky M Y. Lagrangian simulation and tracking of the mesoscale eddies contaminated by Fukushima-derived radionuclides. Ocean Science Discussions, 2017, 13(3): 1–13.
48. Mcdowell S E, Rossby H T. Mediterranean water: an intense mesoscale eddy off the bahamas. Science, 1978, 202(4372): 1084–1087.
49. Flierl G R, Stern M E, Jr J A W. The physical significance of modons: Laboratory experiments and general integral constraints. Dynamics of Atmospheres & Oceans, 1983, 7(4): 233–263.
50. Boyd J P. Equatorial solitary waves. Part 3: Westward-traveling modons. Journal of Physical Oceanography, 1985, 15(1): 46–54.
51. Killworth P D. On the propagation of isolated multilayer and continuously stratified eddies. Journal of Physical Oceanography, 1986, 16(4): 709–716.
52. Zhang Z G, Wang W, Qiu B. Oceanic mass transport by mesoscale eddies. Science, 2014, 345(6194): 322–325.
53. Zhang Z, Tian J, Qiu B, Chang P, Wu D. Observed 3D structure, generation, and dissipation of oceanic mesoscale eddies in the South China Sea. Scientific Reports, 2016, 6: 24349.
54. Shi Y, Yang D, Feng X, Qi, J, Yang H, Yin B. One possible mechanism for eddy distribution in zonal current with meridional shear. Scientific Reports, 2018, 8: 10106.
55. Mana P G P, Zanna L. Toward a stochastic parameterization of ocean mesoscale eddies. Ocean Modelling, 2014, 79(1): 1–20.
56. Anstey J A, Zanna L. A deformation-based parametrization of ocean mesoscale eddy reynolds stresses. Ocean Modelling, 2017, 112: 99–111.
57. Yang J. A statistical analysis of mesoscale eddies in the Bay of Bengal from 22–year altimetry data. Acta Oceanologica Sinica, 2016, 35(11): 16–27.
58. Ponomarev V I, Fayman P A, Prants S V, Budyansky M V, Uleysky M Y. Simulation of mesoscale circulation in the Tatar Strait of the Japan Sea. Ocean Modelling, 2018, 126: 43–55.
59. Kourafalou V H, Androulidakis Y S, Kang H, Smith R H, Valle-Levinson A. Physical connectivity between Pulley Ridge and Dry Tortugas coral reefs under the influence of the Loop Current/Florida Current system. Progress in Oceanography, 2018, 165: 75–99.
60. Jacobs G A, Hogan P J, Whitmer K R. Effects of eddy variability on the circulation of the Japan/East Sea. Journal of Oceanography, 1999, 55(2): 247–256.
61. Zhang D X, Lee T N, Johns W E, Liu C T, Zantopp R. The Kuroshio East of Taiwan: modes of variability and relationship to interior ocean mesoscale eddies. Journal of Physical Oceanography, 2001, 31(4): 1054–1074.

62. Qiu B, Chen S. Variability of the kuroshio extension jet, recirculation gyre, and mesoscale eddies on decadal time scales. Journal of Physical Oceanography, 2005, 35(11): 2090–2103.
63. Liu Q, Kaneko A, Su J. Recent progress in studies of the South China Sea circulation. Journal of Oceanography, 2008, 64(5): 753–762.
64. Lin X, Dong C, Chen D. Three-dimensional properties of mesoscale eddies in the South China Sea based on eddy-resolving model output. Deep Sea Research Part I Oceanographic Research Papers, 2015, 99: 46–64.
65. Huang X, Zhang Z, Zhang X, Qian H, Zhao W, Tian J. Impacts of a mesoscale eddy pair on internal solitary waves in the northern South China Sea revealed by mooring array observations. Journal of Physical Oceanography, 2017, 47(7): 1539–1554.
66. Zhao J, Bower A, Yang J, Lin X. Meridional heat transport variability induced by mesoscale processes in the subpolar North Atlantic. Nature Communications, 2018, 9(1): 1124.
67. Benitez-Nelson C R, Bidigare R R, Dickey T D. Mesoscale eddies drive increased silica export in the subtropical pacific ocean. Science, 2007, 316(5827): 1017–1021.
68. Holloway G. Estimation of oceanic eddy transports from satellite altimetry. Nature, 1986, 323(6085): 243–244.
69. Volkov D L, Tong L, Fu L. Eddy-induced meridional heat transport in the ocean. Geophysical Research Letters, 2008, 35(20): 295–296.
70. Beronvera F J, Olascoaga M J, Yan W, Trinanes J, Pérez-Brunius P. Enduring Lagrangian coherence of a Loop Current ring assessed using independent observations. Scientific Reports, 2018, 8(1): 529.
71. Laiolo L, Matear R, Baird M E. Information content of in situ and remotely sensed chlorophyll-a: Learning from size-structured phytoplankton model. Journal of Marine Systems, 2018, 183: 1–12.
72. Dengler M, Schott F A, Eden C, Brandt, P, Fischer J, Zantopp R J. Break-up of the Atlantic deep western boundary current into eddies at 8 degrees. Nature, 2004, 432(7020): 1018–1020.
73. Adams D K, Jr M G D, Zamudio L, Thurnherr A M, Liang X, Rouxel O, Mullineaux L S. Surface-generated mesoscale eddies transport deep-sea products from hydrothermal vents. Science, 2011, 332(6029): 580–593.
74. Siegel D A, Peterson P, Maritorena S, Nelson N B. Bio-optical footprints created by mesoscale eddies in the Sargasso Sea. Geophysical Research Letters, 2011, 38(13): 392–398.
75. Guo M, Chai F, Xiu P, Li S, Rao S. Impacts of mesoscale eddies in the South China Sea on biogeochemical cycles. Ocean Dynamics, 2015, 65(9–10): 1335–1352.
76. Gruber N, Lachkar Z, Frenzel H, Marchesiello P, Munnich M, Mcwilliams, J C, Plattner, G. Eddy-induced reduction of biological production in eastern boundary upwelling systems. Nature Geoscience, 2011, 4(11): 787–792.
77. Falkowski P G, Ziemann D, Kolber Z, Kolber Z, Bienfang P K. Role of eddy pumping in enhancing primary production in the ocean. Nature, 1991, 352(6330): 55–58.
78. Mcgillicuddy D J, Robinson A R, Siegel D A, Jannasch H W, Johnson R J, Dickey, T D, Knap A H. Influence of mesoscale eddies on new production in the Sargasso Sea. Nature, 1998, 394(6690): 263–266.
79. Williams R G. Oceanography: Ocean eddies and plankton blooms. Nature Geoscience, 2011, 4(11): 739–740.
80. Armi L, Hebert D, Oakey N S, Price F, Richardson L, Rossby H T, Rubbick B. The history and decay of a Mediterranean salt lens. Nature, 1988, 333(6174): 649–651.
81. Pedram H, Philip S, Marcus, Patrice L G. The universal aspect ratio of vortices in rotating stratified flows: Theory and simulation. Journal of Fluid Mechanics, 2012, 706: 46–57.
82. Barranco J A, Marcus P S. Three-dimensional vortices in stratified protoplanetary disks. The Astrophysical Journal, 2005, 623(2): 1157–1701.
83. Barranco J A, Marcus P S. A 3D spectral anelastic hydrodynamic code for shearing, stratified flows. Journal of Computational Physics, 2006, 219(1): 21–46.
84. Beckers, M. Dynamics of Vortices in a Stratified Fluid. Technische Universittndhoven, 1999.
85. Beckers M, Verzicco R, Clercx H J H. Dynamics of pancake-like vortices in a stratified fluid: Experiments, model and numerical simulations. Journal of Fluid Mechanics, 2001, 433: 1–27.

86. Allen J T, Smeed D A, Nurser A J G, Zhang J W, Rixen M. Diagnosis of vertical velocities with the QG omega equation: An examination of the errors due to sampling strategy. Deep Sea Research Part I: Oceanographic Research Papers, 2001, 48(2): 315–346.
87. Emerson S, Quay P D, Stump C, Wilbur D, Schudlich R.Chemical tracers of productivity and respiration in the subtropical pacific-ocean. Journal of Geophysical Research-Oceans, 1995, 100(C8): 15873–15887.
88. Ezer T. On the seasonal mixed layer simulated by a basin-scale ocean model and the Mellor-Yamada turbulence scheme. Journal of Geophysical Research, 2000, 105: 16843–16855.
89. Fasham M J R, Sarmiento J L, Slater R D, Ducklow, H W, Williams, R. Ecosystem behavior at bermuda station "s" and ocean weather station "India": a general circulation model and observational analysis. Global Biogeochem Cycles, 1993, 7: 379–415.
90. Aref H. Point vortex motions with a center of symmetry. Physics of Fluids, 1982, 25: 2183–2187.
91. Aref H. Integrable, chaotic, and turbulent vortex motion in two-dimensional flows. Annual Review of Fluid Mechanics, 1983, 15: 345–389.
92. Berloff P, Dewar W, Kravtsov S, Mcwilliams J. Ocean eddy dynamics in a coupled ocean-Atmosphere model. Journal of Physical Oceanography, 2007, 37(5): 1103–1121.
93. Beckers M M, Clercx H H, Heijst V G, Verzicco R. Evolution and instability of monopolar vortices in a stratified fluid. Physics of Fluids, 2003, 15(4): 1033–1045.
94. Beron-Vera F J, Wang Y, Olascoaga M J, Goni G J, Haller G. Objective detection of oceanic eddies and the Agulhas leakage. Journal of Physical Oceanography, 2013, 43(7): 1426–1438.
95. Babiano A, Provenzale A. Coherent vortices and tracer cascades in two-dimensional turbulence. Journal of Fluid Mechanics, 2007, 574: 429–448.
96. Barcilon V, Pedlosky J. On the steady motions produced by a stable stratification in a rapidly rotating fluid. Journal of Fluid Mechanics Digital Archive, 1967, 29(4): 673–690.
97. Bryden H L, Brady E C. Eddy momentum and heat fluxes and their effects on the circulation of the equatorial Pacific Ocean. Journal of Marine Research, 1989, 47(1): 55–79.
98. Capet X, Klein P, Hua B L, Lapeyre G, McWilliams J C. Surface kinetic and potential energy transfer in SQG dynamics. Journal of Fluid Mechanics, 604 (2008): 165–174.
99. Chelton D B, Schlax M G. The resolution capability of an irregularly sampled dataset: With application to geosat altimeter data. Journal of Atmospheric and Oceanic Technology, 1994, 11(2): 534–550.
100. Chelton D B, Schlax M G. Global observations of oceanic Rossby waves. Science, 1996, 272(5259): 234–238.
101. Chelton D B, Schlax M G. The accuracies of smoothed sea surface height fields constructed from tandem satellite altimeter datasets. Journal of Atmospheric and Oceanic Technology, 2003, 20(9): 1276–1302.
102. Chelton D B, Schlax M G, Samelson R M, De Szoeke R A. Global observations of large oceanic eddies. Geophysical Research Letters, 2007, 34(15): 87–101.
103. Capet, X, McWilliams J C, Molemaker M J, Shchepetkin A F. Mesoscale to submesoscale transition in the California Current System. Part I: Flow structure, eddy flux, and observational tests. Journal of Physical Oceanography 2008, 38(1): 29–43.
104. Bibby T S, Gorbunov M Y, Wyman K W, Falkowski P G. Photosynthetic community responses to upwelling in mesoscale eddies in the subtropical North Atlantic and Pacific Oceans. Deep-sea Research Part II-topical Studies in Oceanography, 2008, 55(10): 1310–1320.
105. Boyd J P. Equatorial solitary waves. Part 3: Westward-travelling modons. Journal of Physical Oceanography, 1985, 15(1): 46–54.
106. Bryden H L, Poleward heat flux and conversion of potential energy in Drake Passage. JMar. Res, 1979, 37: 1–22.
107. Capet X, McWilliams J C, Molemaker M J, Shchepetkin A F. Mesoscale to submesoscale transition in the Califomia Current System. Part II: Frontal processes. Journal of Physical Oceanography. 2008, 38 (1): 44–64.
108. Chaigneau A, Gizolme A, Grados C. Mesoscale eddies off Peru in altimeter records: Identification algorithms and eddy spatio-temporal patterns. Progress in Oceanography, 2008, 79(2–4): 106–119.

109. Chaigneau A, Gérard E, Dewitte B. Eddy activity in the four major upwelling systems from satellite altimetry (1992–2007). Progress in Oceanography, 2009, 83(1–4): 117–123.
110. Chaigneau A, Texier M L, Eldin G, Pizarro O. Vertical structure of mesoscale eddies in the eastern South Pacific Ocean: A composite analysis from altimetry and Argo profiling floats. Journal of Geophysical Research: Oceans, 2011, 116: C11025.
111. Carnevale G F, Kloosterziel R C, Van Heijst G J F. Propagation of barotropic vortices over topography in a rotating tank. Journal of Fluid Mechanics, 1991, 233: 119–139.
112. Claudia C. Laboratory experiments on mesoscale vortices colliding with a seamount. Journal of Geophysical Research: Oceans, 2002, C107(C6): 3053.
113. Chaigneau A, Pizarro O. Eddy characteristics in the eastern South Pacific. Journal of Geophysical Research, 2005, 110: C06005.
114. Chomaz J M, Ortiz S, Gallaire F, Billant P. Stability of quasi-two-dimensional vortices. Lecture Notes in Physics, 2010, 805: 35–59.
115. Cianca A, Helmke P, Mourino B, Rueda M J, Llinas O, Neuer S. Decadal analysis of hydrography and in situ nutrient budgets in the western and eastern North Atlantic subtropical gyre. Journal of Geophysical Research, 2007, 112(7). C07025.
116. Chelton D B, Schlax M G, Samelson R M. Global observations of large oceanic eddies. Geophysical Research Letters, 2007, 34(15): 87–101.
117. Danabasoglu G, Mcwilliams J C, Gent P R. The role of mesoscale tracer transports in the global ocean circulation. Science, 1994, 264(5162): 1123–126.
118. Dritschel, David G. On the persistence of non-axisymmetric vortices in inviscid two-dimensional flows. Journal of Fluid Mechanics, 1998, 371: 141–155.
119. Flierl, G R. A simple model for the structure of warm and cold core rings. Journal of Geophysical Research: Oceans, 1979, 84(C2): 781–785.
120. Flierl G R. Particle motions in large-amplitude wave fields. Geophysical and Astrophysical Fluid Dynamics, 1981, 18: 39–74.
121. Flierl G R, Stern M E, Whitehead J. The physical significance of modons: Laboratory experiments and general integral constraints. Dynamics of Atmospheres and Oceans, 1983, 7(4): 233–263.
122. Flierl G R. Isolated Eddy Models in Geophysics. Annual Review of Fluid Mechanics, 1987, 19(1): 493–530.
123. Flierl G R. On the instability of geostrophic vortices. Journal of Fluid Mechanics, 1988, 197(1): 349–388.
124. Flór, J B. Coherent Vortex Structures in Stratified Fluids. Eindhoven University of Technology, 1994.
125. Fu L, Qiu B. Low-frequency variability of the North Pacific Ocean: The roles of boundary- and wind-driven baroclinic Rossby waves. Journal of Geophysical Research, 2002, 107(C12): 3220.
126. Danioux E, Klein P, Riviere P. Propagation of wind energy into the deep ocean through a fully turbulent mesoscale eddy field. Journal of Physical Oceanography, 2008, 38(10): 2224–2241.
127. Davey M K, Killworth P D. Isolated waves and eddies in a shallow water model. Journal of Physical Oceanography, 1984, 14(6): 1047–1064.
128. Morel Y, Mcwilliams J C. Evolution of isolated interior vortices in the ocean. Journal of Physical Oceanography, 1997, 27(5): 727–748.
129. Morrow R, Birol F, Griffin D, Sudre J. Divergent pathways of cyclonic and anti-cyclonic ocean eddies. Geophysical Research Letters, 2004, 31(24): L24311.
130. Mourio, B, Fernández E, Etienne H, Hernández F, Giraud S. Significance of cyclonic Sub-Tropical Oceanic Rings of Magnitude (STORM) eddies for the carbon budget of the euphotic layer in the subtropical northeast Atlantic. Journal of Geophysical Research: Oceans, 2003, 108(C12): 3383.
131. De Jong M F, Bower A S, Furey H H. Two years of observations of warm-core anticyclones in the labrador sea and their seasonal cycle in heat and salt stratification. Journal of Physical Oceanography, 2014, 44(2): 427–444.

132. Pater I D, Wong M H, Marcus P, Luszcz-Cook S, Conrad A. Persistent rings in and around jupiter's anticyclones-observations and theory. Icarus, 2010, 210(2): 742–762.
133. Dritschel D G, Saravanan R. Three-dimensional quasi-geostrophic contour dynamics, with an application to stratospheric vortex dynamics. Quarterly Journal of the Royal Meteorological Society, 120(519): 1267–1297.
134. Véronique C. G, Oschlies A, Doney S C. The role of mesoscale variability on plankton dynamics in the North Atlantic. Deep-Sea Research II, 2001, 48(10): 2199–2226.
135. Gent P R, Willebrand J, McDougall T J, McWilliams J C. Parameterizing eddy-induced tracer transports in ocean circulation models. Journal of Physical Oceanography, 1995, 25(6): 463–474.
136. Adrian G. Atmosphere-Ocean Dynamics. Academic Press, 1982.
137. Gill A E, Green J S A, Simmons A J. Energy partition in the large-scale ocean circulation and the production of mid-ocean eddies. Deep Sea Research & Oceanographic Abstracts, 1974, 21(7): 499–528.
138. Griffiths R W. The stability of vortices in a rotating stratified fluid. Journal of Fluid Mechanics, 1981, 105: 283–306.
139. Holland, William R. The role of mesoscale eddies in the general circulation of the ocean-numerical experiments using a wind-driven quasi-geostrophic model. Journal of Physical Oceanography, 1978, 8(3): 363–392.
140. James Holton. An Introduction to Dynamic Meteorology. Academic Press, San Diego, 1992.
141. Hopfinger E J, Van Heijst G J. Vortices in rotating fluids. Annual Review of Fluid Mechanics, 1993, 25(1): 241–289.
142. Hu J Y, Gan J P, Sun Z Y, Dai M. Observed three-dimensional structure of a cold eddy in the southwestern South China Sea. Journal of Geophysical Research: Oceans, 2011, 116, C05016.
143. Humphreys T, Marcus P. Vortex street dynamics: The selection mechanism for the areas and locations of Jupiter's vortices. Journal of the Atmospheric Sciences, 2007, 64(4): 1318–1333.
144. Isernfontanet J, Garcialadona E, Font J. Identification of marine eddies from altimetric maps. Journal of Atmospheric and Oceanic Technology, 2003, 20(5): 772–778.
145. Isernfontanet J, Garcialadona E, Font J. Vortices of the Mediterranean Sea: An Altimetric Perspective. Journal of Physical Oceanography, 2006, 36(1): 87–103.
146. Isern-Fontanet J, Lapeyre G, Klein P, Chapron, B, Hecht M W. Three-dimensional reconstruction of oceanic mesoscale currents from surface information. Journal of Geophysical Research, 2008.
147. Jayne S R, Marotzke J. The oceanic eddy heat transport. Journal of Physical Oceanography, 2002, 32(12): 3328–3345.
148. Jenkins W J. Nitrate flux into the euphotic zone near Bermuda. Nature, 1988, 331(6156): 521–523.
149. Johannessen A, Svendsen E, Sandven S, Johannessen M, Lygre K. Three-dimensional structure of mesoscale eddies in the norwegian coastal current. Journal of Physical Oceanography, 1989, 19: 3–19.
150. Johnson K S, Riser S C, Karl D M. Nitrate supply from deep to near-surface waters of the North Pacific subtropical gyre. Nature, 2010, 465(7301): 1062–1065.
151. Kantha L H, Clayson C A. An improved mixed layer model for geophysical applications. Journal of Geophysical Research, 1994, 99(C12): 25235–25266.
152. Killworth P D. On the motion of isolated lenses on a beta-plane. Journal of Physical Oceanography, 1983, 13(3): 368–376.
153. Killworth P D. On the propagation of isolated multilayer and continuously stratified eddies. Journal of Physical Oceanography, 1986, 16(4): 709–716.
154. Killworth P D, Chelton D B, De Szoeke R A. The speed of observed and theoretical long extratropical planetary waves. Journal of Physical Oceanography, 1997, 27(9): 1946–1966.
155. Killworth P D, Blundell J R. The dispersion relation for planetary waves in the presence of mean flow and topography. Part I: analytical theory and one-dimensional examples. Journal of Physical Oceanography, 2004, 34(12): 2692–2711.

156. Killworth P D, Blundell J R. The dispersion relation for planetary waves in the presence of mean flow and topography. Part II: Two-dimensional examples and global results. Journal of Physical Oceanography, 2005, 35(11): 2110–2133.
157. Killworth P D, Blundell J R. Planetary wave response to surface forcing and instability in the presence of mean flow and topography. Journal of Physical Oceanography, 2007, 37(5): 1297–1320.
158. King F D, Devol A H. Estimates of vertical eddy diffusion through the thermocline from phytoplankton nitrate uptake rates in the mixed layer of the eastern tropical Pacific. Limnology & Oceanography, 1979, 24(4): 645–651.
159. Klein P, Lapeyre G. The oceanic vertical pump induced by mesoscale and submesoscale turbulence. Annual Review of Marine Science, 2009, 1(1): 351–375.
160. Kloosterziel R C. On the large-time asymptotics of the diffusion equation on infinite domains. Journal of Engineering Mathematics, 1990, 24(3): 213–236.
161. Kloosterziel R C, vanHeijst G J. An experimental study of unstable barotropic vortices in a rotating fluid. Journal of Fluid Mechanics, 1991, 223: 1–24.
162. Kloosterziel R C, Carnevale G F. On the evolution and saturation of instabilities of two-dimensional isolated circular vortices. Journal of Fluid Mechanics, 1999, 388: 217–257.
163. Koszalka I, Bracco A, Mcwilliams J C, Provenzale A. Dynamics of wind-forced coherent anticyclones in the open ocean. Journal of Geophysical Research: Oceans, 2009, 114, C08011.
164. Koszalka I, Ceballos L, Bracco A. Vertical mixing and coherent anticyclones in the ocean: The role of stratification. Nonlinear Processes in Geophysics, 2010, 17(1): 37–47.
165. Lee M M, Marshall D P, Williams R G. On the eddy transfer of tracers: Advective of diffusive? Journal of Marine Research, 1997, 55(3): 483–505.
166. Ledwell J R, Bratkovich A. A tracer study of mixing in the Santa Cruz Basin. Journal of Geophysical Research, 1995, 100(C10): 20681.
167. Ledwell J R, Watson A J, Law C S. Mixing of a tracer in the pycnocline. Journal of Geophysical Research: Oceans, 1998, 103(C10): 21499–21529.
168. Ledwell J R, Mcgillicuddy D J, Anderson L A. Nutrient flux into an intense deep chlorophyll layer in a mode-water eddy. Deep-sea Research Part II-topical Studies in Oceanography, 2008, 55(10): 1139–1160.
169. Levy M, Ferrari R, Franks P, Martin A P, Riviere P. Bringing physics to life at the submesoscale. Geophysical Research Letters, 2012, 39(14): L14602.
170. Lu Z M, Huang R X. The three-dimensional steady circulation in a homogenous ocean induced by a stationary hurricane. Journal of Physical Oceanography, 2010, 40(7): 1441–1457.
171. Luyten J, Pedlosky R J, Stommel H M. The ventilated thermocline. Journal of Physical Oceanography. 1983, 13: 292–309.
172. Maas L R M. Nonlinear and free-surface effects on the spin-down of barotropic axisymmetric vortices. Journal of Fluid Mechanics, 1993, 246(1): 117–141.
173. Mahadevan A, Archer D. Modeling the impact of fronts and mesoscale circulation on the nutrient supply and biogeochemistry of the upper ocean. Journal of Geophysical Research, 2000: 1209–1225.
174. Mahadevan A, Campbell J. Biogeochemical patchiness at the sea surface. Geophysical Research Letters, 2002, 29(19): 321–324.
175. Marcus P S. Jupiter's Great Red Spot and other vortices. Annual Review of Astronomy & Astrophysics, 1993, 31: 523–573.
176. Mcgillicuddy D J, Robinson A R, Siegel D, Jannasch H W, Johnson R J, Dickey T D, Knap A H. Influence of mesoscale eddies on new production in the Sargasso Sea. Nature, 1998, 394(6690): 263–266.
177. Mcgillicuddy D J, Anderson L A, Doney S C, Maltrud M E. Eddy-driven sources and sinks of nutrients in the upper ocean: Results from a 0.1° resolution model of the North Atlantic. Global Biogeochemical Cycles, 2003, 17(2): 1035.
178. McGillicuddy D J, Anderson L A, Bates N R, Bibby T, Buesseler KO, Carlson C A, Davis C S, Ewart C, Falkowski P G, Goldthwait S A, Hansell D A, Jenkins W J, Johnson R, Kosnyrev V K, Ledwell J R, Li Q P, Siegel D A. Steinberg D K. Eddy/wind interactions stimulate extraordinary mid-ocean plankton blooms. Science, 2007, 316 (5827): 1021–1026.

179. McWilliams J C, Brown E D, Bryden H L, Ebbesmeyer C C, Elioiot B A, Heimmiller R H, Lien H B, Leaman K D, Lindtrom E J, Luyten J R, McDowell S E, Owens W B, Perkins H, Price J F, Regier L, Riser S C, Rossby H T, Sanford T B, Shen C Y, Taft B A, Van Leer J C. Eddies in Marine Science. Springer-Verlag, Berlin, 1983.
180. Mcwilliams J C. The emergence of isolated coherent vortices in turbulent flow. Journal of Fluid Mechanics, 1984, 146, 21–43.
181. Mcwilliams J C. Submesoscale, coherent vortices in the ocean. Reviews of Geophysics, 1985, 23(2), 165–182.
182. Mcwilliams, James C. The vortices of two-dimensional turbulence. Journal of Fluid Mechanics, 1990, 219(1): 361–385.
183. Kunze E, Firing E, Hummon J M, Chereskin T K, Thurnherr A M. Global abyssal mixing inferred from lowered ADCP shear and CTD strain profiles. Journal of Physical Oceanography, 2006, 36(8): 1553–1576.
184. Lapeyre G, Klein, P. Dynamics of the upper oceanic layers in terms of surface quasigeostrophy theory. Journal of Physical Oceanography, 2008, 36(2): 165–176.
185. Lapeyre G. What vertical mode does the altimeter reflect? On the decomposition in baroclinic modes and on a surface-trapped mode. Journal of Physical Oceanography, 2009, 39(11): 2857–2874.
186. Law C S, Abraham E R, Watson A J, Liddicoat M I. Vertical eddy diffusion and nutrient supply to the surface mixed layer of the Antarctic Circumpolar Current. Journal of Geophysical Research: Oceans, 2003, 108(C8): 3272.
187. Mcwilliams J C, Graves L P, Montgomery M T. A formal theory for vortex Rossby waves and vortex evolution. Geophysical and Astrophysical Fluid Dynamics, 2003, 97(4): 275–309.
188. Mcgillicuddy D J, Robinson A R, Siegel D, Jannasch H W, Johnson R J, Dickey T D, Knap A H. Influence of mesoscale eddies on new production in the Sargasso Sea. Nature, 1998, 394(6690): 263–266.
189. Mellor, George L. One-dimensional, ocean surface layer modeling: A problem and a solution. Journal of Physical Oceanography, 2010, 31: 790–809.
190. Nof D. On the beta-induced movement of isolated baroclinic eddies. Journal of Physical Oceanography, 1981, 11 (12): 1662–1672.
191. Zhang Z G, Wang W, Qiu B. Oceanic mass transport by mesoscale eddies. Science, 2014, 345(6194): 322–325.
192. Ye Y Q. Limit Cycle Theory. Shanghai Scientific & Technical Publishers, 1965. (In Chinese)
193. Zhang Z F, Ding T R, Huang W Z. Qualitative theory of differential equations. Science Press, Beijing, 1985. (In Chinese)
194. Qin Y X. Semi stable limit cycles with monotone approximation. Acta Mathematica Sinica, 1954, 4(4): 445–465. (In Chinese)
195. Mandelbrot B B. Fractals: Form, Chance, and Dimension. W. H. Freeman, 1977.
196. Mandelbrot B B. The Fractal Geometry of Nature. Freeman, 1982.
197. Mandelbrot B B. Renewal sets and random cutouts. Zeitschrift Für Wahrscheinlichkeitstheorie Und Verwandte Gebiete, 1972, 22(2): 145–157.
198. Mandelbrot B B. Intermittent turbulence in self-similar cascades: divergence of high moments and dimension of the carrier. Journal of Fluid Mechanics, 1974, 62: 331–358.
199. Mandelbrot B B, Van Ness J W. Fractional Brownian motion, fractional noises and applications. SIAM Review, 1968, 10(4): 422–436.
200. Benoit B. Mandelbrot. Fractal aspects of the iteration of $z \leftarrow \lambda z(1-z)$ for complex λ and z. Annals of the New York Academy of Sciences, 1980, 357: 249–259.
201. Mandelbrot B B. Random Multifractals: Negative dimensions and the resulting limitations of the thermodynamic formalism. Proceedings of the Royal Society A Mathematical Physical & Engineering Sciences, 1991, 434(1890): 79–88.
202. Mandelbrot B B. Fractals and Scaling in Finance: Discontinuity, Concentration, Risk. Springer-Verlag, New York, 1997.
203. Benoit B. Mandelbrot. Multifractals and 1/f Noise: Wild Self-Affinity in Physics. Springer Verlag, New York, 1999.

204. Mandelbrot B B. Gaussian Self-Affinity and Fractals: Globality, the Earth, 1/F Noise, and R/S. Springer Science & Business Media, 2002.
205. Riedi R H, Mandelbrot B B. Multifractal formalism for infinite multinomial measures. Advances in Applied Mathematics, 1995, 16(2): 132–150.
206. Marstrand J M. Some fundamental geometrical properties of plane sets of fractional dimensions. Proceedings of the London Mathematical Society, 1954, (1): 257–302.
207. Marstrand, J. M. The dimension of Cartesian product sets. Mathematical Proceedings of the Cambridge Philosophical Society, 1954, 50(2): 198–202.
208. Aguiar M A M D, Baranger M, Baptestini E M, Kaufman L, Bar-Yam Y. Global patterns of speciation and diversity. Nature, 2009, 460(7253): 384–387.
209. Barth J, Costantini G, Kern K. Engineering atomic and molecular nanostructures at surfaces. Nature, 2005, 437(7059): 671–679.
210. Barthelemy P, Bertolotti J, Wiersma D S. A Lévy flight for light. Nature, 2008, 453(7194): 495–498.
211. Batty M. The size, scale, and shape of cities. Science, 2008, 319(5864): 769–771.
212. Bell T W. Molecular trees: A new branch of chemistry. Science, 1996, 271(5252): 1077–1078.
213. Diemand J, Kuhlen M, Madau P, Moore B, Potter D. Clumps and streams in the local dark matter distribution. Nature, 2008, 454(7205): 735–738.
214. Falconer K. Fractal Geometry: Mathematical Foundations and Applications. John Wiley & Sons, 2004.
215. Fatou P. Sur les équations fonctionnelles. Bulletin de la Société mathématique de France, 47 (1919): 161–271.
216. Haskell J P, Ritchie M. Fractal geometry predicts varying body size scaling relationships for mammal and bird home ranges. Nature, 2002, 418: 527–530.
217. Hermanson K D, Lumsdon S O, Williams J P, Kaler E W, Velev O O. Dielectrophoretic assembly of electrically functional microwires from nanoparticle suspensions. Science, 2001, 294(5544): 1082–1086.
218. Hillier A C, Ward M D. Atomic force microscopy of the electrochemical nucleation and growth of molecular crystals. Science, 1994, 263: 1261–1264.
219. Julia G. Mémoire sur l'itération des fonctions rationnelles. Journal des Mathématiques Pures et Appliquées, 1918, 9(3): 47–245.
220. Meakin P. Models for material failure and deformation. Science, 1991, 252(5003): 226–234.
221. Metzger R J, Krasnow M A. Genetic control of branching morphogenesis. Science, 1999, 284(5420): 1635–1639.
222. Montgomery D R, Dietrich W E. Channel initiation and the problem of landscape scale. Science, 1992, 255(5046): 826–830.
223. Nowak M A, Sigmund K. Evolutionary dynamics of biological games. Science, 2004, 303(5659): 793–799.
224. Perron J T, Kirchner J W, Dietrich W E. Formation of evenly spaced ridges and valleys. Nature, 2009, 460(7254): 502–505.
225. Ridderinkhof H, Zimmerman J T. Chaotic stirring in a tidal system. Science, 1992, 258(5085): 1107–1111.
226. Robinson A L. Fractal fingers in viscuous fluids. Science, 228 (1985): 1077–1081.
227. Schaefer D W. Polymers, fractals, and ceramic materials. Science, 1989, 243(4894): 1023–1027.
228. Sommerer J C, Ott E. Particles Floating on a moving fluid: a dynamically comprehensible physical fractal. Science, 1993, 259(5093): 335–339.
229. Stolum H H. River meandering as a self-organization process. Science, 1996, 271: 1710–1713.
230. Tegmark M. Measuring spacetime: From the big bang to black holes. Science, 2002, 296: 1427.
231. Zaanen J. High-temperature superconductivity: The benefit of fractal dirt. Nature, 2010, 466(7308): 825–827.
232. Zhang Y P, Liu S T. Gradient control and synchronization of Julia sets. Chinese Physics B, 2008, 17(2): 543–549.

233. Zhang Y, Liu S, Sun W. Feedback control in fractal mechanics. Chaos Solitons & Fractals, 2009, 40(1): 337–343.
234. Zhang Y P, Liu S T, Shen S L. Fractals control in particle's velocity. Chaos Solitons & Fractals, 2009, 39(4): 1811–1816.
235. Mandelbrot B B. How long is the coast of Britain? Statistical self-similarity and fractional dimension. Science, 1967, 156(3775): 636–638.
236. Burrough P A. Fractal dimensions of landscapes and other environmental data. Nature, 1981, 294(5838): 240–242.
237. Mandelbrot B B, Passoja D E, Paullay A J. Fractal character of fracture surfaces of metals. Nature, 1984, 308(5961): 721–722.
238. Avnir D, Farin D, Pfeifer P. Molecular fractal surfaces. Nature, 1984, 308(5956): 261–263.
239. Lewis M, Rees D C. Fractal surfaces of proteins. Science, 1985, 230(4730): 1163–1165.
240. Nittmann J, Daccord G, Stanley H E. Fractal growth viscous fingers: quantitative characterization of a fluid instability phenomenon. Nature, 1985, 314(6007): 141–144.
241. Sander L M. Fractal growth processes. Nature, 1986, 322(322): 789–793.
242. Tsonis A A, Elsner J B. Nonlinear prediction as a way of distinguishing chaos from random fractal sequences. Nature, 1992, 358(6383): 217–220.
243. Avnir D, Biham O, Lidar D, Malcai O. Is the geometry of nature fractal? Science, 1998, 279(5347): 39–40.
244. Tsonis A A, Biham O, Malcai O, Avnir D. Fractality in nature. Science, 1998, 279(5357): 1614–1616.
245. Jarvis P, Jefferson B, Parsons S A. Breakage, regrowth, and fractal nature of natural organic matter flocs. Environmental Science & Technology, 2005, 39(7): 2307–2314.
246. Mandelbrot B B, Aizenman M. Fractals: Form, chance, and dimension. Leonardo, 1979, 32(5): 65–66.
247. Chengke Z. SIRS epidemic model under real complex circumstances and its control strategy. Computer Simulation, 2010, 22(1): 195–200.
248. Misra A K, Mishra S, Pathak A L, Srivastava P, Chandra P. A mathematical model for the control of carrier-dependent infectious diseases with direct transmission and time delay. Chaos Solitons & Fractals, 2013: 41–53.
249. Chen L, Sun J. Global stability and optimal control of an SIRS epidemic model on heterogeneous networks. Physica A: Statistical Mechanics and Its Applications, 2014: 196–204.
250. Hu Z, Teng Z, Zhang L. Stability and bifurcation analysis in a discrete SIR epidemic model. Mathematics and Computers in Simulation, 2014, 97: 80–93.
251. Guang P, Jing W. Design of an adaptive sliding mode controller for synchronization of fractional-order chaotic systems. Acta Physica Sinica, 2015, 64(4): 40505.
252. Rao P R, Kumar M N. A dynamic model for infectious diseases: The role of vaccination and treatment. Chaos Solitons & Fractals, 2015, 75: 34–49.
253. Figueiredo T C, Vivas J, Pena N, Miranda J G. Fractal measures of video-recorded trajectories can classify motor subtypes in Parkinson's disease. Physica A: Statistical Mechanics and Its Applications, 2016, 462: 12–20.
254. Li Z, Zhang Y P, Liu J. Fractal dimension of the controlled Julia sets of the output duopoly competing evolution model. Journal of Mathematics and Computer Science-JMCS, 2016, 16(3): 461–471.
255. Zhang N, Liu C, Yang P. Flow of top coal and roof rock and loss of top coal in fully mechanized top coal caving mining of extra thick coal seams. Arabian Journal of Geosciences, 2016, 9(6): 465.
256. Gao W, Yan L, Saeedi M, Nik H S. Ultimate bound estimation set and chaos synchronization for a financial risk system. Mathematics and Computers in Simulation, 2018, 154: 19–33.
257. Zirkohi MM, Khorashadizadeh S. Chaos synchronization using higher-order adaptive PID controller. AEU-International Journal of Electronics and Communications, 2018, 94: 157–167.
258. Metzler R, Klafter J. The random walk's s guide to anomalous diffusion: A fractional dynamics approach. Physics Reports, 2000, 339(1): 1–77.

259. Dzielinski A, Sierociuk D. Stability of discrete fractional order state-space systems. Journal of Vibration & Control, 2008, 14(9–10): 1543–1556.
260. Ahmed E, El-Saka H A. On fractional order models for Hepatitis C. Nonlinear Biomedical Physics, 2010, 4(1): 1–4.
261. Guermah S, Djennoune S, Bettayeb M. A New Approach for Stability Analysis of Linear Discrete-Time Fractional-Order Systems. Springer Netherlands, 2010.
262. Matallana L G, Aníbal M. Blanco, Bandoni JA. Nonlinear dynamic systems design based on the optimization of the domain of attraction. Mathematical & Computer Modelling, 2011, 53(5–6): 731–745.
263. Ding Y, Wang Z, Ye H. Optimal control of a Fractional-Order HIV-Immune system with memory. IEEE Transactions on Control Systems Technology, 20(3): 763–769.
264. Yin C, Dadras S, Zhong S M, Chen Y Q. Control of a novel class of fractional-order chaotic systems via adaptive sliding mode control approach. Applied Mathematical Modelling, 2013, 37(4): 2469–2483.
265. Muthukumar P, Balasubramaniam P. Feedback synchronization of the fractional order reverse butterfly-shaped chaotic system and its application to digital cryptography. Nonlinear Dynamics, 2013, 74(4): 1169–1181.
266. Tang Y G, Zhang X Y, Zhang D L, Zhao G, Guan X P. Fractional order sliding mode controller design for antilock braking systems. Neurocomputing, 2013, 111: 122–130.
267. Wu G C, Baleanu D. Discrete fractional logistic map and its chaos. Nonlinear Dynamics, 2014, 75(1–2): 283–287.
268. Xi H, Li Y, Huang X. Generation and nonlinear dynamical analyses of fractional-order memristor-based Lorenz systems. Entropy, 2014, 16(12): 6240–6253.
269. He J H. A tutorial review on fractal spacetime and fractional calculus. International Journal of Theoretical Physics, 2014, 53(11): 3698–3718.
270. Machado J A. Numerical analysis of the initial conditions in fractional systems. Communications in Nonlinear Science and Numerical Simulation, 2014, 19(9): 2935–2941.
271. Xu B, Chen D, Zhang H, Zhou R. Dynamic analysis and modeling of a novel fractional-order hydro-turbine-generator unit. Nonlinear Dynamics, 2015, 81(3): 1263–1274.
272. Area I, Batarfi H, Losada J, Nieto J J, Shammakh W, Torres A. On a fractional order Ebola epidemic model. Advances in Difference Equations, 2015, 278(1): 1–12.
273. Goufo D, Franc E. A biomathematical view on the fractional dynamics of cellulose degradation. Fractional Calculus & Applied Analysis, 2015, 18(3): 554–564.
274. Witten T A, Sander L M. Diffusion-limited aggregation, a kinetic critical phenomenon. Physical Review Letters, 1981, 47(19): 1400–1403.
275. Argoul F, Arneodo A, Grasseau G, Swinney H L. Self-similarity of diffusion-limited aggregates and electrodeposition clusters. Physical Review Letters, 1988, 61(22): 2558–2561.
276. Wang L, Zheng D Z. A kind of chaotic neural network optimization algorithm based on annealing strategy. Control Theory and Applications, 2000, 17(1): 139–142.
277. Yu S M, Lin Q H, Qiu S S. A family of multiple-folded torus chaotic attractors. Acta Physica Sinica, 2004, 53(7): 2084–2088.
278. Evans P E, Pemble A M E, Sheel D W. Precursor-directed control of crystalline type in atmospheric pressure CVD growth of TiO2 on stainless steel. Chemistry of Materials, 2006, 18(24): 5750–5755.
279. He Q J, Huang Z L. Controlled growth and kinetics of porous hydroxyapatite spheres by a template-directed method. Journal of Crystal Growth, 2007, 300(2): 460–466.
280. Kolesnikov N N, Borisenko E B, Borisenko D N, Bozhko S I. Fractal structures of dendrites in GaSe crystals. Journal of Crystal Growth, 2008, 310(14): 3287–3289.
281. Yu S, Yu Z. A novel fifth-order hyperchaotic circuit and its research. Acta Physica Sinica, 2008, 57(11): 6859–6867.
282. Wang X Y, Wang Y X, Yun J J. An improved fast fractal image compression using spatial texture correlation. Chinese Physics B, 2011, 20(10): 104202.
283. Dias, Eduardo O. Viscous-fingering minimization in uniform three-dimensional porous media. Physical Review E, 2013, 88(6): 1–5.

284. Katyal N, Banerjee V, Puri S. Fractal signatures analogs of interplanetary dust particles. Journal of Quantitative Spectroscopy and Radiative Transfer, 2014, 146: 290–294.
285. Pecora L M, Carroll T L. Synchronization in chaotic systems. Controlling Chaos, 1996, 6(8): 142–145.
286. Fu J C, Jing Y W, Zhang Z H, Zhang S Y. The Study on the SIRS Epidemic Models with Vertical Infection and Continuous Vaccination. Journal of Biomathematics, 2008, 23(2): 273–278.
287. Yin H X, Liang W. Transmission approaches for services of variable bit-rate and differentiated QoS using OCDMA. IEEE/LEOS Summer Topical Meeting, 2009: 11–12.
288. King R D, Brown B, Hwang M, Jeon T, George A T. Fractal dimension analysis of the cortical ribbon in mild Alzheimer's disease. Neuroimage, 2010, 53(2): 471–479.
289. Melesse D Y, Gumel A B. Global asymptotic properties of an SEIRS model with multiple infectious stages. Journal of Mathematical Analysis and Application 2010, 366(1): 202–217.
290. Huo J J, Zhao H Y, Zhu L H. The effect of vaccines on backward bifurcation in a fractional order HIV model. Nonlinear Analysis-real World Applications, 2015: 289–305.
291. Atangana A, Koca I. Chaos in a simple nonlinear system with Atangana-Baleanu derivatives with fractional order. Chaos Solitons & Fractals, 2016, 89: 447–454.
292. Atangana A, Baleanu D. New fractional derivatives with nonlocal and non-singular kernel: theory and application to heat transfer model. Thermal Science, 2016, 20(2): 763–769.
293. Deng Y Q, Liu X X, Zhang Y P. Fractal dimension analysis of the Julia sets of controlled Brusselator model. Discrete Dynamics in Nature and Society, 2016, 2016: 1–13.
294. Rădulescu A, Pignatelli A. Symbolic template iterations of complex quadratic maps. Nonlinear Dynamics, 2016, 84(4): 2025–2042.
295. Yang X J, Machado J A, Cattani C, Gao F. On a fractal LC-electric circuit modeled by local fractional calculus. Communications in Nonlinear Science and Numerical Simulation, 2017, 47: 200–206.
296. Abdelaziz M A, Ismail A I, Abdullah F A, Mohd M H. Bifurcations and chaos in a discrete SI epidemic model with fractional order. Advances in Difference Equations, 2018, 2018(1): 1–19.
297. Lefever R, Turner J W. Dissipative structures in a soluble nonlinear reaction-diffusion system. Physics Letters A, 1977, 60(5): 389–391.
298. Lewis M, Rees D C. Fractal surfaces of proteins. Science, 1985, 230(4730): 1163–1165.
299. Vani P K, Ramanujam G A, Kaliappan P. Painleve analysis and particular solutions of a coupled nonlinear reaction diffusion system. Journal of Physics A, 1993, 26(3): L97–L99.
300. Larsen A L. Weiss approach to a pair of coupled nonlinear reaction-diffusion equations. Physics Letters A, 1993, 179(4–5): 284–290.
301. Ndayirinde I, Malfliet W. New special solutions of the 'Brusselator' reaction model. Journal of Physics A, 1997, 30(14): 5151–5157.
302. Rocco A, West B J. Fractional calculus and the evolution of fractal phenomena. Physica A: Statistical Mechanics and Its Applications, 1999, 265(3): 535–546.
303. Saito Y. Permanence and global stability for general Lotka-Volterra predator-prey systems with distributed delays. Nonlinear Analysis-theory Methods & Applications, 2001, 47(9): 6157–6168.
304. Xu R, Chen LS. Persistence and global stability for a delayed nonautonomous Predator-Prey system without dominating instantaneous negative feedback. Journal of Mathematical Analysis and Applications, 2001, 262(1): 50–61.
305. Zamir M. Fractal dimensions and multifractility in vascular branching. Journal of Theoretical Biology, 2001, 212(2): 183–190.
306. Shu J P, Tan D Q, Wu J. Fractal structure of China stockmarket. Journal Southwest Jiaotong University, 2003, 38(2): 212–215. (In Chinese)
307. Zhao J D, Jiang J F, Lazer A C. The permanence and global attractivity in a nonautonomous Lotka-Volterra system. Nonlinear Analysis: Real World Application, 2004, 5(2): 265–276.
308. Bentley P J. Fractal proteins. Genetic Programming and Evolvable Machines, 2004, 5(1): 71–101.

309. Lin S Q, Lu Z Y. Permanence for two-species Lotka-Volterra systems with delays. Mathematical Bioences and Engineering, 2006, 3(1): 137–144.
310. Jumarie G. Modified Riemann-Liouville derivative and fractional Taylor series of nondifferentiable functions further results. Computers & mathematics with applications, 2006, 51(9–10): 1367–1376.
311. Yonaiguchi N, Ida Y, Hayakawa M, Masuda S. Fractal analysis for VHF electromagnetic noises and the identification of preseismic signature of an earthquake. Journal of Atmospheric and Solar-Terrestrial Physics, 2007, 69(15): 1825–1832.
312. Isvoran A, Pitulice L, Craescu C T, Chiriac A. Fractal aspects of calcium binding protein structures. Chaos Solitons & Fractals, 2008, 35(5): 960–966.
313. Movahed M S, Hermanis E. Fractal analysis of river flow fluctuations. Physica A: Statistical Mechanics and Its Applications, 2008, 387(4): 915–932.
314. Lancia M R, Mosco U, Vivaldi M A. Homogenization for conductive thin layers of pre-fractal type. Journal of Mathematical Analysis and Applications, 2008, 347(1): 354–369.
315. Jiang H, Gu Z. Stokes' theorem for domains with fractal boundary. Journal of Mathematical Analysis and Applications, 2009, 355(1): 164–169.
316. Schindler I, Tintarev K. Semilinear equations on fractal blowups. Journal of Mathematical Analysis and Applications, 2009, 352(1): 57–64.
317. Chen J, Ngai S. Eigenvalues and eigenfunctions of one-dimensional fractal Laplacians defined by iterated function systems with overlaps. Journal of Mathematical Analysis and Applications, 2010, 364(1): 222–241.
318. Liu X, Hong L. Chaos and adaptive synchronizations in fractional-order systems. International Journal of Bifurcation and Chaos, 2013, 23(11): 1350175.
319. Zhang Y P. Feedback control and synchronization of Mandelbrot sets. Chinese Physics B, 2013, 22(1): 0105021.
320. Pitulice L, Vilaseca E, Pastor I, Madurga S, Garces J L, Isvoran A, Mas F. Monte Carlo simulations of enzymatic reactions in crowded media. Effect of the enzyme-obstacle relative size. Bellman Prize in Mathematical Biosciences, 2014, 251: 72–82.
321. Fernandezmartinez M, Sanchezgranero M A. Fractal dimension for fractal structures: A Hausdorff approach revisited. Journal of Mathematical Analysis and Applications, 2014, 409(1): 321–330.
322. Pitulice L, Craciun D, Vilaseca E, Isvoran A. Fractal dimension of the trajectory of a single particle diffusing in crowded media. Romanian Journal of Physics, 2016, 61(7–8): 1276–1286.
323. Sun Y Y, Hou K, Zhao X D, Lu Z X. Research on characteristics of noise-perturbed M-J sets based on equipotential point algorithm. Communications in Nonlinear Science and Numerical Simulation, 2016, 30(1): 284–298.
324. Li C B, Sprott J C, Mei Y. An infinite 2–D lattice of strange attractors. Nonlinear Dynamics, 2017, 89(4): 2629–2639.
325. Wang D, Liu S T. A preliminary study on the fractal phenomenon: "Disconnected plus disconnected = connected". Fractals 2017, 25(1): 1750004.
326. Jiang C M, Zhang F F, Li T X. Synchronization and antisynchronization of N-coupled fractional-order complex chaotic systemswith ring connection. Mathematical Methods in the Applied Sciences, 2018, 41(7): 2625–2638.
327. Tikhomirov G, Petersen P, Qian L L. Fractal assembly of micrometre-scale DNA origami arrays with arbitrary patterns. Nature, 2017, 552(7683): 67–71.
328. Jiang Z Q, Gao X L, Zhou W X. Multifractal cross wavelet analysis. Fractals, 2017, 25(6): 1750054.
329. Namazi H, Farid A A, Chang T S. Fractal-based analysis of the variations of cutting forces along different axes in end milling operation. Fractals, 2018, 26(6): 1850089.
330. Namazi H, Ala T S, Backardjian H. Decoding of steady-state visual evoked potentials by fractal analysis of the electroencephalographic (eeg) signal. Fractals, 2018, 26(6): 1850092.
331. Tian D, Zhou C J, He J H. Hall-petch effect and inverse hall-petch effect: a fractal unification. Fractals, 2018, 26(6): 180083.

332. Wang Q L, Shi X Y, He J H, Li Z B. Fractal calculus and its application to explanation of biomechanism of polar bear hairs. Fractals, 2018, 26(6): 1850086.
333. Wang X X, Hou J G, Song S H, Wang D M, Gong L, Ma K, Liu Y M, Li Y Q, Yan L. Combining pressure-controlled porosimetry and rate-controlled porosimetry to investigate the fractal characteristics of full-range pores in tight oil reservoirs. Journal of Petroleum Science and Engineering, 2018: 353–361.
334. Cai J C, Lin D L, Singh H, Wei W, Zhou S W. Shale gas transport model in 3D fractal porous media with variable pore sizes. Marine and Petroleum Geology, 2018: 437–447.
335. Ahmadipajouh M A, Ala T S, Zamanian F, Namazi H, Jafari S. Fractal-based classification of human brain response to living and non-living visual stimuli. Fractals, 2018, 26(5): 1850069.
336. Namazi H, Ala T S, Kulish V V. Decoding of upper limb movement by fractal analysis of electroencephalogram (eeg) signal. Fractals, 2018, 26(5): 1850080.
337. Alipour Z M, Khosrowabadi R, Namazi H. Fractal-based analysis of the influence of variations of rhythmic patterns of music on human brain response. Fractals, 2018, 26(5): 1850080.
338. Namazi H, Farid A A, Seng C T. Fractal-based analysis of the influence of cutting depth on complex structure of cutting forces in rough end milling. Fractals, 26(5): 1850068.
339. Namazi H, Farid A A, Seng C T. Complexity-based analysis of the influence of tool geometry on cutting forces in rough end milling. Fractals, 2018, 26(5): 18500785.
340. Guariglia E. Harmonic Sierpinski gasket and applications. Entropy, 2018, 20(9): 7149.
341. Wang F, Chen W, Hua Q. A simple empirical formula of origin intensity factor in singular boundary method for two-dimensional Hausdorff derivative Laplace equations with Dirichlet boundary. Computers & mathematics with applications, 2018, 76(5): 1075–1084.
342. He J H. Fractal calculus and its geometrical explanation. Results Physics, 2018, 10: 272–276.
343. MozaffarileghaM, Namazi H, Ahadi M, Jafari S. Complexity-based analysis of the difference in speech-evoked auditory brainstem responses (s-abrs) between binaural and monaural listening conditions. Fractals, 2018, 26(4): 18500524.
344. Namazi H, Jafari S. Age-based variations of fractal structure of EEG signal in patients with epilepsy. Fractals, 2018, 26(4): 1850051.
345. Dawes T, Cai J, Quinlan M, Simoes Monteiro de Marvao A, Ostowski P, Tokarczuk P, Watson G, Wharton J, Howard L, Gibbs J, Cook S, Wilkins M, O'Regan, D P. Fractal analysis of right ventricular trabeculae in pulmonary hypertension. Radiology, 2018, 288(2): 386–395.
346. Han Y Z, Deng Y. An evidential fractal analytic hierarchy process target recognition method. Defence Science Journal, 2018, 68(4): 367–373.
347. Namazi H, Daneshi A, Azarnoush H, Jafari S, Towhidkhah F. Fractal-based analysis of the influence of auditory stimuli on eye movements. Fractals, 2018, 26(3): 1850040.
348. Li X L, Li Z H, Wang E Y, Liang Y P, Li B L, Chen P, Liu Y J. Pattern recognition of mine microseismic and blasting events based on wave fractal features. Fractals, 2018, 26(3): 1850029.
349. Liu C, Chung C E, Yin Y, Schnaiter, M. The absorption Ångström exponent of black carbon: from numerical aspects. Atmospheric Chemistry and Physics, 2017, 18(9): 6259–6273.
350. Reznikov N, Bilton M, Lari L, Stevens, M M, Kroger R. Fractal-like hierarchical organization of bone begins at the nanoscale. Science, 2018, 360(6388): 507.
351. Li Y Y, Zhang S C, Zhang X. Classification and fractal characteristics of coal rock fragments under uniaxial cyclic loading conditions. Arabian Journal of Geosciences, 2018, 11(9): 201.
352. Wang F Y, Yang K, Cai J C. Fractal characterization of tight oil reservoir pore structure using nuclear magnetic resonance and mercury intrusion porosimetry. Fractals, 2018, 26(2): 1840017.
353. Zhang S K, Huang Z W, Huang P P, Wu X G, Xiong C, Zhang C C. Numerical and experimental analysis of hot dry rock fracturing stimulation with high-pressure abrasive liquid nitrogen jet. Journal of Petroleum Science and Engineering, 2018, 163: 156–165.
354. Zhang K, Ma X P, Li Y L, Wu H Y, Zhang X M, Zhang H, Yao K. Parameter prediction of hydraulic fracture for tight reservoir based on micro-seismic and history matching. Fractals, 2018, 26(2): 1840009.

355. Kuznetsov N V, Leonov G A, Mokaev T N, Prasad A, Shrimali M D. Finite-time Lyapunov dimension and hidden attractor of the Rabinovich system. Nonlinear Dynamics, 2018, 92(2): 267–285.
356. Xiao B Q, Zhang X, Wang W, Long G B, Chen H X, Kang H, Ren W. A fractal model for water flow through unsaturated porous rocks. Fractals, 2018, 26(2): 1840015.
357. Laubach S E, Lamarche J, Gauthier B D, Dunne, W M, Sanderson D J. Spatial arrangement of faults and opening-mode fractures. Journal of Structural Geology, 2018, 108: 2–15.
358. Lai J, Wang G W, Cao J T, Xiao C W, Wang S, Pang X J, Dai Q Q, He Z B, Fan X Q, Yang L, Qin Z Q. Investigation of pore structure and petrophysical property in tight sandstones. Marine and Petroleum Geology, 2018, 91: 179–189.
359. Poplin R, Varadarajan A V, Blumer K, Liu Y, McConnell M V, Corrado G S, Peng L, Webster D R. Prediction of cardiovascular risk factors from retinal fundus photographs via deep learning. Nature Biomedical Engineering, 2018, 2(3): 158–164.
360. Taubert F, Fischer R, Groeneveld J, Lehmann S, Muller S M, Rodig E, Wiegand T, Huth A. Global patterns of tropical forest fragmentation. Nature, 2018, 554(7693): 519–522.
361. Corvi F, Pellegrini M, Erba S, Cozzi M, Staurenghi G, Giani A. Reproducibility of vessel density, fractal dimension, and foveal avascular zone using 7 different optical coherence tomography angiography devices. American Journal of Ophthalmology, 2018, 186: 25–31.
362. Xia Y X, Cai J C, Wei W, Hu X Y, Wang X, Ge X M. A new method for calculating fractal dimensions of porous media based on pore size distribution. Fractals, 2018, 26(1): 1850006.
363. Lai J, Wang G W, Fan Z, Zhou Z L, Chen J, Wang S C. Fractal analysis of tight shaly sandstones using nuclear magnetic resonance measurements. AAPG Bulletin, 2018, 102(2): 175–193.
364. Shi X H, Pan J N, Hou Q L, Jin Y, Wang Z Z, Niu Q H, Li M. Micrometer-scale fractures in coal related to coal rank based on micro-CT scanning and fractal theory. Fuel, 2018, 212(15): 162–172.
365. Ghai S, Ghai I, Schmitz G, Effenberg A O. Effect of rhythmic auditory cueing on parkinsonian gait: A systematic review and meta-analysis. Scientific Reports, 2018, 8(1): 1–19.
366. Tao S, Chen S D, Tang D Z, Zhao X, Xu H, Li S. Material composition, pore structure and adsorption capacity of low-rank coals around the first coalification jump: A case of eastern Junggar Basin, China. Fuel, 2018, 211(1): 804–815.
367. Zhu F, Cheng Q Y, Xue S G, Li C X, Hartley W, Wu C, Tian T. Influence of natural regeneration on fractal features of residue microaggregates in bauxite residue disposal areas. Land Degradation & Development, 2018, 29(1): 138–149.
368. Tian D, Li X, He J. Self-assembly of macromolecules in a long and narrow tube. Thermal Science, 2018, 22(4): 1659–1664.
369. Furui W, Michael H S C, Linsheng H, Song G. A novel fractal contact-electromechanical impedance model for quantitative monitoring of bolted joint looseness. IEEE Access, 2018, 6: 40212–40220.
370. Zhang P F, Lu S F, Li J Q, Chen C, Xue H T, Zhang J. Petrophysical characterization of oil-bearing shales by low-field nuclear magnetic resonance (NMR). Marine and Petroleum Geology, 2018, 89: 775–785.
371. Lahmiri S, Bekiros S. Chaos, randomness and multi-fractality in Bitcoinmarket. Chaos Solitons & Fractals, 2018, 106: 28–34.
372. Wang Z Z, Pan J N, Hou Q L, Yu B S, Li M, Niu Q H. Anisotropic characteristics of low-rank coal fractures in the Fukang mining area, China. Fuel, 2018, 211(1): 182–193.
373. Mozaffarilegha M, Namazi H, Tahaei A A, Jafari S. Complexity-based analysis of the difference between normal subjects and subjects with stuttering in speech evoked auditory brainstem response. Journal of Medical and Biological Engineering, 2019, 39(4): 490–497.
374. Tian D, Li X X, He J H. Geometrical potential and nanofiber membrane's highly selective adsorption property. Adsorption Science & Technology, 2019: 367–388.
375. Namazi H, Jafari S. Decoding of simple hand movements by fractal analysis of electromyography (EMG) signal. Fractals, 2019, 27(4): 1950042.
376. Letu H, Nagao T M, Nakajima T Y. Ice cloud properties from Himawari-8/AHI next-generation geostationary satellite: Capability of the AHI to monitor the DC cloud generation process. IEEE Transactions on Geoscience and Remote Sensing, 2019, 57(6): 3229–3239.

377. Wei B, Deng Y. A cluster-growing dimension of complex networks: From the view of node closeness centrality. Physica A: Statistical Mechanics & Its Applications, 2019, 522: 80–87.
378. Ma D, Duan H Y, Liu J F, Li J F, Zhou Z L. The role of gangue on the mitigation of mining-induced hazards and environmental pollution: An experimental investigation. Science of the Total Environment, 2019, 664(10): 436–448.
379. Alipour H, Namazi H, Azarnoush H, Azarnoush H, Jafari S. Complexity-based analysis of the relation between moving visual stimuli and human eye movement. Fractals, 2019, 27(3): 19500243.
380. Namazi H. Fractal-based classification of electromyography (emg) signal in response to basic movements of the fingers. Fractals, 2019, 27(3): 1950037.
381. Namazi H, Jafari S. Estimating of brain development in newborns by fractal analysis of sleep electroencephalographic (EEG) signal. Fractals, 2018, 27(3): 19500213.
382. Namazi H. Decoding of hand gestures by fractal analysis of electromyography (emg) signal. Fractals, 2019, 27(3): 19500223.
383. Xue S, Ye Y, Zhu F, Wang Q L, Jiang J, Hartley W. Changes in distribution and microstructure of bauxite residue aggregates following amendments addition. Journal of Environmental Sciences, 2019, 78: 276–286.
384. Singh J, Kumar D, Baleanu D, Rathore S. On the local fractional wave equation in fractal strings. Mathematical Methods in the Applied Sciences, 2019, 42(5): 1588–1595.
385. Guariglia E. Primality, fractality, and image analysis. Entropy-Switz, 2019, 21(3): 304.
386. Li H, Shi S L, Lu J X, Lu Y. Pore structure and multifractal analysis of coal subjected to microwave heating. Powder Technology, 2019, 346: 97–108.
387. Seyler K L, Rivera P, Yu H Y, Wilson N P, Ray E L, Mandrus D G, Yan J Q, Yao W, Xu X D. Signatures of moiré-trapped valley excitons in MoSe2/WSe2 heterobilayers. Nature, 2019, 567(7746): 66–70.
388. Alipour H, Namazi H, Azarmoush H, Jafari S. Complexity-based analysis of the influence of visual stimulus color on human eye movement. Fractals, 2019, 27(2): 1950002.
389. Xiao B Q, Wang W, Zhang X, Long G B, Chen H X, Cai H T, Deng L. A novel fractal model for relative permeability of gas diffusion layer in proton exchange membrane fuel cell with capillary pressure effect. Fractals, 2018, 27(22): 195001.
390. Fang D Q, Tang F Y, Huang H F, Cheung C Y, Chen H Y. Repeatability, interocular correlation and agreement of quantitative swept-source optical coherence tomography angiography macular metrics in healthy subjects. British Journal of Ophthalmology, 2018, 103(3): 415–420.
391. Namazi H, Jafari S. Decoding of wrist movements' direction by fractal analysis of magnetoencephalography (MEG) signal. Fractals, 2018, 27(12): 195000.
392. Thasthakeer AT, Farid A A, Seng C T, Namazi H. Analysis of the correlation between fractal structure of cutting force signal and surface roughness of machined workpiece in end milling operation. Fractals, 2019, 27(2): 1950013.
393. Wang Y, Deng Q. Fractal derivative model for tsunami traveling. Fractals, 2019, 27(2): 65–68.
394. Ye Q Q, Gu J W, Xi L F. Eigentime identities of fractal flower networks. Fractals, 2019, 27(2): 1950008.
395. Li X X, Tian D, He C H, He J H. A fractal modification of the surface coverage model for an electrochemical arsenic sensor. Electrochimica Acta, 2019, 296: 491–493.
396. Tao G L, Wu X K, Xiao H L, Chen Q S, Cai J C. A unified fractal model for permeability coefficient of unsaturated soil. Fractals, 2019, 27(1): 1940012.
397. Hirano T, Kitahara J, Toriyama Y, Kasamatsu H, Murata T, Sadda S. Quantifying vascular density and morphology using different sweptsource optical coherence tomography angiographic scan patterns in diabetic retinopathy. British Journal of Ophthalmology, 2019, 103(2): 216–221.
398. Kempkes S N, Slot M R, Freeney S E, Zevenhuizen S J M, Vanmaekelbergh D, Swart I. Design and characterization of electrons in a fractal geometry. Nature Physics, 2019, 2018(2): 127–131.
399. Dutkay D E, Haussermann J, Lai C K. Hadamard triples generate self-affine spectral measures. Transactions of the American Mathematical Society, 2018, 371(2): 1439–1481.

400. Alibakhshikenari M, Khalily M, Virdee B S, et al. Mutual coupling suppression between two closely placed microstrip patches using EM-bandgap metamaterial fractal loading. IEEE Access, 2019, 7: 23606–23614.
401. Botero ML, Eaves N, Dreyer A H, Sheng Y, Akroyd J, Yang W, Kraft M. Experimental and numerical study of the evolution of soot primary particles in a diffusion flame. Cambridge Centre for Computational Chemical Engineering, 2019, 37(2): 2047–2055.
402. Harmer G P, Abbott D. Losing strategies can win by Parrondo's paradox. Nature, 1999, 402(6764): 864.
403. Harmer G P, Abbott D. Parrondo's paradox. Statistical Science, 1999, 14(2): 206–213.
404. Minor D P. Parrondo's Paradox-Hope for Losers! College Mathematics Journal, 2003, 34(1): 15–20.
405. Chen L, Li C F, Gong M, Guo G C. Quantum Parrondo game based on a quantum ratchet effect. Physica A: Statistical Mechanics and Its Applications, 2010, 389(19): 4071–4074.
406. Almeida J, Peralta-Salas D, Romera M. Can two chaotic systems give rise to order? physica d nonlinear phenomena, 2005, 200(1–2): 124–132.
407. Danca M, Feckan M, Romera M. Generalized form of parrondo's paradoxical game with applications to chaos control. International Journal of Bifurcation and Chaos, 2014, 24(1): 1450008.
408. Danca MF, Tang WKS. Parrondos paradox for chaos control and anticontrol of fractional-order systems. Chin Physics B, 2015, 25(1): 507–513.
409. Mandelbrot BB. Fractals: Form, Chance, and Dimension. Wiley, San Francisco, 1977.
410. Mojica N S, Navarro J, Marijuan P C, Lahozbeltra. Cellular "bauplans": evolving unicellular forms by means of Julia sets and Pickover biomorphs. BioSystems, 2009, 98(1): 19–30.
411. Dhurandhar S, Bhavsar V C, Gujar U G. Analysis of z-plane fractal images from $z \leftarrow z^\alpha + c$ for $\alpha<0$. Computers & Graphics, 1993, 17(1): 89–94.
412. Wang X, Liu X, Zhu W Y, Gu S S. Analysis of C-plane fractal images from $z \leftarrow z^\alpha + c$ for $(\alpha<0)$. Fractals, 2000, 8(3): 307–314.
413. Rani M, Kumar V. Superior Julia set. Res Math Educ, 2004, 8(4): 261–277.
414. Wang X, Gu L. Research fractal structures of generalized M-J sets using three algorithms. Fractals, 2008, 16(1): 79–88.
415. Nazeer W, Kang S M, Tanveer M, Shahid A A. Fixed point results in the generation of Julia and Mandelbrot sets. Journal of Inequalities and Applications, 2015, 2015, 2015: 298.
416. Wang X Y, Wang Z, Lang Y H, Zhang Z F. Noise perturbed generalized Mandelbrot sets. Journal of Mathematical Analysis and Applications, 2008, 347(1): 179–187.
417. Andreadis I, Karakasidis T E. On a closeness of the Julia sets of noise-perturbed complex quadratic maps. International Journal of Bifurcation and Chaos, 2012, 22(9): 1577–1588.
418. Wang X Y, Ge F D. Quasi-sine Fibonacci M set with perturbation. Nonlinear Dynamics, 2012, 69(4): 1765–1779.
419. Beck C. Physical meaning for Mandelbrot and Julia sets. Physica D: Nonlinear Phenomena, 1999, 125(3): 171–182.
420. Wang X Y, Meng Q Y. Study on the physical meaning for generalized Mandelbrot-Julia sets based on the Langevin problem. Acta Physica Sinica, 2004, 53(2): 388–395.
421. Wang X Y, Chang P J, Gu N N. Additive perturbed generalized Mandelbrot-Julia sets. Applied Mathematics & Computation, 2007, 189(1): 754–765.
422. Wang X Y, Liu W, Yu X J. Research on Brownian movement based on generalized Mandelbrot-Julia sets from a class complex mapping system. Modern Physics Letters B, 2007, 21(20): 1321–1341.
423. Wang X Y, Jia H R, Zhang Z F. The generalized Mandelbrot set perturbed by composing noise of additive and multiplicative. Applied Mathematics and Computation, 2009, 210(1): 107–118.
424. Sun W H, Zhang Y P, Zhang X. Fractal analysis and control in the predator-prey model. International Journal of Computer Mathematics, 2017, 94(4): 737–746.
425. Rochon D. A generalized Mandelbrot set for biocomplex numbers. Fractals, 2000, 8(4): 355–368.

426. Wang X Y, Sun Y Y. The general quaternionic M-J sets on the mapping $z \leftarrow z^{\alpha} + c(\alpha \in N)$. Computers & Mathematics with Applications, 2007, 53(11): 1718–1732.
427. Wang X Y, Song W J. The generalized M-J sets for bicomplex numbers. Nonlinear Dynamics, 2013, 72(1–2): 17–26.
428. Wang X Y, Jin T. Hyperdimensional generalized M-J sets in hypercomplex number space. Nonlinear Dynamics, 2013, 73(1–2): 843–852.
429. Rădulescu A, Pignatelli A. Real and complex behavior for networks of coupled logistic maps. Nonlinear Dynamics, 2017, 87(2): 1295–1313.
430. Danca M, Romera M, Pastor G. Alternated Julia sets and connectivity properties. International Journal of Bifurcation and Chaos, 2009, 19(6): 2123–2129.
431. Danca M, Bourke P, Romera M. Graphical exploration of the connectivity sets of alternated Julia sets. Nonlinear Dynamics, 2013, 73(1): 1155–1163.
432. Wang D, Liu S T. On the boundedness and symmetry properties of the fractal sets generated from alternated complex map. Symmetry, 2016, 8(2): 1–7.
433. Qiu W Y, Yin Y C. Proof of the Branner-Hubbard conjecture on Cantor Julia sets. Science in China (Series A: Mathematics), 2009, 52(1): 45–65. (In Chinese)
434. Julia G. Mémoire sur l'itération des fonctions rationnelles. Journal de Mathématiques Pures et Appliquées, 1918, 9(3): 47–246.
435. Lakhtakia A, Varadan V V, Messier R. On the symmetries of the Julia sets for the process $z \Rightarrow z^{\alpha} + c$. Journal of Physics A General Physics, 1999, 20(11): 3533–3535.
436. Branner B, Hubbard J H. The iteration of cubic polynomials part II: patterns and parapatterns. Acta Mathematica, 1992, 169(3–4): 229–325.
437. Wang X Y, Yu X J. Julia sets for the standard Newton's method, Halley's method, and Schroder's method. Applied Mathematics and Computation, 2007, 189(2): 1186–1195.
438. Rani M, Kumar V. Superior Mandelbrot set. Journalof the Korea Society of Mathematical Education Series D: Research in Mathematical Education, 2004, 8(8): 279–291.
439. Negi A, Rani M. Midgets of superior Mandelbrot set. Chaos Solitons & Fractals, 2008, 36(2): 237–245.
440. Sun Y Y, Lu Z X, Li P. Complex time-delay dynamical systems of quadratic polynomials mapping. Nonlinear Dynamics, 2015, 79(1): 369–375.
441. Liu S T, Zhang Y P. Synchronization of Julia sets of complex systems. Acta Physica Sinica, 2008, 57(2): 737–742.
442. Liu P, Liu C A. Linear generalized synchronization of spatial Julia sets. International Journal of Bifurcation and Chaos, 2011, 21(5): 1281–1291.
443. Levin M. Morphogenetic fields in embryogenesis, regeneration, and cancer: non-local control of complex patterning. BioSystems, 2012, 109(3): 243–261.
444. Sun Y Y, Xu R D, Chen L N, Hu X P. Image compression and encryption scheme using fractal dictionary and Julia set. IET Image Processing, 2015, 9(3): 173–183.
445. Romera M, Small M, Danca M. Deterministic and random synthesis of discrete chaos. Applied Mathematics and Computation, 2007, 192(1): 283–297.
446. Wang X Y, Chang P J. Research on fractal structure of generalized M-J sets utilized Lyapunov exponents and periodic scanning techniques. Applied Mathematics and Computation, 2006, 175(2): 1007–1025.
447. Herzfeld U C. The science of fractal images. Computers & Geoences, 1991, 17(3): 469–470.
448. Bogush A A, Gazizov A Z, Kurochkin Y A, Stosui V T. On symmetry properties of quaternionic analogs of Julia sets. Physics, 2012 (4): 295–299.
449. Uspensky J V. Theory of Equations. McGraw-Hill Paperbacks, New York, 1964.
450. Schwarzenberger R, Falconer K. Mathematical foundations and applications. The Mathematical Gazette, 1990, 74(469): 316.
451. Zhang Y P, Liu S T. Gradient control and synchronization of Julia sets. Chinese Physics B, 2008, 17(2): 543–549.
452. Zhang Y P, Sun W H, Liu C A. Control and synchronization of second Julia sets. Chinese Physics B, 2010, 19(5): 146–153.

453. Brandt M E, Chen G R. Bifurcation control of two nonlinear models of cardiac activity. IEEE Transactions on Circuits and Systems I-regular Papers, 1997, 44(10): 1031–1034.
454. Saitoh N, Saito S, Shimizu A. An analysis of a family of rational maps containing integrable and non-integrable difference analogue of the logistic equation. Journal of Physics A General Physics, 1996, 29(8): 1831–1840.
455. Fornaess J E. The Julia Sset of Hénon maps. Mathematische Annalen, 2006, 334(2): 457–464.
456. Blanchard P, Devaney R L, Garijo A, Russell E. A generalized version of the McMullen domain. International Journal of Bifurcation and Chaos, 2008, 18(8): 2309–2318.
457. Blanchard P, Devaney R, Garijo A, Marotta S M, Russell E D. The rabbit and other Julia sets wrapped in Sierpiński carpets. Complex Dynamics, Families and Friends, 2009, 277–296.
458. Qiu W Y, Wang X G, Yin Y C. Dynamics of McMullen maps. Advances in Mathematics, 2012, 229(4): 2525–2577.
459. Morabito M, Devaney R L. Limiting Julia sets for singularly perturbed rational maps. International Journal of Bifurcation and Chaos, 2008, 18(10): 3175–3181.
460. Zhang Y P. Control and synchronization of Julia sets of the complex perturbed rational maps. International Journal of Bifurcation and Chaos, 2013, 23(5): 1350083.
461. Schwarzenberger R, Falconer K. Mathematical Foundations and Applications. The Mathematical Gazette, 1990, 74(469): 316.
462. Andreadis I, Karakasidis T E. On a topological closeness of perturbed Julia sets. Applied Mathematics & Computation, 2010, 215(10): 3674–3683.
463. Luo X S, Chen G R, Wang B H, Fang J Q. Hybrid control of period-doubling bifurcation and chaos in discrete nonlinear dynamical systems. Chaos Solitons & Fractals, 2003, 18(4): 775–783.
464. Wang D, Liu S T. Synchronization between the spatial Julia sets of complex Lorenz system and complex Henon map. Nonlinear Dynamics, 2015, 81(3): 1197–1205.
465. Liu P, Liu S T. Control and synchronization of Julia sets in coupled map lattice. Communications in Nonlinear Science and Numerical Simulation, 2011, 16(8): 3344–3355.
466. Norton A. Generation and display of geometric fractals in 3–D. International Conference on Computer Graphics and Interactive Techniques, 1982, 16(3): 61–67.
467. Hart J, Sandin D J, Kauffman L H. Ray tracing deterministic 3–D fractals. International Conference on Computer Graphics and Interactive Techniques, 1989, 23(3): 289–296.
468. Liu Y Z, Chen L Q. Nonlinear Oscillation. Shanghai Jiao Tong University Press, 2001. (In Chinese)
469. Wang B H. Introduction to Chaos Dynamics. University of Science and Technology of China Press, 1992. (In Chinese)
470. Wang G R, Chen G Z, Chaotic Motion of Nonlinear Systems.Guangxi science and technology Press, 1995. (In Chinese)
471. Hao B L, Zhang S Y. Subharmonic stroboscopic sampling method for study of period-doubling bifurcation and chaotic phenomena in forced nonlinear oscillators. Acta Physica Sinica, 1983, 32(2): 198–208. (In Chinese)
472. Falconer K J. Fractal Geometry: Mathematical Foundations and Applications. Biometrics, 1990, 46(3).
473. Liu S D, Liu S S. Fractal in Physics. Peking University Press, 2014. (In Chinese)
474. Hao B L, Introduction to Fractal. Inner Mongolia People's Publishing House, 1996. (In Chinese)
475. Steinfeld J I, Francisco J S, Hase W L. Chemical Kinetics and Dynamics. Englewood Cliffs (New Jersey): Prentice Hall, 1989.
476. Wilkinson F. Chemical Kinetics and Reaction Mechanism. Van Nostrand Reinhold Company, 1980.
477. Wu J Z, Introduction to Vortex Dynamics. Higher Education Press, 1993. (In Chinese)
478. Pan C D, Pan C B. Concise Number Theory. Peking University Press, 1998. (In Chinese)
479. Deb K, Pratap A, Agarwal S. A fast and elitist multiobjective genetic algorithm: NSGA-II. IEEE Transactions on Evolutionary Computation, 2002, 6(2): 182–197.
480. Dong C M. Detection and Analysis of Ocean Vortex. Science Press, 2016. (In Chinese)

481. Qing W. Curve fitting by curve fitting toolbox of matlab. Computer Knowledge and Technology, 2010, 6(21): 5822–5823. (In Chinese)
482. Shi L X, Nie X T, Ji M. List curve fitting based on MATLAB curve fitting toolbox. New Technology & New Process, 2007(7): 39–41. (In Chinese)
483. Jin M. Application of Matlab toolbox. Publishing House of Electronics Industry, 2004. (In Chinese)
484. Qin R F, Lin L Z, Kuang C P, Su T C, Mao X D, Zhou Y S. A GIS-based software for forecasting pollutant drift on coastal water surfaces using fractional Brownian motion: A case study on red tide drift. Environmental Modelling & Software, 2017, 92(6): 252–260.
485. Wang B. Introduction to Geophysical Fluid Mechanics. Ocean Press, 1981. (In Chinese)
486. Ma T, Wang S H. Stability and Bifurcation of Nonlinear Evolution Equations. Science Press, 2007. (In Chinese)
487. Lorenz E N. Deterministic nonperiodic flow. Journal of the Atmospheric Sciences, 1963, 20(2): 130–141.
488. Lin, X Y, Dong C M, Chen, D K, Liu Y, Yang J S, Zou B, Guan Y P, Three dimensional properties of mesoscale eddies in the south china sea based on eddy-resolving model output. Deep Sea Research Part I Oceanographic Research Papers, 2015, (99): 46–64.
489. Heaney K D, Campbell R L, Three-dimensional parabolic equation modeling of mesoscale eddy deflection. Journal of the Acoustical Society of America, 2016, 139(2): 918.
490. Anstey J A, Zanna L. A deformation-based parametrization of ocean mesoscale eddy Reynolds stresses. Ocean Modelling, 2017, 112: 99–111.
491. Li Q Y, Wang N C, Yi D Y. Numerical Analysis (5th Edition). Tsinghua University Press, 2006. (In Chinese)
492. Zhang L J, Wang L, Tan L Y, Su J L. Python Data Analysis and Mining. Machinery Industry Press, 2015. (In Chinese)
493. Zhan H M. Numerical Calculation of MATLAB. Machinery Industry Press, 2017. (In Chinese)
494. Hassan K. Khalil. Nonlinear Systems (3rd Edition). Pearson, 2001.
495. Lilly J M, Sykulski A M, Early J J, Olhede S C. Fractional brownian motion, the Matérn process, and stochastic modeling of turbulent dispersion. Nonlinear Processes in Geophysics, 2017, 24(3): 481–514.
496. Huang Y X, Wang L P, Schmitt F G, Zheng X B, Jiang N, Liu Y L. Extremal-point density of scaling processes: From fractional brownian motion to turbulence in one dimension. Physical Review E, 2017, 96(1): 012215.
497. Zhang M M, Zhang Y P. Fractal analysis and control of the competition model. International Journal of Biomathematics, 2016, 9(3): 1–16.
498. Ye W A, Li F Q. Physical Oceanography. China Ocean University Press, 1992. (In Chinese)
499. Stoker J J. Water waves: The mathematical theory with applications. John Wiley & Sons, 2011.
500. Hess J L, Smith A M O. Calculation of non-lifting potential flow about arbitrary three-dimensional bodies. Journal of Ship Research, 1964, 8(4): 22–44.
501. Bai L P. Numerical simulation of three-dimensional random wave. The Ocean Engineering, 2000, 12(2): 26–28. (In Chinese)
502. Teschl G. Ordinary differential equations and dynamical systems. Providence: American Mathematical Society, 2012.
503. Lorenz E N. Dimension of weather and climate attractors. Nature, 1991, 353(6341): 241–244.
504. Broucke R A. On the use of Poincare surfaces of section in rigid-body motion. Journal of the Astronautical Sciences, 1993, 41(4): 593–601.
505. Zhao H, Zhang F Z, Yan J, Wang Y H. Nonlinear differential delay equations using the Poincare section technique. Physical Review E, 1996, 54(6): 6925–6928.
506. Solari H G, Natiello M A, Vázquez M. Braids on the Poincaré section: A laser example. Physical Review E Statistical Physics Plasmas Fluids & Related Interdisciplinary Topics, 1996, 54(4): 3185–3195.
507. Wu W. Invariant manifolds on Poincaré section and its application in Lorenz system. Lanzhou University, 1997. (In Chinese)

508. Jin Y X, He K F. Quantum Poincaré section of a two-dimensional Hamiltonian in a coherent state representation. Chinese Physics Letters, 2002, 19(9): 1264–1267.
509. Liu S Y, Zhu S J, Yang Q C, Yang A B. Application of Poincar é section method in chaotic vibration identification. Applied Physics, 2011, 1(3): 108–115. (In Chinese)
510. Zabihi M, Kiranyaz S, Rad A B, et al. Analysis of high-dimensional phase space via Poincaré section for patient-specific seizure setection. IEEE Transactions on Neural Systems & Rehabilitation Engineering A Publication of the IEEE Engineering in Medicine & Biology Society, 2015, 24(3): 386–398.
511. Burrough P A. Fractal dimensions of landscapes and other environmental data. Nature, 1981, 294(5838): 240–242.
512. Ren F A. Ocean Dynamics. Dalian Maritime University Press, 2001. (In Chinese)
513. Avnir D, Farin D, Pfeifer P. Molecular fractal surfaces. Nature, 1984, 308(5956): 261–263.
514. Lewis M, Rees D C. Fractal surfaces of proteins. Science, 1985, 230(4730): 1163–1165.
515. Nittmann J, Daccord G, Stanley H E. Fractal growth viscous fingers: quantitative characterization of a fluid instability phenomenon. Nature, 1985, 314(6007): 141–144.
516. Sander L M. Fractal growth processes. Nature, 1986, 322(322): 789–793.
517. Tsonis A A, Elsner J B. Nonlinear prediction as a way of disting uishing chaos from random fractal sequences. Nature, 1992, 358(6383): 217–220.
518. Avnir D, Biham O, Lidar D, et al. Is the geometry of nature fractal? Science, 1998, 279(5347): 39–40.
519. Jarvis P, Jefferson B, Parsons S A. Breakage, regrowth, and fractal nature of natural organic matter flocs. Environmental Science & Technology, 2005, 39(7): 2307–14.
520. Qi D X. Fractal and Its Computer Generation. Science Press, 1994. (In Chinese)
521. Liu S D, Liu S S. Introduction to Fractal and Fractal Dimension. China Meteorological Press, 1993. (In Chinese)

The manufacturer's authorised representative in the EU is Springer Nature Customer Service Centre GmbH, Europaplatz 3, 69115 Heidelberg, Germany. If you have any concerns regarding our products, please contact ProductSafety@springernature.com

Printed and bound by CPI Group (UK) Ltd, Croydon, CR0 4YY
30/07/2026
02182030-0002